SPECIAL MATRICES AND THEIR APPLICATIONS IN NUMERICAL MATHEMATICS

Second Edition

Miroslav Fiedler

Academy of Sciences of the Czech Republic

DOVER PUBLICATIONS, INC.
Mineola, New York

Copyright

Copyright © 2008 by Miroslav Fiedler
All rights reserved.

Bibliographical Note

This Dover edition, first published in 2008, is a revised and corrected second edition of the work originally published by Martinus Nijhoff Publishers, Dordrecht, The Netherlands, in 1986.

Library of Congress Cataloging-in-Publication Data

Fiedler, Miroslav.
    [Speciální matice a jejich použití v numerické matematice. English]
    Special matrices and their applications in numerical mathematics / Miroslav Fiedler.
        p. cm.
    Originally published: Dordrecht ; Boston : Martinus Nijhoff Publishers ; Hingham, MA, USA : Distributors for the U.S. and Canada, Kluwer Boston, 1986.
    Includes bibliographical references and index.
    ISBN-13: 978-0-486-46675-0 (pbk.)
    ISBN-10: 0-486-46675-2 (pbk.)
    1. Matrices. 2. Numerical analysis. I. Title.

QA188.F5313 2008
512.9'.434—dc22

2008021479

www.doverpublications.com

Preface

This is the second English edition of the book. The first edition appeared in 1986, published by Martinus Nijhoff Publishers as translation from the Czech original. Only terminological and linguistic changes were performed in the Dover edition.

It was my intention to consider the topic of special matrices in the general sense and to include some more or less auxiliary topics that made it possible to present some facts or processes more intuitively. An example is using the graph theory.

Chapter 1 contains definitions of basic concepts of the theory of matrices, and fundamental theorems. The Schur complement is defined here in full generality and using its properties we prove the theorem on the factorization of a partitioned matrix into the product of a lower block triangular matrix with identity diagonal blocks, a block diagonal matrix, and an upper block triangular matrix with identity diagonal blocks. The theorem on the Jordan normal form of a matrix is given without proof.

Symmetric and Hermitian matrices form the content of Chapter 2. We prove Schur's theorem and, using it, we establish the fundamental theorem describing the factorization of symmetric or Hermitian matrices. Further, the properties of positive definite and positive semidefinite matrices are studied. In the conclusion, Sylvester's law of inertia of quadratic forms and theorems on the singular value decomposition and polar decomposition are proved.

Chapter 3 treats the mutual connections between graphs and matrices. In addition to basic definitions and properties of directed and undirected graphs, it contains e.g. the theorem on the condensation of a digraph with respect to the partition given by its strong components and the related theorem describing the transformation of a square matrix into the upper block triangular form with irreducible diagonal blocks.

The theory of entrywise nonnegative matrices is developed in Chapter 4. The Perron-Frobenius theorem on square nonnegative matrices is proved and its generalizations and extensions are studied. Further, stochastic and doubly stochastic matrices are discussed.

In Chapter 5, we introduce and study M-matrices, or, matrices of class K. Moreover, matrices of classes K_0 and P as well as diagonally dominant and uniformly diagonally dominant matrices are treated.

Important properties of tensor products of matrices and compound matrices including their generalizations form the content of Chapter 6. Spectral properties of such matrices and their significance in solving linear matrix equations are also studied.

In Chapter 7, a matricial representation of polynomials is introduced and its relation to divisibility is discussed. Further, the classes of Bézout, Hankel, Löwner, and (positive) stable matrices are studied. The Routh–Hurwitz theorem and some other criteria for the investigation of the roots of polynomials are given without proof.

Band matrices and their particular case, tridiagonal matrices, are the subject of Chapter 8. An algorithm is described for the approximate solution of the following important problem: Find a simultaneous permutation of the rows and columns of a symmetric matrix that transforms the matrix into a band matrix with minimum bandwidth.

Norms of vectors and matrices are defined and studied in Chapter 9. The measure of nonsingularity of a square matrix with respect to a vector norm is introduced and the concept of dual norms is discussed. The results are applied to estimating eigenvalues of matrices.

The last five chapters treat selected numerical methods for solving problems from the linear algebra field. They make use of the concepts and results of the preceding chapters. In Chapter 10, the direct methods for solving linear systems are explained. Attention is also paid to overdetermined and underdetermined systems.

Iterative methods for solving nonsingular linear systems form the content of Chapter 11. The convergence of the Jacobi, Gauss–

Seidel, and successive overrelaxation methods is discussed. Moreover, the effect of splitting on the rate of convergence of the methods presented is studied for matrices of class K.

In Chapter 12, inversion of special matrices is the main topic. Sherman-Morrison-Woodbury-type results as well as formulae for inverses of updated matrices and aggregated matrices are given.

Chapter 13 contains a description of methods for computing eigenvalues of matrices. The discussion includes the power method and its application to the inverse iteration, the Krylov method, Householder's method for reducing the matrix to the Hessenberg form, the QR-algorithm, and the Jacobi method.

The last chapter deals with large sparse matrices. The emphasis is on the study of the occurrence of additional non-zeros created during the Gaussian elimination process. To this end we employ both the graph theory and the concept of the envelope of a matrix.

Even though the subject is covered rather extensively, not all the classes of special matrices have been included. Moreover, the study of the properties of the classes treated in the book is far from exhaustive. Preference has been given to the classes and properties which are of importance in numerical analysis.

Other topics are mentioned in the exercises. Those referred to in the text are marked with an asterisk.

The author is pleased to express his appreciation to his colleagues Petr Přikryl and Karel Segeth who have not only prepared the first English edition by translating the Czech original but improved the presentation with many useful suggestions.

February 2008   *Miroslav Fiedler*

Summary of Notation

$\{1, 2, 3\}$	finite set of the elements 1,2,3
$\vert M \vert$	number of elements of a finite set
$P \backslash Q$	complement of Q with respect to P
$M \times N$	Cartesian product of M and N
R	the set of real numbers
$[a_1, a_2]$	closed interval
(a_1, a_2)	open interval
$\bar{z}$	complex conjugate of z
$\operatorname{Re} z$	real part of z
$\vert z \vert$	modulus of a complex number
$P = (k_1, k_2, \ldots, k_n),$	permutation P
or	
$P = \begin{pmatrix} 1 & 2 & \ldots & n \\ k_1 & k_2 & \ldots & k_n \end{pmatrix}$	
$\sigma(P)$	sign of permutation P
$E_k(x_1, \ldots, x_n)$	kth elementary symmetric function of $x_1, \ldots, x_n$
$S_k(x)$	kth degree modified Chebyshev polynomial of second kind
$A = [a_{ik}]$	matrix
$a = \begin{bmatrix} a_1 \\ \vdots \\ a_n \end{bmatrix}$	n-component column vector
$b = [b_1, \ldots, b_m]$	m-component row vector
$\langle a, b \rangle$	inner product of a and b
0	zero matrix, zero vector
I	identity matrix
$A(M_1, N_1)$	submatrix
$A(M_1, M_1),$ or $A(M_1)$	principal submatrix
A^T	transpose of A
$\bar{A}$	complex conjugate of A
A^*	conjugate transpose of A
A^{-1}	inverse of A
$\operatorname{tr} A$	trace of A

$\operatorname{Re} A$	symmetric part of A
$\rho(A)$	spectral radius of A
$\det A$	determinant of A
$S(A)$	spectrum of A
$\operatorname{rank} A$	rank of A
$\operatorname{codet} A(M_1, M_2)$	cofactor of the minor $\det A(M_1, M_2)$ of A
$\operatorname{diag}(d_1, \ldots, d_n)$	diagonal matrix with the diagonal entries $d_1, \ldots, d_n$
$\operatorname{diag}(D_1, \ldots, D_m)$	block diagonal matrix with the diagonal blocks $(D_1, \ldots, D_m)$
$J_m(\sigma)$	Jordan block of order m with the diagonal entry σ
A_J	Jordan normal form of A
$[A/A_{11}]$	Schur complement of the nonsingular submatrix A_{11} in the partitioned matrix $A = \begin{bmatrix} A_{11} & A_{12} \\ A_{21} & A_{22} \end{bmatrix}$
$m(A)$	modulus of A
$\operatorname{In} A$	inertia of a square matrix
A^+	pseudoinverse of A
$\Pi(A)$	profile of A
$\langle Ax, y \rangle$	bilinear form with matrix A
$\langle Ax, x \rangle$	quadratic form with matrix A
$A \geq 0$	matrix A is nonnegative
$A > 0$	matrix A is positive
$A \geq B$	$A - B$ is nonnegative
$A > B$	$A - B$ is positive
A_B	Boolean representation of A
Z	the class of real square matrices whose off-diagonal entries are nonpositive
Z_n	the class of $n \times n$ matrices in Z
K	the class of all nonsingular M-matrices
K_0	the class of all (possibly singular) M-matrices
P	the class of real square matrices whose principal minors are positive

P_0	the class of real square matrices whose principal minors are nonnegative
H	the class of complex Hankel matrices
$V(s_1, \ldots, s_r)$, or V_s	Vandermonde matrix
$B(f, g)$	Bezout matrix for the pair of polynomials $f(x), g(x)$
$\mathscr{L}(y, z)$	the class of complex Löwner matrices with points $(y_i), (z_j)$
$\vec{A}$	matrix written as a column vector
$A \otimes B$	tensor product of A and B
$u_1 \wedge u_2 \wedge \cdots \wedge u_k$	exterior product of the vectors $u_1, u_2, \ldots, u_k$
$A^{(k)}$	kth compound of A
$A^{(k,s)}$	generalized compound of A
$A^{[k]}$	kth additive compound of A
$[a]^{(k)}$	$k \times (n + k)$ matrix associated with the polynomial $a(x)$ of degree n
$L(Y, X)$	linear space of matrices as operators from Y into X where Y and X are spaces of column vectors
$g(x)$	norm of a vector
$g_{YX}(A)$	norm of the matrix $A \in L(Y, X)$ induced by the norms g_X and g_Y
$\hat{g}(A)$	measure of nonsingularity of a square matrix
$N(A)$	Schur (Frobenius) norm of A
$\kappa(A)$	condition number of A
$\langle x, x' \rangle$	bilinear form defined for vectors x and x' in dual spaces
$g'(x')$	dual norm of a dual vector
$R(\alpha)$, or $R_{ik}(\alpha)$	matrices of plane rotation through the angle α
$W(w)$	reflection matrix with the normalized vector w
$\vec{G} = (V, H)$	digraph whose set of vertices is V and set of arcs is H

$G = (V, H)$	graph whose set of vertices is V and set of edges is H
(u, v)	arc going from the vertex u to the vertex v in a digraph, or an edge joining the vertices u and v in a graph
$(u_1, u_2, \ldots, u_k)$	directed walk, path in a digraph or walk, path in a graph
$R = [V_1, V_2, \ldots, V_m]$	partition of the set of vertices of a graph
$\vec{G}_R$	condensation of the digraph $\vec{G}$ with respect to the partition R
$\vec{G}_i[V_1]$	subdigraph of $\vec{G}$ induced by the subset V_1 of the set of vertices of $\vec{G}$
$\vec{G}(A)$	digraph of A
$U(\vec{G})$	adjacency matrix of $\vec{G}$
$\vec{G}[W]$	Schur complement of the set of vertices W in $\vec{G}$
$\vec{B} = (V_1, V_2, H)$	bigraph whose first set of vertices is V_1, second set of vertices is V_2 and set of arcs is H
$\vec{B}(A)$	bigraph of A
$\vec{B}^T$	transpose of $\vec{B}$

Contents

1

Basic concepts of matrix theory

This introductory chapter is essentially a survey of well known definitions and general principles from matrix theory which will be useful in the sequel.

1.1 Matrices

A *matrix of type m-by-n* or, equivalently, an $m \times n$ matrix, is a rectangular array of mn numbers (usually real or complex) arranged in m *rows* and n *columns* (m, n positive integers):

$$
\begin{bmatrix}
a_{11} & a_{12} & a_{13} & \cdots & a_{1n} \\
a_{21} & a_{22} & a_{23} & \cdots & a_{2n} \\
\vdots & \vdots & \vdots & \ddots & \vdots \\
a_{m1} & a_{m2} & a_{m3} & \cdots & a_{mn}
\end{bmatrix}
\tag{1.1}
$$

We call the number a_{ik} the *entry* of the matrix in the ith row and the kth column. The set of $m \times n$ matrices with real entries will be denoted by $R^{m \times n}$. Analogously, the set of $m \times n$ matrices with complex entries will be denoted by $C^{m \times n}$. In some cases, entries can be polynomials, variables, functions, etc.

In this terminology, matrices with only one column (thus, $n = 1$) will be called *column vectors* and matrices with only one row (thus, $m = 1$) *row vectors*. In such case, we write R^m instead of $R^{m \times 1}$ and vectors will always be column vectors, unless we specify otherwise.

It is advantageous to denote the matrix by a single symbol, say A, or $[a_{ik}]$, C, etc.

Two matrices $A = [a_{ik}]$, $B = [b_{ik}]$ are *equal*, written $A = B$, if and only if they are both $m \times n$ matrices for some positive integers m and n, and $a_{ik} = b_{ik}$ for $i = 1, \ldots, m$, $j = 1, \ldots, n$.

The importance of matrices is that one can introduce operations generalizing operations with numbers.

Matrices of the same type can be added: If $A = [a_{ik}]$, $B = [b_{ik}]$, then $A + B$ is the matrix $[a_{ik} + b_{ik}]$. The operation of addition is thus *entrywise*. We admit also multiplication of a matrix by a number (real, complex, a parameter, etc.): If $A = [a_{ik}]$ and if α is a number (also called *scalar*), then αA is the matrix $[\alpha a_{ik}]$ of the same type as A.

An $m \times n$ matrix $A = [a_{ik}]$ can be multiplied by an $n \times p$ matrix $B = [b_{k\ell}]$ as follows: it is the $m \times p$ matrix $C = [c_{i\ell}]$, where

$$c_{i\ell} = a_{i1}b_{1\ell} + a_{i2}b_{2\ell} + \ldots + a_{in}b_{n\ell}. \qquad (1.2)$$

It is important to notice that the matrices A and B can be multiplied (in this order) only if the number of columns of A is the same as the number of rows in B. Also, the entries of A and B should be multiplicable. In general, the product AB is not equal to BA, even if the multiplication of both products is possible. On the other hand, the multiplication fulfils the *associative law*

$$(AB)C = A(BC)$$

as well as (in this case, two) *distributive laws*:

$$(A + B)C = AC + BC$$

and

$$A(B + C) = AB + AC,$$

whenever multiplications are possible.

An $m \times n$ matrix is called *square* if $m = n$. We then say that it is *square of order m*.

Of basic importance are the *zero matrices* of which all entries are zeros, and the *identity matrices*; these are *square* matrices, and have ones in the *main diagonal* and zeros elsewhere. Thus

$$[1], \quad \begin{bmatrix} 1 & 0 \\ 0 & 1 \end{bmatrix}, \quad \begin{bmatrix} 1 & 0 & 0 \\ 0 & 1 & 0 \\ 0 & 0 & 1 \end{bmatrix}$$

are identity matrices of *order* one, two and three. We denote zero matrices simply by *0*, the identity matrices by *I*, sometimes with a subscript denoting the order.

The identity matrices of appropriate orders have the property that

$$AI = A \text{ and } IA = A$$

hold for any matrix *A*.

Suppose now $A = [a_{ik}]$ is an $m \times n$ matrix. Choose a subset $M_1 = \{i_1, i_2, \ldots, i_r\}$ of the set of row indices $M = \{1, 2, \ldots, m\}$, and a subset $N_1 = \{k_1, k_2, \ldots, k_s\}$ of the set of column indices $N = \{1, 2, \ldots, n\}$. In addition, we assume that the integers i_j and k_ℓ are ordered in such a way that $i_1 < i_1 < \cdots < i_r$ and $k_1 < k_2 < \cdots < k_s$. The matrix

$$\begin{bmatrix} a_{i_1 k_1} & a_{i_1 k_2} & \cdots & a_{i_1 k_s} \\ a_{i_2 k_1} & a_{i_2 k_2} & \cdots & a_{i_2 k_s} \\ \vdots & \vdots & \ddots & \vdots \\ a_{i_r k_1} & a_{i_r k_2} & \cdots & a_{i_r k_s} \end{bmatrix}$$

is called a *submatrix* of *A*. We denote this submatrix by $A(M_1, N_1)$. If *A* is a square matrix with $M = N$, then every square submatrix $A(M_1, M_1)$ (with $M_1 = N_1$) is called a *principal submatrix*. The principal submatrix will be sometimes denoted also by $A(M_1)$.

The most important type of submatrices are those whose sets of indices consist of consecutive integers. Such submatrices are called *blocks*. For example, in the matrix

$$\left[\begin{array}{cc|cc} a_{11} & a_{12} \ a_{13} & a_{14} \\ a_{21} & a_{22} \ a_{23} & a_{24} \\ \hline a_{31} & a_{32} \ a_{33} & a_{34} \end{array} \right]$$

the submatrices

$$\begin{bmatrix} a_{11} \\ a_{21} \end{bmatrix}, \begin{bmatrix} a_{12} & a_{13} \\ a_{22} & a_{23} \end{bmatrix}, \text{ etc.,}$$

are blocks. A matrix that is in this way subdivided by "vertical" and "horizontal" lines into blocks is said to be *partitioned* as indicated. The above matrix can thus be written as

$$A = \begin{bmatrix} P_{11} & P_{12} & P_{13} \\ P_{21} & P_{22} & P_{23} \end{bmatrix};$$

the matrices P_{ik} have the property that those with first indices equal have the same number of rows and (those with) second indices equal have the same number of columns.

An important property of partitioned matrices is that they behave to some extent as if the individual blocks were numbers. It is only necessary to preserve order in the calculation of products and to assure that the blocks have proper dimensions. Clearly, the addition of two matrices conformally partitioned into blocks can be performed in this manner. As regards multiplication, the following theorem holds (Exercise 1.1).

Theorem 1.1.1 *Let*

$$
P = \begin{bmatrix} P_{11} & P_{12} & \cdots & P_{1s} \\ P_{21} & P_{22} & \cdots & P_{2s} \\ \vdots & \vdots & \ddots & \vdots \\ P_{r1} & P_{r2} & \cdots & P_{rs} \end{bmatrix}, \qquad
Q = \begin{bmatrix} Q_{11} & Q_{12} & \cdots & Q_{1t} \\ Q_{21} & Q_{22} & \cdots & Q_{2t} \\ \vdots & \vdots & \ddots & \vdots \\ Q_{s1} & Q_{s2} & \cdots & Q_{st} \end{bmatrix}
$$

be matrices partitioned in such a way that for each i, $i = 1, \ldots, r$, and each k, $k = 1, \ldots, t$, the number of columns in P_{ij} equals the number of rows in Q_{jk}, whenever $j = 1, \ldots, s$.

Then the product PQ can be partitioned into blocks R_{ik}, $i = 1, \ldots, r$, $k = 1, \ldots, t$, where

$$ R_{ik} = P_{i1}Q_{1k} + P_{i2}Q_{2k} + \cdots + P_{is}Q_{sk}. $$

Example 1.1 Let us write the product

$$
\begin{bmatrix} 1 & 0 & a_{11} & a_{12} \\ 0 & 1 & a_{21} & a_{22} \\ 0 & 0 & b_{11} & b_{12} \\ 0 & 0 & b_{21} & b_{22} \end{bmatrix}
\begin{bmatrix} c_{11} & c_{12} & -a_{11} & -a_{12} \\ c_{21} & c_{22} & -a_{21} & -a_{22} \\ 0 & 0 & 1 & 0 \\ 0 & 0 & 0 & 1 \end{bmatrix} \quad \text{as}
$$

$$
\begin{bmatrix} I_2 & A \\ O & B \end{bmatrix}
\begin{bmatrix} C & -A \\ O & I_2 \end{bmatrix}
$$

with two-by-two blocks. Then

$$
\begin{bmatrix} I_2 C + AO & I_2 A + A I_2 \\ OC + BO & -OA + B I_2 \end{bmatrix} =
\begin{bmatrix} C & O \\ O & B \end{bmatrix},
$$

i.e.

$$\begin{bmatrix} c_{11} & c_{12} & 0 & 0 \\ c_{21} & c_{22} & 0 & 0 \\ o & o & b_{11} & b_{12} \\ 0 & 0 & b_{21} & b_{22} \end{bmatrix},$$

as can be verified by direct multiplication.

Let $A = [a_{ik}]$ be an $m \times n$ matrix. The $n \times m$ matrix $C = [c_{pq}]$ whose entries are $c_{pq} = a_{qp}$, $q = 1, \ldots, m$, $p = 1, \ldots, n$ is called the *transpose* of A; we denote it A^T.

The operation of transposition has two important properties:

$$(AB)^T = B^T A^T \tag{1.3}$$

whenever AB is defined, and further

$$(A^T)^T = A \tag{1.4}$$

for any matrix A.

This notation will be used rather often, especially for column vectors. A column vector u with entries (components) $u_1, \ldots, u_n$ will usually be written as $[u_1, \ldots, u_n]^T$.

We turn now our attention to special cases of square matrices. A square matrix $A = [a_{ik}]$ of order n is *diagonal* if $a_{ik} = 0$ for $i \neq k$, $i, k = 1, \ldots, n$; the diagonal matrix with the diagonal entries $a_{11}, \ldots, a_{nn}$ will be often written as diag $(a_{11}, \ldots, a_{nn})$. The matrix A is a *lower triangular matrix* if $a_{ik} = 0$ for $i < k$, $i, k = 1, \ldots, n$, and it is an *upper triangular matrix* if $a_{ik} = 0$ for $i > k$, $i, k = 1, \ldots, n$. For example, the first two of the matrices

$$[2], \quad \begin{bmatrix} 2 & 0 \\ 3 & -1 \end{bmatrix}, \quad \begin{bmatrix} 1 & -1 & 0 \\ 0 & 1 & 2 \\ 0 & 0 & 1 \end{bmatrix}, \quad \begin{bmatrix} 2 & 0 & 0 \\ 0 & -1 & 0 \\ 0 & 0 & 3 \end{bmatrix}$$

are lower triangular, the third is upper triangular, and the fourth matrix is a diagonal matrix which may be also written as diag $(2, -1, 3)$. Clearly, the following statement holds.

Theorem 1.1.2 *The product of two lower triangular matrices of the same order is again a lower triangular matrix, and the product of two upper triangular matrices of the same order is upper triangular. The product of two diagonal matrices of the same order is again a diagonal matrix.*

A further important concept is that of a *nonsingular matrix*. We say that a square matrix A is nonsingular if there exists a matrix B such that $AB = BA = I$. The matrix B is then called the *inverse* of A and is denoted by A^{-1}. The following theorems can easily be proved.

Theorem 1.1.3 *Given a nonsingular matrix A, its inverse is determined uniquely. Furthermore, we have $(A^{-1})^{-1} = A$.*

Theorem 1.1.4 *If A and B are nonsingular matrices of the same order, then their product AB is nonsingular as well and*

$$(AB)^{-1} = B^{-1}A^{-1}.$$

Hence, the inverse of the product is the product of the inverses in the reverse order.

Theorem 1.1.5 *If A is nonsingular, then its transpose A^T is nonsingular as well and*

$$(A^T)^{-1} = (A^{-1})^T.$$

Hence, the inverse of the transpose is the transpose of the inverse.

1.2 Determinants

Determinants play a crucial role in distinguishing whether a square matrix is singular or nonsingular.

The *determinant* $\det A$ of an $n \times n$ matrix $A = [a_{ik}]$ is the number

$$\det A = \sum_{P=(k_1,\dots,k_n)} \sigma(P) a_{1k_1} a_{2k_2} \dots a_{nk_n}. \qquad (1.5)$$

The summation is over all permutations P of the integers $1, \dots, n$ and the factor $\sigma(P)$ is the *sign of the permutation P*. It is equal to $+1$ if the permutation P is even, and to -1 if P is odd. The permutation

$$\begin{pmatrix} 1 & 2 & \dots & n \\ k_1 & k_2 & \dots & k_n \end{pmatrix}$$

is *even* if it has an even number of *inversions*, i.e. such pairs i, j for which $i < j$ whereas $k_i > k_j$. An *odd* permutation has an odd number of inversions.

We state two consequences of (1.5). Clearly, if I is an identity matrix, then

$$\det I = 1. \tag{1.6}$$

The properties of permutations imply that for any square matrix A

$$\det A^T = \det A. \tag{1.7}$$

Let $A = [a_{ik}]$ be an $m \times n$ matrix. Let $M = \{1, 2, \ldots, m\}$, $N = \{1, 2, \ldots, n\}$ and let $A(M_1, M_2)$ be a square submatrix of A ($M_1 \subset M$, $M_2 \subset N$, $|M_1| = |M_2|$).[1] Then, we are able to consider the determinant $\det A(M_1, M_2)$ of the submatrix; such a determinant will be called a *minor* of A. If A is a square matrix, then together with the above minor there exists the minor $\det A(M \setminus M_1, N \setminus M_2)$.[2] It is the determinant of the submatrix whose rows consists precisely of the rows of A (in the same order as they occur in A) that are not included in the submatrix $A(M_1, M_2)$, and whose columns consist precisely of the columns of A that are not included in $A(M_1, M_2)$, again in the same order as they are in A. The number $(-1)^{s(M_1)+s(M_2)} \det A(M \setminus M_1, N \setminus M_2)$, where $s(M_1)$ is the sum of the row indices belonging to M_1 and $s(M_2)$ is the sum of the column indices belonging to M_2, is then called the *cofactor* of the minor $\det A(M_1, M_2)$. It is advantageous to denote the cofactor of $\det A(M_1, M_2)$ by codet $A(M_1, M_2)$. If A is a square matrix, then the determinants of principal submatrices play an important role. It is customary to call them *principal minors*.

Example 1.2 Let

$$A = \begin{bmatrix} a_{11} & a_{12} & a_{13} \\ a_{21} & a_{22} & a_{23} \\ a_{31} & a_{32} & a_{33} \end{bmatrix}.$$

---

[1] $|M|$ denotes the number of the elements of the set M; also in the sequel.
[2] Here, and in the sequel, $P \setminus Q$ denotes the complement of the set Q with respect to P, i.e., the set of all those elements of the set P that do not belong to Q.

Then for $M_1 = \{1,3\}$, $M_2 = \{2,3\}$ we have

$$\det A(M_1, M_2) = \det \begin{bmatrix} a_{12} & a_{13} \\ a_{32} & a_{33} \end{bmatrix},$$

$\det A(M \setminus M_1, N \setminus M_2) = \det[a_{21}] = a_{21}$, $s(M_1) = 4$, $s(M_2) = 5$.
Hence, codet $A(M_1, M_2) = -a_{21}$.

The following important theorem gives an expansion of the determinant of A by some of its rows (or columns).[3]

Theorem 1.2.1 (Laplace expansion) *Let $M \subset N = \{1, 2, \ldots, n\}$ be a fixed set of indices which has m elements. Let $A = [a_{ij}]$, $i, j = 1, \ldots, n$, be a square matrix of order n. Then*

$\det A =$
$(-1)^{s(M)} \sum_{M_j} (-1)^{s(M_j)} \det A(M, M_j) \det A(N \setminus M, N \setminus M_j)$,

where the summation is over all the subsets M_j of the set of column indices that have m elements.

Remark. Since

$$(-1)^{s(M)+s(M_j)} \det A(N \setminus M, N \setminus M_j) = \text{codet } A(M, M_j),$$

the statement of the theorem may be rewritten as

$$\det A = \sum_{M_j} \det A(M, M_j) \text{ codet } A(M, M_j), \qquad (1.8)$$

where the summation is over all the sets $M_j \subset N$ that have m elements.

Applying Theorem 1.2.1 to the transpose of A and turning back to the original matrix, we obtain a similar formula giving the expansion of the determinant by its columns:

$$\det A = \sum_{M_j} \det A(M_j, M) \text{ codet } A(M_j, M), \qquad (1.9)$$

where M is an arbitrary subset of N and the summation is over all the sets $M_j \subset N$ that have the same number of elements as M.

---

[3] A proof may be found e.g. in [60].

As a corollary of the Laplace theorem we have the formulae (for $A = [a_{ik}]$ of order n)

$$\det A = \sum_{k=1}^{n} a_{ik} A_{ik}, \qquad i \in \{1, \dots, n\}, \qquad (1.10)$$

where A_{ik} is the cofactor of the entry a_{ik} (the expansion of $\det A$ by its ith row), and

$$\det A = \sum_{i=1}^{n} a_{ik} A_{ik}, \qquad k \in \{1, \dots, n\}, \qquad (1.11)$$

(the expansion of $\det A$ by its kth column).

An immediate consequence of (1.10) and (1.11) is the next theorem.

Theorem 1.2.2 *Let A and B be square matrices of equal orders which differ only in the entries of one row or column. Let C be the matrix whose entries in this row or column are sums of the corresponding entries of A and B, the other entries being the same as in A and B. Then*

$$\det C = \det A + \det B.$$

Assuming that A and B differ only in the first column and denoting the columns of A by $A_1, A_2, \dots, A_n$, and the columns of B by $B_1, A_2, \dots, A_n$, we can write the above formula as

$$\det[A_1 + B_1, A_2, \dots, A_n] =$$
$$\det[A_1, A_2, \dots, A_n] + \det[B_1, A_2, \dots, A_n].$$

Repeatedly using this result, we establish a formula for the *determinant of the sum of two matrices*, A and B:

$$\det[A_1 + B_1, A_2 + B_2, \dots, A_n + B_n] =$$
$$\sum_{i_1, \dots, i_n = 1}^{2} \det[C_1^{(i_1)}, C_2^{(i_2)}, \dots, C_n^{(i_n)}].$$

Here, the summation is over $i_1 = 1, 2$, $i_2 = 1, 2, \dots$, $i_n = 1, 2$, and $C_k^{(1)} = A_k$, $C_k^{(2)} = B_k$, $k = 1, \dots, n$.

If we now use (1.9) to expand each determinant on the right-hand side by all its columns that are from A we arrive at the formula

$$\det(A + B) = \sum_{M_i, M_j} \det A(M_i, M_j) \operatorname{codet} B(M_i, M_j), \qquad (1.12)$$

where the summation is over all pairs of subsets M_i, M_j, $M_i \subset N$, $M_j \subset N$, which have the same number of elements ($|M_i| = |M_j|$).

A further consequence of (1.10) and (1.11) is the following theorem.

Theorem 1.2.3 *If a row or a column of a square matrix A contains only zero entries, then $\det A = 0$.*

If two rows of A are equal, then applying the Laplace expansion 1.2.1 to these rows one obtains the following theorem.

Theorem 1.2.4 *If a square matrix A has two identical rows, then $\det A = 0$. The same is true if A has two identical columns.*

Let us apply the theorem to the matrix B obtained from a square matrix $A = [a_{ik}]$ of order n by replacing (for $j \neq i$) the jth row of A by its ith row (without changing the ith row). Expanding $\det B$ by its jth row we have, according to (1.10)

$$(\det B =)0 = \sum_{k=1}^{n} a_{jk} A_{ik}, \qquad i, j = 1, \ldots, n, \ \ j \neq i. \qquad (1.13)$$

The same procedure applied to columns gives

$$\sum_{i=1}^{n} a_{i\ell} A_{ik} = 0, \qquad k, \ell = 1, \ldots, n, \ \ k \neq \ell. \qquad (1.14)$$

The following important theorem can be proved by induction on r using the expansion (1.8) and Theorem 1.2.3.

Theorem 1.2.5 *Let $A = [A_{ik}]$, $i, k = 1, \ldots, r$, be a partitioned square matrix whose diagonal blocks are square matrices and the* *blocks below diagonal are zero blocks:* $A_{ik} = 0$ *for* $i > k$ *(such a matrix is said to be upper block triangular), i.e.,*

$$A = \begin{bmatrix} A_{11} & A_{12} & \dots & A_{1r} \\ 0 & A_{22} & \dots & A_{2r} \\ 0 & 0 & \dots & A_{3r} \\ \vdots & \vdots & \ddots & \vdots \\ 0 & 0 & \dots & A_{rr} \end{bmatrix}.$$

Then,

$$\det A = \det A_{11} \det A_{22} \dots \det A_{rr}. \tag{1.15}$$

Remark. A similar formula holds for lower block triangular and for block diagonal matrices.

The next theorem presents the celebrated *Binet-Cauchy formula.*

Theorem 1.2.6 *Let* P *be an* $m \times n$ *matrix and let* Q *be an* $n \times m$ *matrix,* $m \leq n$. *Set* $M = \{1, 2, \dots, m\}$, $N = \{1, 2, \dots, n\}$. *Then*

$$\det PQ = \sum_{N_i} \det P(M, N_i) \det Q(N_i, M), \tag{1.16}$$

where the summation is over all the subsets N_i *of* N *that have* m *elements.*

We do not give the proof here[4] but we state an important corollary for $m = n$.

Theorem 1.2.7 *If* A *and* B *are square matrices of equal orders, then*

$$\det AB = \det A \cdot \det B. \tag{1.17}$$

1.3 Nonsingular matrices. Inverse matrices

We now turn to nonsingular matrices. First, we prove the following fundamental theorem.

---

[4]See, for example, [27].

Theorem 1.3.1 *A square matrix $A = [a_{ik}]$ is nonsingular if and only if $\det A \neq 0$. In this case, the inverse $A^{-1} = [\alpha_{ik}]$ has the entries*

$$\alpha_{ik} = \frac{A_{ki}}{\det A},$$

where A_{ki} is the cofactor of a_{ki} in the matrix A.

Proof. Suppose A has the inverse A^{-1}. Then

$$AA^{-1} = I,$$

and by Theorem 1.2.7 and (1.6), we have

$$\det A \cdot \det A^{-1} = 1. \qquad (1.18)$$

Therefore,

$$\det A \neq 0.$$

Conversely, suppose that $\det A \neq 0$. Define the matrix $B = [\alpha_{ik}]$ by $\alpha_{ik} = A_{ki}/\det A$. Then it follows that

$$\sum_{k=1}^{n} a_{ik}\alpha_{ki} = 1, \qquad i = 1, \ldots, n, \qquad (1.19)$$

as well as

$$\sum_{k=1}^{n} \frac{a_{ik}A_{ik}}{\det A} = 1$$

by (1.10).

Further, if $i \neq j$, then

$$\sum_{k=1}^{n} a_{ik}\alpha_{kj} = (\det A)^{-1} \sum_{k=1}^{n} a_{ik}A_{jk} = 0 \qquad (1.20)$$

according to (1.13).

Equations (1.19) and (1.20) may also be written in the form

$$AB = I.$$

A similar argument yields the relation

$$BA = I.$$

Thus, B is the (unique) inverse of A, indeed. $\qquad \square$

Before proceeding further, we restate (1.18) as a theorem.

Theorem 1.3.2 *If A is a nonsingular matrix, then*

$$\det A^{-1} = \frac{1}{\det A}.$$

Theorem 1.3.1 enables us, at least theoretically, to find the solution of a system of linear equations written as

$$Ax = b$$

in matrix notation, provided A is a nonsingular square matrix. Indeed, supposing $x = [x_1, \ldots, x_n]^T$ is the column solution vector and left-multiplying the system by

$$A^{-1} = \left[\frac{A_{ki}}{\det A}\right],$$

we get

$$x_j = \frac{1}{\det A} \sum_i A_{ji} b_i = \frac{\det M_j}{\det A}.$$

Here, M_j is the matrix obtained from A by replacing its jth column by b. This method of solving systems of linear equations is called *Cramer's rule*. A consequence of this formula is the next theorem.

Theorem 1.3.3 *A system of n linear equations in n unknowns with a nonsingular matrix of coefficients has a unique solution.*

Theorem 1.3.4 *Let A be a square matrix. If a matrix B satisfies the relation*

$$AB = I, \tag{1.21}$$

where I is the identity matrix, then $B = A^{-1}$.

Remark. Thus, if we want to prove that B is the inverse of A it is sufficient to verify (1.21). It is not necessary to show also that $BA = I$.

Proof. By Theorems 1.2.7, 1.3.1, and by (1.6), A is nonsingular. The first column of B in (1.21) can be taken for the solution of the system

$$Ax = e_1, \quad e_1 = [1, 0, \ldots, 0]^T.$$

But the first column of A^{-1} satisfies the equation as well. By Theorems 1.3.3, the first columns of B and A^{-1} are identical. The same can be proved for the remaining columns, and thus $B = A^{-1}$. ◻

Theorem 1.3.5 *Let A be a square matrix. If the system $Ax = 0$ has a solution in which not all the unknowns are zero, then* $\det A = 0$.

Proof. The system in question always has a (trivial) solution $x = 0$. If A were nonsingular, then Theorem 1.3.3 would imply that the system has no other solution, contrary to the hypothesis of the theorem. ◻

Square matrices whose determinant is zero are called *singular*. Therefore, Theorem 1.3.5 gives a sufficient condition for A to be a singular matrix. Later we shall see that this condition is necessary as well (Theorem 1.5.3).

The following theorem is an immediate consequence of the definition of a singular matrix and Theorem 1.2.7.

Theorem 1.3.6 *If one of the factors of a product of square matrices is singular, then the product is singular as well.*

Note also the following corollary of Theorem 1.2.5.

Theorem 1.3.7 *A block triangular matrix (lower or upper) is nonsingular if and only if all its diagonal blocks are nonsingular. In particular, a (lower or upper) triangular matrix is nonsingular if and only if all its diagonal entries are different from zero.*

We are now in a position to study the properties of the inverse of a nonsingular lower (or upper) triangular matrix.

Theorem 1.3.8 *If $A = [a_{ik}]$ is a nonsingular lower (upper) triangular matrix, then its inverse A^{-1} is lower (upper) triangular as well.*

Proof. In view of Theorem 1.3.4 it is sufficient to show that there is a lower triangular matrix B satisfying

$$AB = I.$$

The entries of such a matrix B, if it exists, fulfil the condition

$$a_{ii}b_{ii} = 1, \qquad i = 1, \ldots, n, \tag{1.22}$$

as well as

$$\sum_{k=j}^{i} a_{ik}b_{kj} = 0, \quad \text{whenever } i > j.$$

Hence,

$$b_{ij} = -a_{ii}^{-1} \sum_{k=j}^{i-1} a_{ik}b_{kj}, \qquad i > j. \tag{1.23}$$

However, from these equations the entries b_{ij} can be calculated successively by using $i - j = 1, 2, \ldots, n - 1$. Indeed, the right-hand side of (1.23) contains only the b_{kj}'s with $k - j$ less than $i - j$.

If B is the lower triangular matrix whose entries are the calculated b_{ij}'s, then it does hold $AB = I$, i.e. $B = A^{-1}$. The corresponding statement for an upper triangular matrix is obtained by taking transposes and employing Theorem 1.1.5. $\qquad \Box$

Remark. Note also the following two important consequences of (1.22) and (1.23):

(1) If $A = [a_{ik}]$ is a lower triangular matrix satisfying $a_{ii} > 0$ for every i and $a_{ij} \leq 0$ for all i, j, $i \neq j$, then all the entries of A^{-1} are nonnegative. (Naturally, the same is true for upper triangular matrices as well.)

(2) If A is a nonsingular lower block triangular matrix, the proof applies without changes, except for writing

$$A_{ii}B_{ii} = I_i \tag{1.24}$$

(I_i are identity matrices) instead of (1.22) and considering a_{ik} and b_{ik} as blocks. Therefore, the theorem also holds for lower (or upper) block triangular matrices. Another corollary is the next theorem.

Theorem 1.3.9 *Lower triangular matrices of order n constitute a subset of square matrices which is closed[5] with respect to addition, multiplication, scalar multiplication and inversion (of nonsingular matrices). The same holds for upper triangular matrices.*

Moreover, carrying out the operations with triangular matrices we find that the kth diagonal entry of the result (k fixed, $1 \leq k \leq n$) is, according to the matrix operation being performed, the sum, product or reciprocal of the kth diagonal entries of the operands and is not affected by other entries.

1.4 Schur complement. Factorization

We now introduce a further concept. Let

$$A = \begin{bmatrix} A_{11} & A_{12} \\ A_{21} & A_{22} \end{bmatrix}$$

be a partitioned matrix and A_{11} its nonsingular block. The *Schur complement* $[A/A_{11}]$ *of the block* A_{11} *in* A is then defined [34] to be the matrix

$$[A/A_{11}] = A_{22} - A_{21}A_{11}^{-1}A_{12}. \tag{1.25}$$

(Here, A need not be square.)

Theorem 1.4.1 *Let $A = [A_{ij}]$, $i, j = 1, \ldots, r$, be a partitioned matrix. Let the submatrices*

$$\begin{bmatrix} A_{11} & A_{12} & \cdots & A_{1k} \\ \vdots & \vdots & \ddots & \vdots \\ A_{k1} & A_{k2} & \cdots & A_{kk} \end{bmatrix}, \qquad k = 1, \ldots, r-1,$$

be nonsingular. Then there exist a lower block triangular matrix B having identity matrices as diagonal blocks, an upper block

---

[5]A set M is closed with respect to an operation if the result of the operation with elements of M always gives an element of M.

triangular matrix C having identity diagonal blocks, and a block diagonal matrix D such that

$$A = BDC. \tag{1.26}$$

Moreover, the matrices B, D, and C are determined uniquely by the conditions of the theorem; the diagonal blocks $D_1, \ldots, D_r$ of D satisfy

$$D_1 = A_{11}, \tag{1.27}$$

$$D_k = \left[\begin{bmatrix} A_{11} & A_{12} & \cdots & A_{1k} \\ \vdots & \vdots & \ddots & \vdots \\ A_{k1} & A_{k2} & \cdots & A_{kk} \end{bmatrix} \Big/ \begin{bmatrix} A_{11} & \cdots & A_{1,k-1} \\ \vdots & \ddots & \vdots \\ A_{k-1,1} & \cdots & A_{k-1,k-1} \end{bmatrix} \right],$$
$$k = 2, \ldots, r, \tag{1.28}$$

and $D_1, \ldots, D_{r-1}$ are nonsingular.

Proof. First we prove the uniqueness assuming that $D_1, \ldots, D_{r-1}$ are nonsingular. Consider two possible factorizations of A, (1.26) and $A = \hat{B}\hat{D}\hat{C}$, that satisfy the conditions of the theorem. Since $\hat{B}$ and C are nonsingular, we may write

$$\hat{B}^{-1}AC^{-1} = \hat{B}^{-1}BD = \hat{D}\hat{C}C^{-1}.$$

By Theorem 1.3.8, the matrix $\hat{B}^{-1}BD$ is lower block triangular, whereas $\hat{D}\hat{C}C^{-1}$ is upper block triangular. Hence both matrices are block diagonal, the former being equal to D, the latter to $\hat{D}$ (the diagonal blocks of both $\hat{B}^{-1}B$ and $\hat{C}C^{-1}$ are identity matrices). Thus, $D = \hat{D}$ and writing $\hat{B}^{-1}B = [Y_{ik}]$, i.e. $Y_{jj} = I_j$, $Y_{ik} = 0$ for $i < k$, and further $\hat{C}C^{-1} = [Z_{ik}]$, i.e. $Z_{jj} = I_j$, $Z_{ik} = 0$ for $i > k$, we have

$$Y_{ik}D_k = 0 \qquad \text{whenever } i > k,$$
$$D_iZ_{ik} = 0 \qquad \text{whenever } i < k.$$

Since D_i, $i = 1, \ldots, r-1$ are nonsingular, we get $Y_{ik} = 0$ whenever $i > k$, $Z_{ik} = 0$ whenever $i < k$. Hence $\hat{B}^{-1}B = I = \hat{C}C^{-1}$, and thus $B = \hat{B}$, $C = \hat{C}$.

The existence of the factorization (1.26), and the validity of (1.27) and (1.28) will be proved by induction on r. For $r = 1$, both (1.26) and (1.27) hold. Let A have $r > 1$ block rows and assume the existence of (1.26) for matrices having $r - 1$ block rows. The matrix $\tilde{A}$ obtained from A by deleting its last block row and the last block column, has $r - 1$ block rows and satisfies the nonsingularity assumption of the theorem. By the induction hypothesis, there exist a lower block triangular matrix $\tilde{B}$ with identity diagonal blocks, an upper block triangular matrix $\tilde{C}$ with identity diagonal blocks, and a block diagonal magtrix $\tilde{D}$ with the diagonal blocks $D_1, \ldots, D_{r-1}$ given by (1.27), (1.28), such that $\tilde{A} = \tilde{B}\tilde{D}\tilde{C}$.

Now write A in the partitioned form

$$A = \begin{bmatrix} \tilde{A} & \tilde{A}_{1r} \\ \tilde{A}_{r1} & A_{rr} \end{bmatrix}$$

and try to find B, C, and D,

$$B = \begin{bmatrix} \tilde{B} & 0 \\ B_{r1} & I_{r_1} \end{bmatrix}, \quad C = \begin{bmatrix} \tilde{C} & C_{1r} \\ 0 & I_{r_2} \end{bmatrix}, \quad D = \begin{bmatrix} \tilde{D} & 0 \\ 0 & D_r \end{bmatrix}$$

such that $A = BDC$. This requirement gives

$$\begin{bmatrix} \tilde{A} & \tilde{A}_{1r} \\ \tilde{A}_{r1} & A_{rr} \end{bmatrix} = \begin{bmatrix} \tilde{B} & 0 \\ B_{r1} & I_{r_1} \end{bmatrix} \begin{bmatrix} \tilde{D} & 0 \\ 0 & D_r \end{bmatrix} \begin{bmatrix} \tilde{C} & C_{1r} \\ 0 & I_{r_2} \end{bmatrix}.$$

The right-hand side is

$$\begin{bmatrix} \tilde{B}\tilde{D}\tilde{C} & \tilde{B}\tilde{D}C_{1r} \\ B_{r1}\tilde{D}\tilde{C} & B_{r1}\tilde{D}C_{1r} + D_r \end{bmatrix}$$

and it is therefore sufficient to satisfy the equations

$$\tilde{B}\tilde{D}C_{1r} = \tilde{A}_{1r},$$
$$B_{r1}\tilde{D}\tilde{C} = \tilde{A}_{r1},$$
$$B_{r1}\tilde{D}C_{1r} + D_r = A_{rr}.$$

Since $\tilde{A}$ is nonsingular, the matrices $\tilde{B}\tilde{D}$ and $\tilde{C}$ are nonsingular as well and the first two equations may be solved for C_{1r} and B_{r1}. We thus obtain

$$C_{1r} = \tilde{D}^{-1}\tilde{B}^{-1}\tilde{A}_{1r}, \quad B_{r1} = \tilde{A}_{r1}\tilde{C}^{-1}\tilde{D}^{-1}.$$

The last of the equations is then satisfied by

$$D_r = A_{rr} - \tilde{A}_{r1}\tilde{C}^{-1}\tilde{D}^{-1}\tilde{D}\tilde{D}^{-1}\tilde{B}^{-1}\tilde{A}_{1r} = A_{rr} - \tilde{A}_{r1}\tilde{A}^{-1}\tilde{A}_{1r},$$

i.e., by $D_r = [A/\tilde{A}]$. Therefore, the factorization (1.26) exists and (1.28) holds also for $k = r$.

At the same time we showed that $D_1, \ldots, D_{r-1}$ are nonsingular, which is necessary for the uniqueness proof. $\quad\square$

The theorem has several corollaries.

Theorem 1.4.2 *Let A be a square matrix which is partitioned in the form*

$$A = \begin{bmatrix} A_{11} & A_{12} \\ A_{21} & A_{22} \end{bmatrix},$$

where A_{11} is nonsingular. Then

$$\det[A/A_{11}] = \det A/\det A_{11}. \qquad (1.29)$$

Hence A is nonsingular if and only if the Schur complement $[A/A_{11}]$ of the submatrix A_{11} in A is nonsingular. In such case, the inverse of the Schur complement $[A/A_{11}]$ is the submatrix of A^{-1} in the position complementary to that of A_{11}.

In other words, if A^{-1} is partitioned as was A above,

$$A^{-1} = \begin{bmatrix} B_{11} & B_{12} \\ B_{21} & B_{22} \end{bmatrix},$$

then

$$[A/A_{11}] = B_{22}^{-1}. \qquad (1.30)$$

Proof. Using the formulae (1.27) and (1.28) of Theorem 1.4.1 with $r = 2$, we get

$$D_1 = A_{11}, \qquad D_2 = [A/A_{11}].$$

But $\det A = \det B \det D \det C$ by (1.26) and (1.17), where $\det B = \det C = 1$, $\det D = \det D_1 \det D_2$ because of (1.15). Altogether, we have $\det A = \det A_{11} \det[A/A_{11}]$, which proves (1.29).

Now suppose that A, and thus also $[A/A_{11}] = D_2$, is nonsingular. The factorization (1.26) implies $A^{-1} = C^{-1}D^{-1}B^{-1}$. But B^{-1}, C^{-1}, and D^{-1} are of the form

$$B^{-1} = \begin{bmatrix} I_1 & 0 \\ \tilde{B}_{21} & I_2 \end{bmatrix}, C^{-1} = \begin{bmatrix} I_1 & \tilde{C}_{12} \\ 0 & I_2 \end{bmatrix}, D^{-1} = \begin{bmatrix} D_1^{-1} & 0 \\ 0 & D_2^{-1} \end{bmatrix},$$

so that

$$A^{-1} = \begin{bmatrix} I_1 & \tilde{C}_{12} \\ 0 & I_2 \end{bmatrix} \begin{bmatrix} D_1^{-1} & 0 \\ 0 & D_2^{-1} \end{bmatrix} \begin{bmatrix} I_1 & 0 \\ \tilde{B}_{21} & I_2 \end{bmatrix}.$$

This implies $B_{22} = D_2^{-1}$, which is (1.30).   □

If a square matrix $A = [a_{ik}]$ of order n has all principal minors $\det A(N_k)$, $N_k = \{1, 2, \ldots, k\}$, $k = 1, \ldots, n$, different from zero, we call it *strongly nonsingular*. An immediate corollary of Theorem 1.4.1 is the next statement on strongly nonsingular matrices.

Theorem 1.4.3 *If $A = [a_{ik}]$ is a strongly nonsingular matrix of order n, then there is a unique factorization $A = BDC$, where B and C are lower and upper triangular matrices, respectively, with ones along the main diagonal, and D is a diagonal matrix with diagonal entries $d_1, \ldots, d_n$. Moreover, we have $d_1 = a_{11}$, $d_k = \det A(N_k)/\det A(N_{k-1})$, $k = 2, \ldots, n$.*

We shall now show another property of the Schur complement.

Theorem 1.4.4 [8] *Let*

$$A = \begin{bmatrix} A_{11} & A_{12} & A_{13} \\ A_{21} & A_{22} & A_{23} \\ A_{31} & A_{32} & A_{33} \end{bmatrix}$$

be a partitioned matrix whose submatrices A_{11} and

$$\tilde{A} = \begin{bmatrix} A_{11} & A_{12} \\ A_{21} & A_{22} \end{bmatrix}$$

are both nonsingular. Then

$$[A/\tilde{A}] = [[A/A_{11}]/[\tilde{A}/A_{11}]]. \tag{1.31}$$

Proof. By Theorem 1.4.1, there exists a unique factorization $A = BDC$, where B is a lower block triangular matrix having three block rows and identity diagonal blocks, C is upper block triangular with analogous structure, and D is a block diagonal matrix with diagonal blocks D_1, D_2, D_3, where

$$D_3 = [A/\tilde{A}]. \tag{1.32}$$

On the other hand, A may be partitioned as

$$\begin{bmatrix} A_{11} & \tilde{A}_{13} \\ \tilde{A}_{31} & \tilde{A}_{33} \end{bmatrix}$$

and factored into factors with two block rows. According to (1.26) we thus obtain $A = \tilde{B}\tilde{D}\tilde{C}$, where

$$\tilde{D} = \begin{bmatrix} D_1 & 0 \\ 0 & \tilde{D}_2 \end{bmatrix}, \qquad \tilde{D}_2 = [A/A_{11}].$$

Next, $\tilde{D}_2$ may be factored according to (1.26),

$$\tilde{D}_2 = \hat{B} \begin{bmatrix} \hat{D}_2 & 0 \\ 0 & \hat{D}_3 \end{bmatrix} \hat{C},$$

so that

$$A = \tilde{B} \begin{bmatrix} I_1 & 0 \\ 0 & \hat{B} \end{bmatrix} \begin{bmatrix} D_1 & 0 & 0 \\ 0 & \hat{D}_2 & 0 \\ 0 & 0 & \hat{D}_3 \end{bmatrix} \begin{bmatrix} I_1 & 0 \\ 0 & \hat{C} \end{bmatrix} \tilde{C}. \tag{1.33}$$

It is easy to verify that the matrix

$$\tilde{B} \begin{bmatrix} I_1 & 0 \\ 0 & \hat{B} \end{bmatrix}$$

written as a partitioned matrix with three block rows, has identity diagonal blocks and this is true of the matrix

$$\begin{bmatrix} I_1 & 0 \\ 0 & \hat{C} \end{bmatrix} \tilde{C}$$

as well. Hence (1.33) is only another way of writing the factorization $A = BDC$, and thus $\hat{D}_3 = D_3$ of (1.32). Using now (1.28) we obtain

$$D_2 = [\tilde{A}/A_{11}], \quad \hat{D}_3 = [\tilde{D}_2/\hat{D}_2] = [\tilde{D}_2/D_2] = [[A/A_{11}]/[\tilde{A}/A_{11}]]$$

and (1.31) follows. $\qquad\qquad\qquad\qquad\qquad\qquad\qquad\qquad\qquad\square$

1.5   Vector spaces. Rank

We now give a brief survey of the properties of linear vector spaces. A *real linear vector space* L is a set of elements, called *vectors*, for which the *addition of vectors* and *multiplication of a vector by a real number* (a number of R) is defined. In addition, the following rules are required to hold:

VS1: $u + v = v + u$ for all $u, v \in L$;

VS2: $(u + v) + w = u + (v + w)$ for all $u, v, w \in L$;

VS3: there exists a vector 0 (the *zero vector*) such that $u + 0 = u$ for all $u \in L$;

VS4: for each $u \in L$ there exists a vector $-u$ (the *negative* of u) in L satisfying $u + (-u) = 0$;

VS5: $\alpha(u + v) = \alpha u + \alpha v$ for all $u, v \in L$ and all $\alpha \in R$;

VS6: $(\alpha + \beta)u = \alpha u + \beta u$ for all $u \in L$ and all $\alpha, \beta \in R$;

VS7: $(\alpha\beta)u = \alpha(\beta u)$ for all $\alpha, \beta \in R$ and all $u \in L$;

VS8: $-u = (-1)u$ for all $u \in L$.

We sometimes call L a *vector space* or a *linear space* as well. The multiplication of vectors by numbers (scalars) is often called *scalar multiplication*.

The definition of a *complex linear vector space* is analogous (the multiplication of vectors by complex numbers is defined instead of that by real numbers).

In our considerations, the most important example of a linear vector space will be that of the linear space of (real or complex) column vectors. The *n-dimensional linear space of real column vectors* is the set of all real n-component column vectors, where the operation of addition corresponds to that previously defined for $n \times 1$ matrices and the multiplication by a real number is defined as the scalar multiplication of $n \times 1$ matrices. Analogously, the n-dimensional linear space of complex column vectors is defined.

Consider now the abstract vector space again. We introduce the concept of the linear dependence of finite sets of vectors in the following manner. The vectors $u_1, u_2, \ldots, u_s$ are *linearly dependent* if there exist numbers $\alpha_1, \alpha_2, \ldots, \alpha_s$ (real in case the space is real, complex in case the space is complex), not all zero, such that

$$\alpha_1 u_1 + \alpha_2 u_2 + \cdots + \alpha_s u_s = 0.$$

Otherwise, the set of vectors $u_1, \ldots, u_s$ is said to be *linearly independent*.

Let $u_1, \ldots, u_s$ be a set of vectors in a linear vector space and let v be a vector in this space. We say that the vector v is *linearly dependent* on the vectors $u_1, \ldots, u_s$ if there exist numbers $\alpha_1, \ldots, \alpha_s$ such that $v = \alpha_1 u_1 + \cdots + \alpha_s u_s$.

A linear vector space is *finite-dimensional* if there exists a positive integer m such that any set of more than m vectors is linearly dependent. The *dimension* of the space is then that integer n for which there exists a set of n linearly independent vectors, whereas any set of more than n vectors is linearly dependent. Such a system of n linearly independent vectors is called a *basis* of the space.

The n-dimensional linear space of column vectors (real or complex) is of dimension n in fact. One of its bases is constituted by the vectors $e_1 = [1, 0, \ldots, 0]^T$, $e_2 = [0, 1, 0, \ldots, 0]^T$, $\ldots$, $e_n = [0, 0, \ldots, 0, 1]^T$.

If L is a vector space and L_1 some of its subsets, it may happen that L_1 is itself a vector space with respect to the operations of addition and scalar multiplication in L. In that case L_1 is called a *subspace* of L. If $L_1, L_2, \ldots, L_k$ are subspaces of L, then their intersection $L_1 \cap L_2 \cap \cdots \cap L_k$ is itself a subspace of L.

If M is a set of vectors in a finite-dimensional vector space L, then the subspace having the least dimension of all the subspaces containing M is called the *linear hull* of M. Its dimension is called the *rank* of M.

We state without proof[6] an important theorem of the theory of matrices.

---

[6]See, for example, [5].

Theorem 1.5.1 *Let A be an $m \times n$ matrix. Then the rank r of the set of all its m row vectors is the same as the rank of the set of all its n column vectors. Moreover, this rank r is equal to the maximal order of all nonsingular submatrices of A (if A is a zero matrix, then r equals zero).*

The number r is called the *rank* of the matrix A and will be denoted by rank A.

The next statement is evident.

Theorem 1.5.2 *For any matrix A, $\operatorname{rank} A = \operatorname{rank} A^T$.*

As a consequence of Theorem 1.5.1, we also obtain the completion of Theorem 1.3.5.

Theorem 1.5.3 *A square matrix A is singular if and only if there exists a vector $x \neq 0$ such that $Ax = 0$.*

Proof. If there is a vector $x \neq 0$ satisfying $Ax = 0$, then $\det A = 0$ by Theorem 1.3.5.

Conversely, if $\det A = 0$, then by Theorem 1.5.1 the rank of A is at most $n - 1$, where n is the order of A. Therefore, the rank of the set of the column vectors of A is less than n, too, and the columns $A_1, \ldots, A_n$ are thus linearly dependent:

$$c_1 A_1 + c_2 A_2 + \cdots + c_n A_n = 0,$$

where $c_1, \ldots, c_n$ are not all equal to zero. But this may be written as $Ax = 0$, where $x = [c_1, \ldots, c_n]^T \neq 0$. $\square$

We also have the following important result.

Theorem 1.5.4 *If A and B are matrices whose product AB is defined, then*

$$\operatorname{rank} AB \leq \operatorname{rank} A, \qquad \operatorname{rank} AB \leq \operatorname{rank} B. \tag{1.34}$$

If B is nonsingular, then $\operatorname{rank} AB = \operatorname{rank} A$; if A is nonsingular, then $\operatorname{rank} AB = \operatorname{rank} B$.

Proof. The first inequality (1.34) is a consequence of Theorem 1.5.1 since the columns of AB are linear combinations of the columns of A. The second inequality follows similarly using rows.

If B is nonsingular, then in addition to (1.34) we have rank $A = \text{rank}(ABB^{-1}) \leq \text{rank } AB$, i.e., rank $AB = \text{rank } A$. If A is nonsingular, then also rank $B = \text{rank}(A^{-1}AB) \leq \text{rank } AB$, i.e., rank $AB = \text{rank } B$. $\qquad\square$

1.6 Eigenvectors, eigenvalues. Characteristic polynomial

We now introduce an important notion. Let $A = [a_{ik}]$ be a square matrix. An *eigenvector* (or *characteristic vector*) of A is a vector $x \neq 0$ such that $Ax = \lambda x$ for some number λ. This λ is then called an *eigenvalue* (or *characteristic value*, sometimes also *root*) of A *associated* with x.

Theorem 1.6.1 *A number λ is an eigenvalue of the matrix A (i.e., an eigenvalue associated with some eigenvector) if and only if the matrix $A - \lambda I$ is singular; that is, if and only if*

$$\det(A - \lambda I) = 0. \tag{1.35}$$

Proof. This follows directly from Theorem 1.5.3. $\qquad\sqcup$

Theorem 1.6.2 *If A is a square matrix of order n, then $\det(A - xI)$ is a polynomial in x of degree n. Moreover,*

$$\det(A - xI) = (-x)^n + c_1(-x)^{n-1} + \cdots + c_{n-1}(-x) + c_n, \tag{1.36}$$

where the coefficient c_k is the sum of all the principal minors of A of order k,

$$c_k = \sum_{M \subset N, |M|=k} \det A(M, M), \quad N = \{1, 2, \dots, n\}. \tag{1.37}$$

Proof. By (1.12)

$$\det(A + (-x)I) = \sum_{M_1, M_2} \det A(M_1, M_2) \, \text{codet} B(M_1, M_2),$$

where
$$B = (-x)I.$$

But clearly codet $B(M_1, M_2) = 0$ whenever $M_1 \neq M_2$. If $M_1 = M_2$, $|M_1| = k$, then codet $B(M_1, M_2) = \det B(N \setminus M_1, N \setminus M_1) = (-x)^{n-k}$. It follows that

$$\det(A - xI) = \sum_{k=0}^{n} \left[\sum_{|M|=k} \det A(M, M) \right] (-x)^{n-k},$$

so that (1.37) does hold. $\qquad \qquad \square$

Remark. The polynomial (1.36) is called the *characteristic polynomial* of A.

Theorem 1.6.3 *A square matrix A of order n has n eigenvalues (some of them may, however, coincide), namely all the roots of the characteristic polynomial. Denoting them by $\lambda_1, \lambda_2, \ldots, \lambda_n$, we have*

$$\sum_{i=1}^{n} \lambda_i = \sum_{i=1}^{n} a_{ii}, \qquad (1.38)$$

and

$$\lambda_1 \lambda_2 \ldots \lambda_n = \det A. \qquad (1.39)$$

In general,

$$E_k(\lambda_1, \ldots, \lambda_n) = \sum_{|M|=k} \det A(M, M), \qquad (1.40)$$

where $E_k(x_1, \ldots, x_n)$ is the kth elementary symmetric function of n variables, that is,

$$E_k(x_1, \ldots, x_n) = \sum x_1^{\varepsilon_1} x_2^{\varepsilon_2} \ldots x_n^{\varepsilon_n},$$

where the summation is over all $\varepsilon_1 = 0, 1; \varepsilon_2 = 0, 1; \ldots; \varepsilon_n = 0, 1$ satisfying $\sum_{j=1}^{n} \varepsilon_j = k$.

Proof. We have (in the sense of equality for polynomials)

$$\det(A - xI) = \prod_{i=1}^{n} (\lambda_i - x) = \sum_{k=0}^{n} [E_k(\lambda_1, \ldots, \lambda_n)](-x)^{n-k},$$

and comparing this with (1.36) and (1.37) we get (1.40). Formulae (1.38) and (1.39) are particular cases of (1.40) for $k = 1$ and $k = n$, respectively. □

Remark. The number $\sum_{i=1}^{n} a_{ii}$, that is the sum of the diagonal entries of a square matrix $A = [a_{ik}]$ of order n, is called the *trace* of the matrix; it is denoted by tr A. Therefore, by (1.38) the trace of a matrix equals the sum of all its eigenvalues (every eigenvalue taken the number of times equal to its multiplicity). Of course, the eigenvalues need not be real if the matrix is real.

Theorem 1.2.5 and (1.35) imply the next two theorems.

Theorem 1.6.4 *If $A = [a_{ik}]$ is an upper or a lower triangular matrix of order n, then its diagonal entries a_{ii} are just all the eigenvalues of A.*

Theorem 1.6.5 *If $A = [A_{ik}]$ is an upper or a lower block triangular matrix with square diagonal blocks, then the set of its eigenvalues is the union of the sets of eigenvalues of the diagonal blocks.*

Remark. The set of all the eigenvalues of a matrix is called the *spectrum* of the matrix. We shall denote the spectrum by $S(A)$. Hence the preceding theorem may be restated as

$$S(A) = \bigcup_i S(A_{ii}). \tag{1.41}$$

1.7 Similarity. Jordan normal form

In discussing the spectral properties of matrices, i.e. those connected with the eigenvalues, the concept of similarity plays an essential role. We say that a square matrix A is *similar* to the matrix B if there exists a nonsingular matrix T such that

$$A = TBT^{-1}. \tag{1.42}$$

Remark. To formulate it more exactly, we should speak about similarity over the real field or over the complex field, according to whether the matrix T is real or complex, respectively (cf. Theorem 1.7.10).

Theorem 1.7.1 *The relation of similarity possesses the following properties:*

1. *Every square matrix is similar to itself (reflexivity).*

2. *If A is similar to B, then B is similar to A (symmetry).*

3. *If A is similar to B and B is similar to C, then A is similar to C (transitivity).*

Therefore, the (real or complex) similarity is an equivalence relation in the set of (real or complex) square matrices of the same order. This set can thus be partitioned into classes of mutually equivalent (i.e. similar) matrices.

Proof. This follows directly from (1.42). □

Remark. In studying equivalence relations, it is usually useful to pick one element of each equivalence class, which is then called the representative of the respective class. Before doing this for the similarity relation, we establish four theorems.

Theorem 1.7.2 *Similar matrices have the same characteristic polynomial. They have thus the same eigenvalues, and these occur with the same multiplicity.*

If $A = TBT^{-1}$ and if x is an eigenvector of B associated with the eigenvalue λ, then Tx is an eigenvector of A associated with the same eigenvalue.

Proof. The first part of the theorem follows from the fact that $A = TBT^{-1}$ implies

$$\det(A - xI) = \det(T(B - xI)T^{-1})$$
$$= \det T \cdot \det(B - xI) \cdot \det T^{-1}$$
$$= \det(B - xI)$$

according to (1.17).

To prove the second part, observe that $A = TBT^{-1}$ and $Bx = \lambda x$. It is now sufficient to write

$$A(Tx) = (TBT^{-1})(Tx) = TBx = \lambda(Tx).$$

□

The Fundamental Theorem of Algebra[7] stating that every polynomial of degree at least one with complex coefficients has at least one complex root, implies the following theorem.

Theorem 1.7.3 *Every real or complex matrix has at least one complex (possibly real, of course) eigenvalue.*

Theorem 1.7.4 *Similar matrices have the same rank.*

Proof. This is a consequence of Theorem 1.5.4. □

In the next theorem and later as well, we shall need the concept of the substitution of a matrix into a polynomial. If $f(x) = a_0 x^m + a_1 x^{m-1} + \cdots + a_{m-1} x + a_m$ is a polynomial, then the matrix $f(A) = a_0 A^m + a_1 A^{m-1} + \cdots + a_{m-1} A + a_m I$ is called a *matrix polynomial*.

Theorem 1.7.5 *If A and B are similar matrices, and $f(x)$ is a polynomial, then the matrices $f(A)$ and $f(B)$ are similar as well. In fact, if $A = TBT^{-1}$, then $f(A) = Tf(B)T^{-1}$.*

Proof. If $A = TBT^{-1}$, then $A^k = TB^k T^{-1}$ for $k = 0, 1, \ldots$. Hence $f(A) = Tf(B)T^{-1}$, which proves the theorem. □

We saw (Theorem 1.7.2) that similar matrices have the same characteristic polynomial. However, the converse is not true if the order of the matrices is at least two. Really, the matrices

$$\begin{bmatrix} 1 & 0 \\ 0 & 1 \end{bmatrix}, \quad \begin{bmatrix} 1 & 1 \\ 0 & 1 \end{bmatrix}$$

have the same characteristic polynomial $(1 - \lambda)^2$. But the first (identity) matrix is similar only to itself and it forms one similarity class, whereas the second matrix does not belong to this class. In order to find an appropriate representative of a class of mutually similar matrices and to answer in this way the question of characterizing pairs of matrices that are similar to each other, the concept of a Jordan block and the concept of a matrix in Jordan normal form are useful. Let σ be a complex number and

---

[7]See, for example, [5].

let m be a positive integer. By the *Jordan block* $J_m(\sigma)$ is then meant the square matrix

$$J_m(\sigma) = \begin{bmatrix} \sigma & 1 & 0 & \ldots & 0 & 0 \\ 0 & \sigma & 1 & \ldots & 0 & 0 \\ \vdots & \vdots & \vdots & \ddots & \vdots & \vdots \\ 0 & 0 & 0 & \ldots & \sigma & 1 \\ 0 & 0 & 0 & \ldots & 0 & \sigma \end{bmatrix} \tag{1.43}$$

of order m.

We say that a matrix is in *Jordan normal form* if it is block diagonal and each of its diagonal blocks is a Jordan block:

$$J = \begin{bmatrix} J_{m_1}(\sigma_1) & 0 & \ldots & 0 \\ 0 & J_{m_2}(\sigma_2) & \ldots & 0 \\ 0 & 0 & \ldots & 0 \\ 0 & 0 & \ldots & J_{m_r}(\sigma_r) \end{bmatrix} \tag{1.44}$$

(the numbers σ_i which are not necessarily distinct, are the eigenvalues of the matrix J).

At this point we state without proof[8] a fundamental theorem which completely solves the problem of representing classes of similar matrices over the complex field.

Theorem 1.7.6 (Jordan) *Any complex or real matrix is similar over the complex field to a matrix in Jordan normal form.*

Theorem 1.7.7 *If A and B are similar matrices and λ is their eigenvalue (an eigenvalue of both A and B by Theorem 1.7.2), then*

$$\operatorname{rank}(A - \lambda I)^k = \operatorname{rank}(B - \lambda I)^k, \quad k = 1, 2, \ldots. \tag{1.45}$$

Conversely, if

$$\det(A - \lambda I) = \det(B - \lambda I) \tag{1.46}$$

for any complex λ, and if (1.45) holds for all the eigenvalues λ of A and $k = 1, \ldots, n(\lambda)$, where $n(\lambda)$ is the multiplicity of λ as an eigenvalue of A, then the matrices A and B are similar.

---

[8]See, for example, [5].

Proof. If A and B are similar, then $(A - \lambda I)^k$ and $(B - \lambda I)^k$ are similar by Theorem 1.7.5, and thus their ranks are equal by Theorem 1.7.4. Conversely, let A and B satisfy (1.46) and (1.45) with $k = 1, \ldots, n(\lambda)$. By the Jordan Theorem 1.7.6, the matrix A is similar to a matrix A_J, whereas B is similar to B_J, both A_J and B_J being in Jordan normal form. For all the eigenvalues λ of A_J we thus have

$$\operatorname{rank}(A_J - \lambda I)^k = \operatorname{rank}(B_J - \lambda I)^k, \quad k = 1, \ldots, n(\lambda), \quad (1.47)$$

where $n(\lambda)$ is the multiplicity of λ as an eigenvalue of A_J, and the matrix B_J has Jordan blocks with the same numbers λ as A_J. The dimensions of the blocks belonging to λ in the matrix A_J may be determined in the following way. If n is the order of A, there are $n - \operatorname{rank}(A - \lambda I)$ such blocks since for each Jordan block belonging to λ the difference of the order of the block and the rank of the block of $A_J - \lambda I$ equals 1, whereas for the blocks which do not belong to λ this difference is zero. The number of blocks of individual orders is determined from the fact that $\operatorname{rank}(A - \lambda I) - \operatorname{rank}(A - \lambda I)^2$ is the number of blocks whose order is at least 2 (see Exercise 1.3), the number of blocks of order at least 3 equals $\operatorname{rank}(A - \lambda I)^2 - \operatorname{rank}(A - \lambda I)^3$, etc. From this and (1.47) we thus find that the matrices A_J and B_J have the same Jordan normal form, possibly except for the order in which the Jordan blocks are arranged along the diagonal. The rest of the proof follows from an auxiliary statement, which will be of use also later and, therefore, is formulated as a theorem.   □

Theorem 1.7.8 *Let D be a block diagonal (or diagonal) matrix (we write $D = \operatorname{diag}(D_1, D_2, \ldots, D_m)$), and let $(k_1, k_2, \ldots, k_m)$ be a permutation of the integers $1, 2, \ldots, m$. Then the matrix $\tilde{D} = \operatorname{diag}(D_{k_1}, D_{k_2}, \ldots, D_{k_m})$ satisfies*

$$\tilde{D} = P^T D P = P^{-1} D P,$$

where P is the permutation matrix corresponding to the permutation $(k_1, k_2, \ldots, k_m)$. It is a partitioned matrix with m block rows and m block columns. The dimensions of the blocks of rows correspond to the orders of $D_1, \ldots, D_m$. The first column of P contains the identity matrix in the k_1th row and zero matrices

*elsewhere; the second column of P contains the identity matrix
in the k_2th row, zero matrices elsewhere, etc. Finally, the mth
column has the identity matrix in the k_mth row and zero matrices
elsewhere.*

Proof. By the definition, the matrix DP is the block matrix
whose first block column contains D_{k_1} in the k_1th row, the second
block column contains D_{k_2} in the k_2th row, and, finally, the mth
column contains D_{k_m} in the k_mth row. Left-multiplication by P^T
transforms D_{k_1} into the diagonal position in the first column, D_{k_2}
into the diagonal position in the second column, and, finally, D_{k_m}
is transformed into the diagonal position in the mth column. We
thus obtain $\tilde{D}$ as required.

To prove $P^T = P^{-1}$ it is sufficient to calculate $P^T P$, which
gives the identity matrix.   □

Remark. A *permutation matrix* is a square matrix which in each
row and in each column has exactly one nonzero entry, always
equal to 1. Theorem 1.7.8 shows that (if all the blocks $D_1, \ldots, D_m$
consist of only one entry) $P^T AP$ is the matrix whose columns
and rows have been permuted by the permutation P. Such a
transformation will be called a *simultaneous permutation* of the
rows and columns of A. The diagonal entries of the original
matrix remain then on the diagonal but they may be moved to
another diagonal position.

We now state a theorem which follows from Theorems 1.7.7
and 1.7.8 and supplements Theorem 1.7.6.

Theorem 1.7.9 *Two matrices in Jordan normal form are similar if and only if each is obtained from the other by a permutation
of the diagonal blocks.*

The Jordan theorem holds, as we know, over the complex
field. For real matrices we have the following statement.

Theorem 1.7.10 *Let the matrices A and B be real and similar
over the complex field. Then they are similar over the real field
as well.*

Proof. Suppose $B = TAT^{-1}$, where $T = P + iQ$ is a complex
nonsingular matrix and P, Q are real. Then $BT = TA$, and thus

$BP = PA$, $BQ = QB$. Since $\det(P + iQ) \neq 0$, the polynomial $\det(P + xQ)$ (in x) is not identically zero. Such a polynomial has only a finite number of roots and, therefore, we can find a real number x_0 such that $\det(P + x_0 Q) \neq 0$. Setting $T_0 = P + x_0 Q$ we then have $B T_0 = T_0 A$, and finally $B = T_0 A T_0^{-1}$. $\qquad \square$

We shall now discuss several consequences of the Jordan theorem. First of all, we describe the powers of the Jordan block $J_m(\sigma)$.

Theorem 1.7.11 *If k is a positive integer, then*

$$(J_m(\sigma))^k = \begin{bmatrix} \sigma^k & \binom{k}{1}\sigma^{k-1} & \binom{k}{2}\sigma^{k-2} & \cdots & \binom{k}{m-1}\sigma^{k-m+1} \\ 0 & \sigma^k & \binom{k}{1}\sigma^{k-1} & \cdots & \binom{k}{m-2}\sigma^{k-m+2} \\ \vdots & \vdots & \vdots & \ddots & \vdots \\ 0 & 0 & 0 & \cdots & \sigma^k \end{bmatrix}.$$

(1.48)

If $|\sigma| < 1$, then $\lim_{k \to \infty} (J_m(\sigma))^k = 0$.

Proof. The first statement follows by induction on k employing the relation

$$\binom{k}{j+1} + \binom{k}{j} = \binom{k+1}{j+1}.$$

The second statement follows from the relation

$$\lim_{k \to \infty} \binom{k}{j} \sigma^k = 0,$$

which holds for any fixed j and $|\sigma| < 1$. This is a consequence of the fact that the exponential function t^k grows for $t = |\sigma|^{-1} > 1$ more rapidly than $\binom{k}{j}$, which is (with j fixed) a polynomial in k of degree j. $\qquad \square$

This result leads to the following statement.

Theorem 1.7.12 *Let A be a square matrix of order n and let $f(x)$ be a polynomial. If $\lambda_1, \lambda_2, \ldots, \lambda_n$ are the eigenvalues of A, then the eigenvalues of the matrix $f(A)$ are exactly the numbers $f(\lambda_1), f(\lambda_2), \ldots, f(\lambda_n)$. If, in addition, $g(x)$ is a polynomial such that $g(\lambda_i) \neq 0$, $i = 1, \ldots, n$, then $g(A)$ is a nonsingular* *matrix and the eigenvalues of $f(A)(g(A))^{-1}$ are exactly the numbers $f(\lambda_1)/g(\lambda_1), f(\lambda_2)/g(\lambda_2), \ldots, f(\lambda_n)/g(\lambda_n)$. Furthermore, if x is an eigenvector of A associated with an eigenvalue λ, then x is also an eigenvector of $f(A)$ and $f(A)(g(A))^{-1}$ associated with the eigenvalues $f(\lambda)$ and $f(\lambda)/g(\lambda)$, respectively.*

Proof. Let $TAT^{-1} = A_J$, where A_J is in Jordan normal form. By Theorem 1.7.5, $Tf(A)T^{-1} = f(A_J)$, where $f(A_J)$ is an upper triangular matrix with diagonal entries $f(\lambda_1), f(\lambda_2), \ldots, f(\lambda_n)$ by Theorem 1.3.9. Using now Theorem 1.6.4 we find that these entries are exactly all the eigenvalues of $f(A_J)$, and thus those of $f(A)$ as well.

If a polynomial $g(x)$ satisfies $g(\lambda_i) \neq 0$, $i = 1, \ldots, n$, then $g(A)$ is nonsingular as it has no zero eigenvalue. Moreover, $g(A_J)$ is again an upper triangular matrix with diagonal entries $g(\lambda_i)$ and, therefore, using Theorems 1.3.9 and 1.6.4 we find that the eigenvalues of the matrix $f(A)(g(A))^{-1}$ are $f(\lambda_i)/g(\lambda_i)$.

Finally, if $Ax = \lambda x$, then also $A^k x = \lambda^k x$, $k = 0, 1, \ldots$, hence $f(A)x = f(\lambda)x$ and, moreover, $f(A)(g(A))^{-1}x = (f(\lambda)/g(\lambda))x$. $\square$

In order to formulate the next theorem, which is due to Oldenburger [48] and plays a very important part in the study of iterative methods and nonnegative matrices (Chapter 11, and Chapters 4 and 5), we introduce the concept of the *spectral radius*. The spectral radius $\rho(A)$ of a square matrix A is defined to be

$$\rho(A) = \max\{|\lambda|; \lambda \in \mathcal{S}(A)\}.$$

Therefore, $\rho(A)$ is the radius of the smallest disk in the complex plane, with centre at the origin, which contains all the eigenvalues of the matrix A inside or on its boundary.

Theorem 1.7.13 (Oldenburger) *If A is a square matrix, then $\lim_{k \to \infty} A^k = 0$ if and only if $\rho(A) < 1$.*

Proof. By Theorem 1.7.6, there exists a nonsingular complex matrix T such that $TAT^{-1} = A_J$, where A_J is a matrix in Jordan normal form. Furthermore, $TA^k T^{-1} = A_J^k$.

Thus if $\lim_{k \to \infty} A^k = 0$, then also $\lim_{k \to \infty} A_J^k = 0$, and hence $\lim_{k \to \infty} \lambda^k = 0$ for any $\lambda \in \mathcal{S}(A)$ since λ^k is a diagonal entry of A_j^k by (1.48) and Theorem 1.3.9. Therefore, $|\lambda| < 1$ for any $\lambda \in \mathcal{S}(A)$, and thus $\rho(A) < 1$. Conversely, if $\rho(A) < 1$, then $|\lambda| < 1$ for any $\lambda \in \mathcal{S}(A)$, and thus $\lim_{k \to \infty}(J_m(\lambda))^k = 0$ by Theorem 1.7.11. Hence $\lim_{k \to \infty} A_j^k = 0$, and thus $\lim_{k \to \infty} A^k = 0$ as well. $\square$

The following theorem is a corollary of Theorem 1.7.13.

Theorem 1.7.14 *If a square matrix A satisfies the condition $\rho(A) < 1$, then the geometric series $I + A + A^2 + \dots$ converges and its sum is $(I - A)^{-1}$.*

Proof. It follows from $\rho(A) < 1$ that $I - A$ is nonsingular. The relation

$$I + A + A^2 + \dots + A^k = (I - A^{k+1})(I - A)^{-1}$$

and the fact that $\lim_{k \to \infty} A^k = 0$ by Theorem 1.7.13 then imply the required statement. $\square$

The last theorem to be proved in this chapter is *Schur's lemma* providing the possibility to find whether some given eigenvalue is simple or multiple.

Theorem 1.7.15 *Let A be a square matrix and let λ be any eigenvalue of A. Then λ is simple if and only if both the following conditions are fulfilled:*

1. *There exists only one linearly independent eigenvector of A associated with λ, say u, and thus also only one linearly independent eigenvector of A^T associated with λ, say v (this means that $\mathrm{rank}(A - \lambda I) = n - 1$, where n is the order of A).*

2. *The vectors u and v of 1. satisfy $v^T u \neq 0$.*

Proof. First of all, let us show that the conditions 1. and 2. are invariant under similarity.

Indeed, if $B = PAP^{-1}$ and the matrix A satisfies 1., i.e. $\mathrm{rank}(A - \lambda I) = n - 1$, then $\mathrm{rank}(B - \lambda I) = n - 1$ as well. Further, suppose that 1. is satisfied and let $Au = \lambda u$, $A^T v = \lambda v$. Then the only linearly independent eigenvector of the matrix B corresponding to λ is Pu and the only linearly independent eigenvector of B^T is $(P^T)^{-1}v$ since

$$B(Pu) = PAu = \lambda(Pu).$$

$$B^T((P^T)^{-1}v) = (P^T)^{-1}A^T P^T((P^T)^{-1}v)$$
$$= (P^T)^{-1}A^T v = \lambda((P^T)^{-1}v).$$

Therefore, the vectors $\tilde{u} = Pu$ and $\tilde{v} = (P^T)^{-1}v$ satisfy $\tilde{v}^T\tilde{u} \neq 0$ if and only if $v^T u \neq 0$ as $\tilde{v}^T\tilde{u} = v^T((P^T)^{-1})^T Pu = v^T P^{-1}Pu = v^T u$.

Note also that to a Jordan block $J_m(\sigma)$ there corresponds a unique linearly independent eigenvector, namely the (m-component) vector $u_0 = [1, 0, \ldots, 0]^T$. To the transposed block $J_m^T(\sigma)$ there corresponds a unique linearly independent (m-component) eigenvector $v_0 = [0, 0, \ldots, 0, 1]^T$.

Suppose now that λ is a simple eigenvalue of A. Then λ is a simple eigenvalue of the Jordan normal form A_J of A as well. Therefore, there is only one Jordan block $J_m(\lambda)$ that belongs to λ and its order is $m = 1$. The associated eigenvector u of the matrix is then unique except for a nonzero factor and may be obtained from the eigenvector $u_0 = [1]$ of the block $J_1(\lambda)$ by adding zeros for the other components. Similarly, the eigenvector v of A_J^T associated with λ is obtained from the eigenvector $v_0 = [1]$ of $J_1^T(\lambda)$ by adding zeros for the other components. Since $v_0^T u_0 = 1 \neq 0$ we have $v^T u \neq 0$ as well. Conversely, if a matrix A and its eigenvalue λ satisfy the conditions 1. and 2., the conditions are satisfied by the Jordan normal form A_J as well. The condition 1. implies that A_J has only one Jordan block corresponding to the eigenvalue λ. But the condition 2. gives $m = 1$ since if $m > 1$ then the eigenvectors $u_0 = [1, 0, \ldots, 0]^T$ and $v_0 = [0, 0, \ldots, 1]^T$ of the blocks $A_m(\lambda)$ and $A_m^T(\lambda)$ would satisfy $v_0^T u_0 = 0$ and thus we would have $v^T u = 0$ for the eigenvectors u of A_J and v of A_J^T belonging to λ. Therefore, λ is a simple eigenvalue of A_J and hence also of A. $\qquad\square$

Exercises

1.1. Prove Theorem 1.1.1.

1.2.* If A and B are matrices such that both AB and BA are defined, prove that $\operatorname{tr} AB = \operatorname{tr} BA$.

1.3.* Let A be a square matrix and let λ be some of its eigenvalues. Show that the number of blocks of order at least 2 associated with λ in the Jordan normal form of A is equal to $\mathrm{rank}(A - \lambda I) - \mathrm{rank}(A - \lambda I)^2$.

1.4.* Prove: If ε is a nonzero number, then the Jordan block $J_m(\lambda)$ is similar to the $m \times m$ matrix

$$\begin{bmatrix} \lambda & \varepsilon & & & \\ & \lambda & \varepsilon & & \\ & & \ddots & \ddots & \\ & & & \lambda & \varepsilon \\ & & & & \lambda \end{bmatrix}.$$

1.5.* We say that the matrices A and B *commute* if $AB = BA$. Show that if $f(x)$ and $g(x)$ are polynomials, then the matrices $f(A)$ and $g(A)$ commute.

1.6.* Prove that any linearly independent set of vectors of a finite-dimensional vector space L can be extended to a basis of L.

1.7.* Let L_1 and L_2 be subspaces of a finite-dimensional vector space. Denote by $\langle L_1 \cup L_2 \rangle$ the subspace of L having the least dimension among all subspaces containing both L_1 and L_2.

Prove that the dimensions satisfy

$$\dim L_1 + \dim L_2 = \dim\langle L_1 \cup L_2 \rangle + \dim(L_1 \cap L_2).$$

1.8.* The subspaces L_1 and L_2 of a linear vector space L are said to form a *direct decomposition* of L if for any $x \in L$ there exist unique vectors $x_1 \in L_1$ and $x_2 \in L_2$ such that $x = x_1 + x_2$. (We also say that L is the *direct sum* of L_1 and L_2.) Show that L is directly decomposed into L_1 and L_2 if and only if $\langle L_1 \cup L_2 \rangle = L$ and $L_1 \cap L_2 = \{0\}$.

1.9.* Let A and B be two $m \times n$ matrices. Show that

$$\mathrm{rank}(A + B) \leq \mathrm{rank}\, A + \mathrm{rank}\, B.$$

1.10.* Let A and B be two $m \times n$ matrices. Prove that the following conditions are equivalent:

(i) $\operatorname{rank}(A + B) = \operatorname{rank} A + \operatorname{rank} B$;

(ii) there exist nonsingular matrices P and Q, and matrices A_0, B_0 such that

$$A = P \begin{bmatrix} A_0 & 0 \\ 0 & 0 \end{bmatrix} Q, \qquad B = P \begin{bmatrix} 0 & 0 \\ 0 & B_0 \end{bmatrix} Q,$$

where the matrices on the right-hand sides are conformally partitioned;

(iii) if $L(A), L(B)$ and $L(A + B)$ are the linear spaces of vectors of the form Ax, Bx and $(A + B)x$, respectively, then $L(A + B)$ is the direct sum of the subspaces $L(A)$ and $L(B)$.

Remark. If the matrices satisfy one of the conditions (i), (ii), and (iii) (and thus all of them), we say that $A + B$ is the *quasi-direct sum* of A and B.

1.11.* Let A be a partitioned matrix,

$$A = \begin{bmatrix} A_{11} & A_{12} \\ A_{21} & A_{22} \end{bmatrix},$$

and let A_{11} be a nonsingular square matrix of order r. Prove that $\operatorname{rank} A = r$ if and only if $A_{22} = A_{21} A_{11}^{-1} A_{12}$.

1.12.* Let A and A_{11} be the matrices from Exercise 1.11. Show that A is the quasi-direct sum of the matrices B and C, where

$$B = \begin{bmatrix} A_{11} & A_{12} \\ A_{21} & A_{21} A_{11}^{-1} A_{12} \end{bmatrix}, C = \begin{bmatrix} 0 & 0 \\ 0 & A_{22} - A_{21} A_{11}^{-1} A_{12} \end{bmatrix}.$$

1.13.* Show that any $m \times n$ matrix of rank r can be expressed in the form UV where U is an $m \times r$ matrix whose columns are linearly independent, and V is an $r \times n$ matrix whose rows are linearly independent.

1.14.* Prove that all the entries of a matrix of rank one having no zero-row and no zero-column, are nonzero.

1.15.* Suppose that the matrix A can be expressed in the form $A = BDC$, where B is a lower triangular matrix, C is an upper triangular matrix, both with ones along the main diagonal, and D is a diagonal matrix with nonzero diagonal entries. Prove that A is then strongly nonsingular.

1.16.* Let $A = [a_{ik}]$ be a square matris of order n, and let rank $A = n-1$. Show that rank $B = 1$, where $B = [A_{ik}]$ is the matrix whose entries are the cofactors A_{ik} of corresponding entries a_{ik}. (*Hint*: Prove that if $Au = 0$, $u \neq 0$, and $A^T v = 0$, $v \neq 0$, then $B = \rho u v^T$, where ρ is a nonzero number.)

1.17.* Let A be a square matrix. Prove that the derivative $f'(\lambda)$ of the characteristic polynomial $f(\lambda) = \det(A - \lambda I)$ is $f'(\lambda) = -\operatorname{tr} B$, where B is the matrix whose entries are the cofactors of the corresponding entries in $A - \lambda I$.

1.18. Use the results of Exercises 1.16 and 1.17 to give another proof of Theorem 1.7.15.

2

Symmetric Matrices. Positive Definite and Semidefinite Matrices

In this chapter we study properties of a broad and very important class of symmetric matrices. It is advantageous to employ geometrical approach and geometrical terminology. We have explained in Chapter 1 the concept of a linear vector space. In the present chapter, the principal part is played by the Euclidean or unitary space, which, as we shall see, is a linear space where a real or complex inner product $\langle x, y \rangle$ is defined for any pair of vectors x and y. For the sake of simplicity, we shall confine ourselves to linear spaces of column vectors.

2.1 Euclidean and unitary spaces

The *Euclidean space of dimension* n is a linear space of real n-component column vectors with the inner product $\langle x, y \rangle$ of two vectors $x = [x_i]$ and $y = [y_i]$ given as

$$\langle x, y \rangle = \sum_{i=1}^{n} x_i y_i.$$

The *unitary space of dimension* n is a linear space of complex n-component column vectors with the inner product $\langle x, y \rangle$ of two vectors $x = [x_i]$ and $y = [y_i]$ given as

$$\langle x, y \rangle = \sum_{i=1}^{n} x_i \bar{y}_i,$$

41

where here (and in the sequel) the bar denotes the complex conjugate number.

We denote such a unitary or Euclidean space by X_n.

Theorem 2.1.1 *The inner product defined above satisfies (in both cases) the following rules that are part of the general definition:*

1. $\langle y, x \rangle = \overline{\langle x, y \rangle}$ *for all* $x, y \in X_n$.

2. $\langle x_1 + x_2, y \rangle = \langle x_1, y \rangle + \langle x_2, y \rangle$ *for all* $x_1, x_2, y \in X_n$.

3. $\langle \alpha x, y \rangle = \alpha \langle x, y \rangle$ *for all* $x, y \in X_n$ *and all (complex or real) numbers* α.

4. $\langle x, x \rangle$ *is real and* $\langle x, x \rangle \geq 0$ *for all* $x \in X_n$ *where the equality takes place only for* $x = 0$.

Proof. The properties 1. to 3. are clear from the definition of inner product. The property 4. follows from the relation

$$\langle x, x \rangle = \sum_{i=1}^{n} x_i \bar{x}_i = \sum_{i=1}^{n} |x_i|^2$$

which holds also in the case of a real space.   □

Remark. The properties 3. and 1. imply that

3′. $\langle x, \beta y \rangle = \bar{\beta} \langle x, y \rangle$ for all $x, y \in X_n$ and all complex numbers β.

The property 4. of Theorem 2.1.1 allows to introduce the *length of a vector* x as the nonnegative number $\sqrt{\langle x, x \rangle}$. Further, vectors x and y are said to be *orthogonal* if $\langle x, y \rangle = 0$. A set of vectors $u_1, u_2, \ldots, u_k$ is said to be *orthonormal* if

$$\langle u_i, u_j \rangle = \delta_{ij} = \begin{cases} 1 & \text{for} \quad i = 1, \\ 0 & \text{for} \quad i \neq j; \ i, j = 1, \ldots, k. \end{cases}$$

An orthonormal set is always linearly independent (Exercise 2.1). If it has n elements, where n is the dimension, then it is called *an orthonormal basis.*

An example of an orthonormal basis is the set

$$e_1 = [1, 0, \ldots, 0]^T, \quad e_2 = [0, 1, 0, \ldots, 0]^T, \quad \ldots, \quad e_n = [0, 0, 0, \ldots, 1]^T.$$

We define now the operation $*$:

Let $A = [a_{ik}]$ be a complex (or real) $m \times n$ matrix. The matrix A^* with entries $\bar{a}_{ki}$ is then called *the conjugate transpose* of A. Thus $A^* = \bar{A}^T$ (i.e., A^* is the transpose of the complex conjugate of A).

Remark. If the matrix A is real, then clearly $A^* = A$.

Theorem 2.1.2 *Let A and B be matrices. Then the following statements hold:*

1. *If the product of A and B exists, then $(AB)^* = B^*A^*$.*

2. *$(A^*)^* = A$.*

3. *If A is nonsingular, then A^* is nonsingular as well and*
$$(A^{-1})^* = (A^*)^{-1}.$$

Proof. We have $(AB)^* = \overline{(AB)^T} = \overline{B^T A^T} = \bar{B}^T \bar{A}^T = B^*A^*$, which is 1. Further, $(A^*)^* = \overline{(\bar{A}^T)^T} = (A^T)^T = A$, which proves 2. To prove 3., it is sufficient to verify that $(A^{-1})^*$ is an inverse of A^* since the inverse is determined uniquely. But this statement is true as $(A^{-1})^* A^* = (AA^{-1})^* = I^* = I$ by 1. $\qquad\square$

The basic property of the operation $*$ for square matrices is described in the next theorem.

Theorem 2.1.3 *We have*

$$\langle Ax, y \rangle = \langle x, A^*y \rangle$$

for any two vectors x and y, and any matrix A. If $x = [x_i]$, $y = [y_i]$, and $A = [a_{ik}]$, then both the sides are equal to $\sum_{i,k} a_{ik}\bar{y}_i x_k$.

Proof. Let the matrix $A = [a_{ik}]$ be of order n. Set $Ax = z = [z_i]$ so that $z_i = \sum_{k=1}^n a_{ik}x_k$. For $y = [y_i]$, then

$$\langle Ax, y \rangle = \sum_{i=1}^n z_i \bar{y}_i = \sum_{i=1}^n \left(\sum_{k=1}^n a_{ik} x_k \right) \bar{y}_i$$

$$= \sum_{k=1}^n \left(\sum_{i=1}^n a_{ik} \bar{y}_i \right) x_k = \sum_{k=1}^n x_k \bar{u}_k = \langle x, u \rangle,$$

where $u = [u_k], u_k = \sum_{i=1}^{n} y_i \bar{a}_{ik}$. Therefore, $u = A^*y$ and $\langle Ax, y \rangle = \langle x, A^*y \rangle$. The rest is obvious.   □

2.2 Symmetric and Hermitian matrices

A matrix A is called *symmetric* if $A^T = A$, i.e., if $A = [a_{ik}]$ and $a_{ik} = a_{ki}$ for all i and k. A matrix A is said to be *Hermitian* if $A^* = A$, i.e., if $A = [a_{ik}]$ and $a_{ik} = \bar{a}_{ki}$ for all i and k.

With any square matrix A we can associate the Hermitian matrix $\frac{1}{2}(A + A^*)$. This matrix is called the *symmetric part of A* and is usually denoted by Re A. The reason for such a notation will be clear from the next example.

Example 2.1. Given a complex matrix $A = [z]$ of order 1, A is Hermitian if and only if the number z is real. The matrix Re A is then equal to the matrix [Rez] where the real part Rez of the complex number $z = a + bi$ (a and b real) is a, i.e., Re$z = \frac{1}{2}(z + \bar{z})$.

We turn now back to symmetric matrices. They are important, in particular, for studying properties of quadratic forms. A *quadratic form* is a homogeneous polynomial of degree 2. For example, $x^2 + 4xy - 2y^2$ is a quadratic form in two variables, x and y; $2x_1^2 - 3x_1x_2 + x_2^2 + 2x_1x_3 + 6x_2x_3 - 6x_3^2$ is a quadratic form in three variables, x_1, x_2 and x_3. To any real quadratic form, a unique symmetric matrix $A = [a_{ik}]$ can be assigned, such that the given quadratic form equals $\langle Ax, x \rangle$, i.e. $\sum_{i,k} a_{ik}x_ix_k$. Such a matrix for the first example is

$$\begin{bmatrix} 1 & 2 \\ 2 & -2 \end{bmatrix},$$

for the second example

$$\begin{bmatrix} 2 & -\frac{3}{2} & 1 \\ -\frac{3}{2} & 1 & 3 \\ 1 & 3 & -6 \end{bmatrix}.$$

A generalization to quadratic forms $\sum_{i,k} a_{ik}\bar{x}_ix_k$ in complex variables shows that Hermitian matrices play a similar role in this case.

Studying quadratic forms, we are concerned with the important problem of finding a nonsingular linear substitution that transforms the given quadratic form into a simpler quadratic form, i.e. a sum or difference of squares, etc. For example, by substituting $x = x' - 2y'/\sqrt{6}$, $y = y'/\sqrt{6}$ into the quadratic form $x^2 + 4xy - 2y^2$, we obtain the quadratic form $x'^2 - y'^2$ in these new variables. The celebrated Sylvester's law of inertia of quadratic forms (see Theorem 2.6.1) states that if a real nonsingular substitution transforms the given real quadratic form into a linear combination of squares, e.g. $a_1\tilde{x}_1^2 + a_2\tilde{x}_2^2 + \cdots + a_n\tilde{x}_n^2$, then the three numbers, the number of positive coefficients a_i, the number of negative coefficients a_i as well as the number of zero coefficients a_i, are independent of the substitution employed. This fact has also a geometrical meaning. Quadratic forms for $n = 2$ may be represented by conics, forms for $n = 3$ may be represented by quadrics in the three-dimensional space, etc. The number of positive, negative, or zero coefficients expresses the type of this conic, quadric, etc. (for example, an ellipse, hyperbola, ellipsoid, hyperboloid of one sheet or of two sheets, etc.), namely the type in the so-called affine space, where e.g. the centre of a line segment is significant, but orthogonality is not defined.

An even more important problem – and more difficult as well – is the problem of expressing a conic, quadric, etc. in the canonical form, i.e., finding an *orthonormal* coordinate system in the Euclidean space, such that the conic, quadric, etc. has a simple form in this system, e.g. $a_1x_1^2 + a_2x_2^2 + \cdots + a_mx_m^2 = b$. This problem – in its algebraic form – is treated in what follows. We first introduce orthogonal matrices (over the real field) and unitary matrices (over the complex field).

2.3 Orthogonal, unitary matrices

A matrix A is called *orthogonal* if $AA^T = I$. A matrix A is called *unitary* if $AA^* = I$.

We remark that the term "symmetric matrix" means a real symmetric matrix and the term "orthogonal matrix" means a real orthogonal matrix throughout the book. Thus a real Hermitian matrix is symmetric and a real unitary matrix is orthogonal. As a rule, it is then sufficient to prove theorems in the complex field; the corresponding theorems in the real field are their corollaries.

Example 2.2. A square matrix $[z]$ of order 1 is unitary if and only if $|z| = 1$. It is real orthogonal if and only if $z = 1$ or $z = -1$. Thus, unitary matrices are a generalization of complex numbers whose modulus equals 1.

Of course, every unitary matrix U satisfies

$$U^* = U^{-1}. \tag{2.1}$$

The property $U^*U = I$ can also be stated as follows.

Theorem 2.3.1 *The columns of a unitary matrix form an orthonormal basis. Conversely, if the columns of a matrix form an orthonormal basis, then the matrix is unitary.*

Moreover, the following theorem holds.

Theorem 2.3.2 1. *The product of unitary matrices of the same order is a unitary matrix as well.*

2. *A unitary matrix is nonsingular and its inverse is unitary as well.*

3. *If a matrix U is unitary, then U^* is unitary as well.*

4. *The identity matrix I is unitary.*

Corollary 1 *Unitary matrices of order n with the operation of matrix multiplication form a group.*[1]

Corollary 2 *Orthogonal matrices of order n with the operation of matrix multiplication form a group.*

Theorem 2.3.3 (Fundamental Theorem on Unitary Matrices) *Let U be a complex square matrix of order n. Then the following properties are equivalent:*

---

[1] A set G together with an associative operation $*$ constitutes a *group* if the following statements hold:

1. If $a \in G$ and $b \in G$, then $a * b \in G$.

2. There exists an element $e \in G$ such that $a * e = e * a = a$ for all $a \in G$.

3. For each $a \in G$, there exists an element $\hat{a} \in G$ such that $a * \hat{a} = \hat{a} * a = e$.

1. U is unitary.

2. For all (complex n-component column) vectors x we have

$$|Ux| = |x|.$$

3. For any two vectors x and y the equality

$$\langle Ux, Uy \rangle = \langle x, y \rangle$$

   holds.

4. Whenever the vectors $u_1, u_2, \ldots, u_n$ form an orthonormal basis, then $Uu_1, Uu_2, \ldots, Uu_n$ form also an orthonormal basis.

5. There exists an orthonormal basis $f_1, f_2, \ldots, f_n$ such that Uf_1, $Uf_2, \ldots, Uf_n$ is again an orthonormal basis.

Proof. We shall prove the implications $1. \to 2., 2. \to 3., 3. \to 4., 4. \to 5.$, and $5. \to 1.$

Suppose 1.; if x is a vector, then $\langle Ux, Uy \rangle = \langle x, U^*Ux \rangle = \langle x, x \rangle$ by Theorem 2.1.3, therefore $|Ux| = |x|$ and 2. holds.

Let 2. be true and x and y be vectors. We shall distinguish two cases separately: a) the space is real; b) the space is complex. In the first case,

$$\langle U(x + y), U(x + y) \rangle = \langle x + y, x + y \rangle,$$
$$\langle U(x - y), U(x - y) \rangle = \langle x - y, x - y \rangle.$$

Subtracting the second equality from the first and using the fact that $\langle Ux, Uy \rangle = \langle Uy, Ux \rangle$ and $\langle x, y \rangle = \langle y, x \rangle$, we have $\langle Ux, Uy \rangle = \langle x, y \rangle$, which is 3. In the case b) we use the relations

$$\langle U(x + y), U(x + y) \rangle = \langle x + y, x + y \rangle,$$
$$\langle U(x - y), U(x - y) \rangle = \langle x - y, x - y \rangle,$$
$$\langle U(x + iy), U(x + iy) \rangle = \langle x + iy, x + iy \rangle,$$
$$\langle U(x - iy), U(x - iy) \rangle = \langle x - iy, x - iy \rangle.$$

From the first two equations we obtain by subtraction

$$\langle Ux, Uy \rangle + \langle Uy, Ux \rangle = \langle x, y \rangle + \langle y, x \rangle,$$

while from the third and fourth equations we similarly have

$$\langle Ux, Uy \rangle - \langle Uy, Ux \rangle = \langle x, y \rangle - \langle y, x \rangle.$$

This now implies $\langle Ux, Uy \rangle = \langle x, y \rangle$ which is again 3.

In order to prove the implication 3. $\rightarrow$ 4., suppose that $u_1, \ldots,$ u_n form an orthonormal basis. For $v_i = Uu_i$, $i = 1, \ldots, n$, we then have by 3. that

$$\langle v_i, v_k \rangle = \langle Uu_i, Uu_k \rangle = \langle u_i, u_k \rangle = \delta_{ik}$$

for all $i, k = 1, \ldots, n$. Therefore Uu_i is an orthonormal basis as 4. asserts.

The implication 4. $\rightarrow$ 5. follows from the fact that there exists an orthonormal basis. It remains to prove that 5. implies 1.

Let $f_1, f_2, \ldots, f_n$ be an orthonormal basis, and $Uf_1, Uf_2, \ldots,$ Uf_n also. Then $\langle Uf_i, Uf_k \rangle = \delta_{ik} = \langle f_i, f_k \rangle$. From this it follows by Theorem 2.1.3 that $\langle f_i, U^*Uf_k \rangle = \langle f_i, f_k \rangle$, i.e.

$$\langle f_i, (U^*U - I)f_k) \rangle = 0 \qquad \text{for all } i \text{ and } k. \tag{2.2}$$

Let k be fixed. Set $(U^*U - I)f_k = z_k$. Expressing z_k in the basis $f_1, f_2, \ldots, f_n$ as $z = \alpha_1 f_1 + \alpha_2 f_2 + \cdots + \alpha_n f_n$, we obtain by (2.2) that

$$\langle z_k, z_k \rangle = \sum_{i=1}^{n} \alpha_i \langle f_i, z_k \rangle = \sum_{i=1}^{n} \alpha_i \langle f_i, (U^*U - I)f_k \rangle = 0.$$

Thus $z_k = 0$ by the property 4. of the inner product, i.e.

$$(U^*U - I)f_k = 0.$$

But this is true for $k = 1, \ldots, n$. As $f_1, f_2, \ldots, f_n$ form a basis, it holds $(U^*U - I)x = 0$ for any vector x, which means that

$$U^*U = I.$$

Therefore U is unitary. $\square$

Remark. A completely analogous theorem holds also for orthogonal matrices in a real space.

We shall now show examples of unitary and orthogonal matrices. The identity matrix I and any permutation matrix (a matrix whose one entry in each row and column equals 1 and all the other entries are zeros) is clearly orthogonal as well as unitary. Any diagonal matrix whose diagonal entries are 1 or -1 is orthogonal as well. Any diagonal matrix all of whose diagonal entries are complex units (i.e. numbers the modulus of which is 1) is unitary.

Further important examples of orthogonal and unitary matrices are treated in the next theorem.

Theorem 2.3.4 *The matrix*

$$R(\alpha) = \begin{bmatrix} \cos\alpha & \sin\alpha \\ -\sin\alpha & \cos\alpha \end{bmatrix}$$

with real α is orthogonal. If w is an (in general complex) n-component vector of length 1 (i.e., $|w| = 1$), then the matrix

$$W(w) = I - 2ww^*$$

is unitary (and even Hermitian).

Proof. Verifying that $R(\alpha)(R(\alpha))^T = I$, immediately proves the first statement. In order to prove the second statement, we first note that $(W(w))^* = W(w)$. Further

$$\begin{aligned} W(w)(W(w))^* &= (I - 2ww^*)(I - 2ww^*) \\ &= I - 4ww^* + 4ww^*ww^* \\ &= I - 4ww^* + 4ww^* = I \end{aligned}$$

(since $w^*w = 1$). □

Remark. The matrix $R(\alpha)$ corresponds to the plane rotation with angle α. The matrix $W(w)$ is called the *reflection matrix* corresponding to the vector w. The image $W(w)$ of a vector x and the vector x are symmetric with respect to the hyperplane orthogonal to w. The matrices $R(\alpha)$ and $W(w)$ will be also used later (Chapter 13).

To complete our discussion, we present the following theorem.

Theorem 2.3.5 *The determinant of an orthogonal matrix equals 1 or −1. The modulus of the determinant of a unitary matrix equals 1.*

Proof. If C is an orthogonal matrix, then $CC^T = I$ and, further, by Theorem 1.2.7, $\det C \det C^T = 1$. But $(\det C)^2 = 1$ since $\det C^T = \det C$. Thus $\det C$ equals either 1 or −1.

If U is a unitary matrix, then $UU^* = I$ and $\det U \det U^* = 1$. But $\det U^* = \det \overline{U^T} = \det \bar{U} = \overline{\det U}$, and thus $|\det U| = 1$. □

We shall now study the behaviour of a Hermitian (or symmetric) matrix A under a similarity transformation PAP^{-1} with a unitary (or orthogonal) matrix P.

Theorem 2.3.6 *If the matrix A is Hermitian and P unitary, then the matrix PAP^{-1} is Hermitian as well. If A is symmetric and P orthogonal, then PAP^{-1} is again symmetric.*

Proof. In the first case, we have $(PAP^{-1})^* = (PAP^*)^* = PA^*P^* = PAP^* = PAP^{-1}$ by (2.1) and Theorem 2.1.3, and the matrix PAP^{-1} is thus Hermitian. Similarly, $(PAP^{-1})^T = (PAP^T)^T = PA^TP^T = PAP^T = PAP^{-1}$ in the second case and PAP^{-1} is thus symmetric.   □

Therefore, there is a question of characterizing the class of matrices that are unitarily or orthogonally similar to a given Hermitian or symmetric matrix A. Another problem is to find a suitable representation of this class of matrices. These problems are completely solved by Theorem 2.4.7 (or, 2.4.6), where we show that there is essentially just one diagonal matrix in each such class and that this matrix has real diagonal entries.

First, however, we prove a more general theorem due to Schur, which states that an upper triangular matrix is always contained in the class of complex matrices unitarily similar to a given complex matrix.

2.4   Gram-Schmidt orthonormalization

Theorem 2.4.1 *Let $u_1, u_2, \ldots, u_r$ be n-component column (in general complex) vectors of length 1, any two of which are orthogonal. Then $r \leq n$ and there exist vectors $u_{r+1}, \ldots, u_n$. such* *that the matrix*

$$U = [u_1, u_2, \ldots, u_n]$$

is unitary, or orthogonal in the case that the vectors $u_1, u_2, \ldots,$ u_r *are real.*

Proof. The vectors $u_1, \ldots, u_r$ are linearly independent. Indeed, if $\sum_{i=1}^{r} \alpha_i u_i = 0$, then, left-multiplying by u_k^*, we obtain $\sum_{i=1}^{r} \alpha_i (u_k^* u_i) = \alpha_k = 0$, $k = 1, \ldots, r$. Thus, $r \leq n$. Some vectors $w_{r+1}, \ldots, w_n$ may be added to the vectors $u_1, \ldots, u_r$ to complete them to a basis of the space of all n-component vectors. In fact, the $n \times r$ matrix $[u_1, \ldots, u_r]$ has rank r by Theorem 1.5.1 and thus there are r rows of this matrix, say with indices $i_1, \ldots, i_r$, in which a nonsingular submatrix of order r is contained. It is then sufficient to choose as the w_k's the vectors $e_k = [0, \ldots, 0, 1, 0, \ldots, 0]^T$ having 1 as their kth component and such that k is different from $i_1, \ldots, i_r$. The resulting matrix of order n is nonsingular, e.g. by the Laplace Theorem 1.2.1.

We now construct a recurrence relation for the vectors $u_{r+1}, \ldots, u_n$: Suppose that we have constructed an orthonormal set $u_1, \ldots, u_s$ such that the vector u_k, $r + 1 \leq k \leq s$, is linearly dependent on $u_1, \ldots, u_{k-1}$, and w_k. Set $\tilde{u}_{s+1} = w_{s+1} - \sum_{k=1}^{s} u_k^* w_{s+1}) u_k$. The vector constructed in this way is orthogonal to $u_1, \ldots, u_s$ since $u_k^* \tilde{u}_{s+1} = 0$ for $k = 1, \ldots, s$. Moreover, $\tilde{u}_{s+1}$ is nonzero since the vector w_{s+1} would otherwise be linearly dependent on $u_1, \ldots, u_s$, i.e. on $u_1, \ldots, u_r$, $w_{r+1}, \ldots, w_s$, which contradicts the fact that $u_1, \ldots, u_r$, $w_{r+1}, \ldots, w_{s+1}$ constitute a part of the basis. Setting now $u_{s+1} = \tilde{u}_{s+1}/(\tilde{u}_{s+1}^* \tilde{u}_{s+1})^{1/2}$, we obtain a vector u_{s+1} of length 1 and such that it is linearly dependent on $u_1, \ldots, u_s$ and w_{s+1}. We continue in this way until we obtain all the vectors $u_1, \ldots, u_n$. Clearly, the resulting matrix is unitary by Theorem 2.3.1. $\square$

Remark. The method presented in the proof of Theorem 2.4.1 is called the *Gram-Schmidt orthonormalization process*.

Theorem 2.4.2 *Let A be a complex square matrix of order n. Then there exist a unitary matrix V and a complex number λ such that the first column of the matrix $V^* A V$ is $[\lambda, 0, \ldots, 0]^T$.*

Proof. By the Fundamental Theorem of Algebra, there exists at least one (in general complex) root λ of the characteristic equation $\det(A - xI) = 0$. Denote by $\hat{u}$ some of the eigenvectors associated with the eigenvalue λ. Then the length of the vector $u = \hat{u}/|\hat{u}|$ is 1. By Theorem 2.4.1, there exists a unitary matrix V whose first column is u. Let us show that V^*AV satisfies the statement of the theorem. The unitary matrix V can be written in partitioned form

$$V = [u, V_1],$$

where V_1 is an $n \times (n - 1)$ matrix which satisfies

$$V_1^* u = 0.$$

Calculate now V^*AV employing the relation $Au = \lambda u$. We obtain

$$V^*AV = V^*A[u, V_1] = V^*[Au, AV_1] = V^*[\lambda u, AV_1]$$
$$= \begin{bmatrix} u^* \\ V_1^* \end{bmatrix} [\lambda u, AV_1] = \begin{bmatrix} \lambda u^* u & u^* AV_1 \\ \lambda V_1^* u & V_1^* AV_1 \end{bmatrix}.$$

Since $u^* u = 1$ and $V_1^* u = 0$, the first column has the form required. $\square$

Remark. If a square real matrix A has a real eigenvalue λ, then, by Theorem 2.4.2, there exists an orthogonal matrix C such that the first column of the real matrix $C^T AC$ is $[\lambda, 0, \ldots, 0]^T$.

Now we formulate *Schur's theorem* [55].

Theorem 2.4.3 (Schur). *Any complex square matrix A can be expressed in the form*

$$A = UTU^*,$$

where U is a unitary matrix and T is an (in general complex) upper triangular matrix.

Proof. We use induction on the order n of A. If $n = 1$, the theorem holds. Let $n > 1$ and suppose that the theorem is true for matrices of order $n - 1$.

By Theorem 2.4.2, there exist a unitary matrix V and a complex number λ such that

$$V^*AV = \begin{bmatrix} \lambda & b \\ 0 & A_1 \end{bmatrix},$$

where A_1 is a square complex matrix of order $n - 1$ and b is a row $(n - 1)$-component vector. By the induction hypothesis, A_1 can be expressed in the form $A_1 = U_1 T_1 U_1^*$ where U_1 is a unitary matrix of order $n-1$ and T_1 is a complex upper triangular matrix. Then

$$W = \begin{bmatrix} 1 & 0 \\ 0 & U_1 \end{bmatrix}$$

is a unitary matrix of order n and

$$W^* V^* A V W = \begin{bmatrix} 1 & 0 \\ 0 & U_1^* \end{bmatrix} \begin{bmatrix} \lambda & b \\ 0 & A_1 \end{bmatrix} \begin{bmatrix} 1 & 0 \\ 0 & U_1 \end{bmatrix}$$

$$= \begin{bmatrix} \lambda & bU_1 \\ 0 & U_1^* A_1 U_1 \end{bmatrix} = \begin{bmatrix} \lambda & bU_1 \\ 0 & T_1 \end{bmatrix}.$$

Denoting this matrix by T and the unitary matrix VW by U, we have $A = UTU^*$. □

In the real field, a slightly modified theorem holds. Its proof is left to the reader (see Exercise 2.3).

Theorem 2.4.4 *If all the eigenvalues of a real square matrix A are real, then there exist a (real) orthogonal matrix C and a real upper triangular matrix T such that $A = CTC^T$.*

Now we turn back to the study of symmetric and Hermitian matrices. First of all, we prove an important theorem.

Theorem 2.4.5 *All eigenvalues of a Hermitian matrix (and thus also those of a symmetric matrix) are real.*

Proof. Let λ be an eigenvalue of a Hermitian matrix A and u one of the associated eigenvectors. Then $\langle Au, u \rangle$ is real since $\overline{\langle Au, u \rangle} = \overline{\langle u, A^*u \rangle} = \overline{\langle u, Au \rangle} = \langle Au, u \rangle$ by a property of the inner product. As $Au = \lambda u$, this real number is equal to $\langle \lambda u, u \rangle$, i.e. to $\lambda \langle u, u \rangle$. But $\langle u, u \rangle$ is real (even positive), and thus λ is real. □

Theorem 2.4.6 (Fundamental Theorem on Symmetric Matrices) *Any (real) symmetric matrix A can be expressed in the form*

$$A = CDC^T.$$

where C is an orthogonal matrix and D a real diagonal matrix. The diagonal entries of D are eigenvalues of A and the columns of C are eigenvectors of A (the kth column of C corresponds to the kth diagonal entry of D).

Proof. All the eigenvalues of the matrix A are real by Theorem 2.4.5. Using Theorem 2.4.4, we can thus write $A = CTC^T$, where C is an orthogonal matrix and T is a real upper triangular matrix. Since A is symmetric, the matrix T which is orthogonally similar to A is also symmetric by Theorem 2.3.6. Thus T is diagonal (and real); let us write $D = \mathrm{diag}(\lambda_1, \lambda_2, \ldots, \lambda_n)$ instead of T.

We can express the matrix C in partitioned form

$$C = [C_1, C_2, \ldots, C_n],$$

where n is the order of C and C_k is the kth column. The relation

$$A = CDC^T$$

implies

$$AC = CD,$$

or

$$A[C_1, C_2, \ldots, C_n] = [C_1, C_2, \ldots, C_n] \begin{bmatrix} \lambda_1 & 0 & \ldots & 0 \\ 0 & \lambda_2 & \ldots & 0 \\ \vdots & \vdots & \ddots & \vdots \\ 0 & 0 & \ldots & \lambda_n \end{bmatrix}.$$

By Theorem 1.1.1 on the multiplication of partitioned matrices, we thus have

$$[AC_1, AC_2, \ldots, AC_n] = [\lambda_1 C_1, \lambda_2 C_2, \ldots, \lambda_n C_n),$$

or

$$\begin{aligned} AC_1 &= \lambda_1 C_1, \\ AC_2 &= \lambda_2 C_2, \\ &\vdots \\ AC_n &= \lambda_n C_n. \end{aligned}$$

Therefore, the kth column C_k of C (which is distinct from the zero vector) is really an eigenvector of A associated with the kth diagonal entry of D. □

Remark. Conversely, the product CDC^T is a symmetric matrix if D is a real diagonal matrix and C is an orthogonal matrix.

Theorem 2.4.7 (Fundamental Theorem on Hermitian Matrices) *Any Hermitian matrix A can be expressed in the form*

$$A = UDU^*,$$

where U is a unitary matrix and D is a real diagonal matrix. The diagonal entries of D are eigenvalues of A (which are real by Theorem 2.4.5) and the columns of U are eigenvectors of A (the kth column of U corresponds to the kth diagonal entry of D).

Proof is analogous to that of Theorem 2.4.6 and is left to the reader (Exercise 2.4).

Remark. Conversely, if D is a real diagonal matrix and U is a unitary matrix, then the matrix UDU^* is Hermitian.

2.5 Positive definite matrices

Given a Hermitian matrix A, the corresponding quadratic form $\langle Ax, x \rangle = \sum a_{ik}\bar{x}_i x_k$ assumes only real values which follows from the relation $\langle Ax, x \rangle = \langle x, A^*x \rangle = \langle x, Ax \rangle = \overline{\langle Ax, x \rangle}$. This certainly holds also for a real symmetric matrix even when the vector x is not real.

In physics and engineering applications, the quadratic from $\langle Ax, x \rangle$ often assumes only nonnegative values and, moreover, it frequently assumes only positive values for vectors different from the zero vector. This happens e.g. if the quadratic form represents the energy of some system of mass points. We will study now matrices connected with such quadratic forms.

A Hermitian or symmetric matrix $A = [a_{ik}]$ is called *positive definite* if

$$\langle Ax, x \rangle > 0,$$

or
$$\sum_{i,k} a_{ik}\bar{x}_i x_k > 0,$$

for any complex (real, when A is symmetric) vector $x \neq 0$.

A Hermitian or symmetric matrix $A = [a_{ik}]$ is called *positive semidefinite* if
$$\langle Ax, x \rangle \geq 0,$$

or
$$\sum_{i,k} a_{ik}\bar{x}_i x_k \geq 0,$$

for any complex (real, when A is symmetric) vector x.

Example 2.3. The identity matrix is clearly positive definite. A Hermitian or symmetric matrix $[z]$ of order 1 (where z is real as in Example 2.1) is positive definite if and only if z is positive, and it is positive semidefinite if and only if z is nonnegative. Positive definite and positive semidefinite matrices are thus a generalization of positive and nonnegative numbers, respectively, in the sense of Examples 2.1 and 2.2.

In Theorems 2.5.1 and 2.5.2, we summarize the most important properties of positive definite and positive semidefinite matrices.

Theorem 2.5.1 (Fundamental Theorem on Positive Definite Matrices) *Let A be a square Hermitian matrix (a square symmetric matrix, if A is real) of order n. Then the following properties are equivalent:*

1. *A is positive definite.*

2. *All the eigenvalues of all the principal submatrices of A are positive.*

3. *All the principal minors of A are positive.*

4. *$\det A(N_k, N_k) > 0$ for $k = 1, \ldots, n$, where $N_k = \{1, 2, \ldots, k\}$, i.e.*
$$a_{11} > 0, \ \det \begin{bmatrix} a_{11} & a_{12} \\ a_{21} & a_{22} \end{bmatrix} > 0,$$

$$\det \begin{bmatrix} a_{11} & a_{12} & a_{13} \\ a_{21} & a_{22} & a_{23} \\ a_{31} & a_{32} & a_{33} \end{bmatrix} > 0, \ldots, \det A > 0.$$

5. There exists a nonsingular lower triangular matrix T such that $A = TT^*$.

6. There exists a nonsingular matrix C such that $A = CC^*$.

7. For $k = 1, \ldots, n$, the sums of all the principal minors of A of order k are positive.

8. All the eigenvalues of A are positive.

9. There exist a unitary matrix U (an orthogonal matrix U, if A is real) and a diagonal matrix D with positive diagonal entries, such that $A = UDU^*$.

Proof. We prove the implications 1. $\to$ 2., 2. $\to$ 3., 3. $\to$ 4., 4. $\to$ 5., 5. $\to$ 6., 6. $\to$ 1., 3. $\to$ 7., 7. $\to$ 8., 8. $\to$ 9., and 9. $\to$ 6. The equivalence of the conditions will then be established.

1. $\to$ 2. Let $\emptyset \neq M \subset N = \{1, 2, \ldots, n\}$, let λ be an eigenvalue of the principal submatrix $A(M, M)$. Denote by $x(M)$ an eigenvector of $A(M, M)$ associated with λ (the components of $x(M)$ are defined for the indices in M), that is

$$A(M, M)x(M) = \lambda x(M).$$

Denote by x the n-component column vector obtained from $x(M)$ by adding zeros for the components not belonging to M. Then $\langle Ax, x \rangle = \langle A(M, M)x(M), x(M) \rangle$. Since $x(M) \neq 0$, $x \neq 0$ as well and the left-hand side of the last equality is positive by 1. The right-hand side can be modified to read $\langle \lambda x(M), x(M) \rangle = \lambda \langle x(M), x(M) \rangle$. As $\langle x(M), x(M) \rangle$ is positive, λ is positive as well.

2. $\to$ 3. If $M \subset N$, then $\det A(M, M)$ equals the product of all the eigenvalues of $A(M, M)$ by (1.39). Since these eigenvalues are positive by 2., we have $\det A(M, M) > 0$ for any $M \subset N$.

3. $\to$ 4. Clear.

4. $\to$ 5. The matrix A is strongly nonsingular by the assumption 4. By Theorem 1.4.3, there exists a unique factorization $A = BDC$, where B is a lower triangular matrix with ones along

the main diagonal, C is an upper triangular matrix with ones along the main diagonal, and D is a diagonal matrix. By Theorem 2.1.2, $A^* = C^* D^* B^*$. But $A^* = A$, and thus $B = C^*, D = D^*$, and $C = B^*$. Moreover, the diagonal entries of D are positive by Theorem 1.4.2 and the assumption 4., and thus there exists a diagonal matrix W with positive diagonal entries such that $W^2 = D$. The matrix $T = BW$ is therefore lower triangular and $A = TT^*$.

5. → 6. Obvious.

6. → 1. Let $A = CC^*$, where C is a nonsingular matrix. If x is a column vector, $x \neq 0$, then also $C^* x = y \neq 0$, and hence

$$\langle Ax, x \rangle = \langle CC^*x, x \rangle = \langle C^*x, C^*x \rangle = \langle y, y \rangle > 0.$$

3. → 7. Obvious.

7. → 8. By Theorem 2.4.5, all the eigenvalues of A are real. It is therefore sufficient to obtain a contradiction from the assumption that some of the eigenvalues of A, say, λ, is nonpositive. By (1.36), we have then

$$(-\lambda)^n + c_1(-\lambda)^{n-1} + c_2(-\lambda)^{n-2} + \cdots + c_n = 0, \qquad (2.3)$$

where c_k is the sum of all the principal minors of A. All the terms on the left-hand side are thus nonnnegative and the last term is even positive, which contradicts the fact that the right-hand side is zero.

8. → 9. By Theorem 2.4.7 (by Theorem 2.4.6, if A is real), A can be expressed as $A = UDU^*$, where U is a unitary matrix (an orthogonal matrix, if A is real) and D is a diagonal matrix. Moreover, the diagonal entries of D are exactly all the eigenvalues of A. If 8. holds, then these eigenvalues are positive and 9. holds as well.

9. → 6. Let $A = UDU^*$, where $D = \mathrm{diag}(d_1, \ldots, d_n)$, $d_i > 0$, $i = 1, \ldots, n$. Set $W = \mathrm{diag}(\sqrt{d_1}, \ldots, \sqrt{d_n})$; then $D = WW^*$. We thus have $A = CC^*$, where $C = UW$ is indeed nonsingular.   □

Remark. The property 4. is called *Sylvester's criterion for positive definiteness of a matrix.*

In Chapter 10 we will see how to decide, with the help of this criterion, whether a numerically given Hermitian or symmetric matrix is or is not positive definite.

Theorem 2.5.2 (Fundamental Theorem on Positive Semi-definite Matrices) *Let* A *be a square Hermitian matrix (a square symmetric matrix, if* A *is real) of order* n. *Then the following properties are equivalent:*

1. A *is positive semidefinite.*

2. *The matrix* $A + \varepsilon I$ *is positive definite for any* $\varepsilon > 0$.

3. *All the eigenvalues of all the principal submatrices of* A *are nonnegative.*

4. *All the principal minors of* A *are nonnegative.*

5. *For* $k = 1, \ldots, n$, *the sums of all the principal minors of* A *of order* k *are nonnegative.*

6. *All the eigenvalues of* A *are nonnegative.*

7. *There exist a unitary matrix* U *(an orthogonal matrix* U, *if* A *is real) and a diagonal matrix* D *with nonnegative diagonal entries, such that* $A = UDU^*$.

8. *There exists a square matrix* C, *such that* $A = CC^*$.

Proof. We leave to the reader (Exercise 2.5) to prove the implications 1. → 2., 2. → 3., 3. → 4., 4. → 5., 5. → 6., 6. → 7., 7. → 8., and 8. → 1. □

Clearly, a positive definite matrix is also positive semidefinite. The next theorem states when a positive semidefinite matrix is positive definite.

Theorem 2.5.3 *A positive semidefinite matrix is positive definite if and only if it is nonsingular.*

Proof. By the property 3. of Theorem 2.5.1, the determinant of a positive definite matrix A is positive and thus A is nonsingular. Conversely, if a positive semidefinite matrix A is nonsingular, then $\det A \neq 0$. By (1.39), all the eigenvalues of A are nonzero. As they are nonegative by 6. of Theorem 2.5.2, they have to be positive. Therefore, by 8. of Theorem 2.5.1, A is positive definite. □

We present some further properties of positive definite and positive semidefinite matrices.

Theorem 2.5.4 *If A is a positive definite matrix, then its inverse A^{-1} exists and is positive definite.*

Proof. Let $x \neq 0$. Since the positive definite matrix A is nonsingular by the property 4. of Theorem 2.5.1, the vector $y = A^{-1}x$ exists and $y \neq 0$. Then

$$\langle A^{-1}x, x \rangle = \langle y, Ay \rangle = \langle Ay, y \rangle > 0.$$

Therefore, A^{-1} is positive definite.   □

Theorem 2.5.5 *Let a matrix A be positive definite. Then every principal submatrix of A is positive definite. If A_{11} is a principal submatrix of A, then the Schur complement $[A/A_{11}]$ of A_{11} in A is also positive definite.*

Proof. The first statement follows easily from the property 3. of Theorem 2.5.1.

To prove the second statement, we employ the relation (1.30) and Theorem 2.5.4. Since the matrix A^{-1} is positive sefinite, its principal submatrix B_{22} is positive definite as well. Thus also the matrix $B_{22}^{-1} = [A/A_{11}]$ is positive definite.   □

Theorem 2.5.6 *A necessary and sufficient condition that a symmetric (or Hermitian) matrix (with a square block A_{11})*

$$A = \begin{bmatrix} A_{11} & A_{12} \\ A_{21} & A_{22} \end{bmatrix}$$

be positive definite is that both the submatrix A_{11} and its Schur complement $[A/A_{11}] = A_{22} - A_{21}A_{11}^{-1}A_{12}$ be positive definite.

Proof. We prove it for the unitary space, i.e. for the complex inner product, which is the more general case. If A is positive definite, then both A_{11} and $[A/A_{11}]$ are positive definite by Theorem 2.5.5.

Conversely, let both A_{11} and $[A/A_{11}]$ be positive definite. Suppose that

$$x = \begin{bmatrix} x_1 \\ x_2 \end{bmatrix}$$

is a non-zero vector partitioned in the same way as A. Then

$$\langle Ax, x \rangle = [x_1^*, x_2^*] \begin{bmatrix} A_{11} & A_{12} \\ A_{21} & A_{22} \end{bmatrix} \begin{bmatrix} x_1 \\ x_2 \end{bmatrix}$$
$$= x_1^* A_{11} x_1 + x_1^* A_{12} x_2 + x_2^* A_{21} x_1 + x_2^* A_{22} x_2$$
$$= (x_1^* + x_2^* A_{21} A_{11}^{-1}) A_{11} (x_1 + A_{11}^{-1} A_{12} x_2 + $$
$$+ x_2^* (A_{22} - A_{21} A_{11}^{-1} A_{12}) x_2.$$

This sum is nonnegative by the assumption that A_{11} as well as $[A/A_{11}] = A_{22} - A_{21} A_{11}^{-1} A_{12}$ are positive definite matrices. If the sum were equal to zero we would have $x_2 = 0$ (from the second term), but then also $x_1 = 0$ (from the first term), which is a contradiction. Hence A is positive definite. □

The next theorems follow immediately from the definition of positive definiteness and positive semidefiniteness.

Theorem 2.5.7 *If A is a positive semidefinite matrix and α is a nonnegative number, then the matrix αA is also positive semidefinite. If A is a positive definite matrix and α is a positive number, then the matrix αA is also positive definite.*

Theorem 2.5.8 *If A and B are positive semidefinite matrices of the same order, then the matrix $A + B$ is also positive semidefinite. If, moreover, A is a positive definite matrix, then the matrix $A + B$ is also positive definite.*

Both these theorems imply that the set S_p of all real positive semidefinite matrices of order n forms a convex cone in the real linear space S of all symmetric matrices of order n (cf. Exercise 2.2). We recall that a set M of a real linear space L is called a *convex cone* if the following two statements hold:

1. If $x \in M$ and $\alpha \geq 0$, then $\alpha x \in M$.
2. If $x \in M$ and $y \in M$, then $x + y \in M$.

Further, the set S_p is a convex hull of the zero matrix and the set of all positive semidefinite matrices of rank 1 (i.e., S_p is the smallest convex set containing the zero matrix and all positive semidefinite matrices of rank 1; a set is said to be *convex set* if together with any two elements x and y it contains also all the elements $\lambda x + (1 - \lambda) y$, $0 < \lambda < 1$).

In the following chapters a few more theorems will be needed.

Theorem 2.5.9 *If A is a positive definite matrix and G is a nonsingular matrix, then the matrix GAG^* is again positive definite. If A is a positive semidefinite matrix and G is a (possibly rectangular) matrix such that GA does exist, then GAG^* is again positive semidefinite.*

Proof. Let x be a vector, $x \neq 0$. By Theorem 2.1.3,

$$\langle GAG^*x, x \rangle = \langle AG^*x, G^*x \rangle = \langle Ay, y \rangle > 0$$

since $y = G^*x \neq 0$ as well.

In the second case it is sufficient to use the fact that $\langle Ay, y \rangle \geq 0$. □

2.6   Sylvester's law of inertia

We introduce now the concept of the *inertia* of a square (in general complex) matrix A as an ordered triple

$$\text{In } A = (p^+, p^-, p^{(0)}), \tag{2.4}$$

where p^+ is the number of eigenvalues of A with positive real part, p^- is the number of eigenvalues with a negative real part, and $p^{(0)}$ is the number of eigenvalues with zero real part.

Hence $\text{In } A = (n, 0, 0)$ for a positive definite matrix A of order n, $p^- = 0$ for a positive semidefinite matrix, etc.

We have already mentioned *Sylvester's law of inertia*.

Theorem 2.6.1 *If A is a Hermitian or real symmetric matrix and G is a nonsingular matrix of the same order, then*

$$\text{In } A = \text{In}(GAG^*).$$

Proof. It suffices to prove the assertion only for a Hermitian matrix A.

By Theorem 2.4.7, there exist a unitary matrix U and a real diagonal matrix D such that $A = UDU^*$. Similarly there exist a unitary matrix V and a real diagonal matrix L such that $GAG^* = VLV^*$. Hence

$$R^*DR = L, \tag{2.5}$$

where $R = U^*G^*V$ is a nonsingular matrix; moreover,

$$\text{In } A = \text{In } D, \qquad \text{In}(GAG^*) = \text{In } L. \qquad (2.6)$$

Let n be the order of A. If $d_1, \ldots, d_n$ are the diagonal entries of D and $\ell_1, \ldots, \ell_n$ the diagonal entries of L, we put $M_1 = \{i; d_i > 0\}$, $M_2 = \{i; d_i < 0\}$, $M_3 = \{i; d_i = 0\}$, $N_1 = \{i; \ell_i > 0\}$, $N_2 = \{i; \ell_i < 0\}$, $N_3 = \{i; \ell_i = 0\}$. Hence $\text{In } A = (|M_1|, |M_2|, |M_3|)$ and $\text{In}(GAG^*) = (|N_1|, |N_2|, |N_3|)$.

Clearly $|M_1| + |M_2| = \text{rank } D$ and also $|N_1| + |N_2| = \text{rank } L$. But by (2.5) and Theorem 1.5.4, $\text{rank } D = \text{rank } L$. Since $|M_1| + |M_2| + |M_3| = |N_1| + |N_2| + |N_3|$ we have

$$|M_3| = |N_3|.$$

To prove that also $|M_1| = |N_1|$, and, therefore, also $|M_2| = |N_2|$, suppose that $|M_1| < |N_1|$. (If $|M_1| > |N_1|$ just use the matrices $-A$ and $G(-A)G^*$.) We thus have $|M_1| + |N_2| + |N_3| < n$; therefore, there exists a solution $y = [y_1, \ldots, y_n]^T \neq 0$ of the linear system (with $|M_1| + |N_2| + |N_3|$ equations)

$$x_i = 0, \qquad i \in N_2 \cup N_3,$$

$$\sum_{k=1}^{n} r_{jk} x_k = 0, \qquad j \in M_1,$$

where r_{ik} are the entries of the matrix R from (2.5).

Since $y_i \neq 0$ for some $i \in N_1$, we have

$$\langle Ly, y \rangle = \sum_{i \in N_1} \ell_i |y_i|^2 > 0.$$

But

$$\langle R^*DRy, y \rangle = \langle DRy, Ry \rangle = \sum_{i=1}^{n} d_i \left| \sum_{k=1}^{n} r_{ik} y_k \right|^2$$

$$= \sum_{i \in M_2} d_i \left| \sum_{k=1}^{n} r_{ik} y_k \right|^2 \leq 0,$$

which is a contradiction, since $\langle Ly, y \rangle = \langle R^*DRy, y \rangle$ by (2.5). Hence $|M_1| = |N_1|$ (and also $|M_2| = |N_2|$) and thus $\text{In } A = \text{In}(GAG^*)$ by (2.6). $\qquad \square$

Theorem 2.6.2 (Jacobi) *Let A be a strongly nonsingular Hermitian (or real symmetric) matrix of order n. Then* $\operatorname{In} A = (k_1, k_2, 0)$, *where k_1 is the number of agreements in sign and k_2 is the number of changes of sign in the sequence*

$$1, \det A(N_1), \det A(N_2), \ldots, \det A(N_n),$$

where $N_k = \{1, 2, \ldots, k\}$.

Proof. By Theorem 1.4.3, the matrix A can be uniquely expressed in the form $A = BDC$, where B is a lower and C is an upper triangular matrix with ones on the diagonal, and D is a diagonal matrix. But $A = A^* = C^* D^* B^*$ is also of this form, and hence $B = C^*$, $A = BDB^*$. Moreover, $D = \operatorname{diag}(d_1, \ldots, d_n)$, $d_1 = \det A(N_1)$, and

$$d_k = \det A(N_k) / \det A(N_{k-1}).$$

Clearly, each positive entry d_k corresponds to one agreement in sign (between $\det A(N_{k-1})$ and $\det A(N_k)$) and each negative entry corresponds to one change of sign in the sequence mentioned. As $\operatorname{In} A = \operatorname{In} D$ by Theorem 2.6.1, the proof is complete.   □

2.7   Singular value decomposition

The following theorem on the *singular value decomposition* of a matrix is very important.

Theorem 2.7.1 *Let A be a complex $m \times n$ matrix of rank r. Then there exist a unitary (orthogonal, if A is real) matrix U of order m, a unitary (orthogonal) matrix V of order n, and a diagonal matrix S of order r with positive diagonal entries such that*

$$A = U \begin{bmatrix} S & 0 \\ 0 & 0 \end{bmatrix} V^*; \qquad (2.7)$$

here, the zero blocks complete S to obtain an $m \times n$ matrix. The matrix S is determined uniquely except for the order of its diagonal entries.

Remark. The diagonal entries of S are called the *singular values* of the matrix A (if A is a square matrix, even the remaining zero diagonal entries of the matrix between U and V^* on the right-hand side of (2.7) are, as a rule, considered singular values).

Proof. Without loss of generality, we may assume $m \leq n$; otherwise we employ the matrix A^* instead of A. The matrix A^*A is identical with $\tilde{A}^*\tilde{A}$, where

$$\tilde{A} = \begin{bmatrix} A \\ 0 \end{bmatrix}$$

is a square matrix of order n with $n - m$ zero rows. By the property 8. of Theorem 2.5.2, A^*A is positive semidefinite. By the property 7. of the same theorem, there exist a unitary matrix V and a diagonal matrix D of order n with nonnegative diagonal entries d_i such that $A^*A = VDV^*$. Hence

$$(AV)^*(AV) = D. \tag{2.8}$$

We may assume that the first p diagonal entries of the matrix D (of order n) are positive and the remaining are zero:

$$D = \begin{bmatrix} \tilde{D} & 0 \\ 0 & 0 \end{bmatrix},$$

where $\tilde{D} = \operatorname{diag}(d_1, \ldots, d_p)$ is a nonsingular matrix. Denote by $w_1, \ldots, w_p$ the vectors $d_i^{1/2}(AV)_i$, where $(AV)_i$ means the ith column of the matrix AV, $i = 1, \ldots, p$. These m-component vectors w_i are mutually orthogonal by (2.8) and each of them has length 1. Theorem 2.4.1 implies the existence of vectors $w_{p+1}, \ldots, w_m$ such that $U = [w_1, w_2, \ldots, w_m]$ is a unitary matrix (of order m). Let us show that

$$AV = U\hat{D}, \tag{2.9}$$

where

$$\hat{D} = \begin{bmatrix} S & 0 \\ 0 & 0 \end{bmatrix}$$

is an $m \times n$ matrix, $S = \operatorname{diag}(\sqrt{d_1}, \ldots, \sqrt{d_p})$. The relation (2.9) holds for the first p columns of AV by the definition of $w_1, \ldots, w_p$.

But the other columns of AV are zero since, by (2.8), the length of the vector $(AV)_i$ is equal to the ith diagonal entry of D for $i > p$ and is therefore zero.

From (2.9) it follows that $A = U\hat{D}V^*$. By Theorem 1.5.4, however, rank A = rank $\hat{D} = p$, i.e. $p = r$, and (2.7) is proved. The uniqueness of S follows from the fact that the diagonal entries of S are nonnegative square roots of the eigenvalues of A^*A.

$\square$

The theorem on *polar decomposition* is an easy consequence.

Theorem 2.7.2 *Let A be a square matrix. Then there exist a positive semidefinite matrix P and a unitary matrix U such that $A = UP$.*

Proof. By Theorem 2.7.1, $A = U_1 D_1 V_1^*$ where U_1 and V_1 are unitary matrices and D_1 is a diagonal matrix with nonnegative diagonal entries. The matrix $P = V_1 D_1 V_1^*$ is therefore positive semidefinite by the property 7. of Theorem 2.5.3 and the matrix $U = U_1 V_1^*$ is unitary by Theorem 2.3.2. Moreover,

$$UP = U_1 V_1^* V_1 D_1 V_1^* = U_1 D_1 V_1^* = A.$$

$\square$

Remark. Applying Theorem 2.7.1 to the matrix A^*, we obtain that there also exists the decomposition $A = QV$ where Q is a positive semidefinite matrix and V is a unitary matrix.

In the conclusion of this chapter, let us mention an important property of the smallest and largest eigenvalues of a symmetric or Hermitian matrix.

Theorem 2.7.3 *Let A be a Hermitian (or symmetric) matrix of order n with the eigenvalues $\lambda_1 \geq \lambda_2 \geq \cdots \geq \lambda_n$. Then λ_1 is the maximal and λ_n the minimal value of the quadratic form $\langle Ax, x \rangle$ for normalized vectors x (i.e., $\langle x, x \rangle = 1$). Moreover, the maximal value is attained for any normalized eigenvector associated with the eigenvalue λ_1, and the minimal value is attained for any normalized eigenvector associated with the eigenvalue λ_n.*

Proof. By Theorem 2.4.7 (by Theorem 2.4.6, if A is a real symmetric matrix), there exist a unitary matrix U (an orthogonal matrix, if A is real) and a real diagonal matrix D such that $A = UDU^*$. The diagonal entries $d_1, \ldots, d_n$ of D are the eigenvalues of A, i.e., the numbers $\lambda_1, \ldots, \lambda_n$ in some order. By Theorem 2.1.3, we have $\langle Ax, x \rangle = \langle UDU^*x, x \rangle = \langle DU^*x, U^*x \rangle = \langle Dy, y \rangle$, where $y = U^*x$, and, moreover, $\langle y, y \rangle = \langle U^*x, U^*x \rangle = \langle UU^*x, x \rangle = \langle x, x \rangle$. Hence $\langle Dy, y \rangle$ assumes the same values for $\langle y, y \rangle = 1$ as $\langle Ax, x \rangle$ does for $\langle x, x \rangle = 1$.

Examine now what values the expression

$$\langle Dy, y \rangle = \sum_{i=1}^{n} d_i |y_i|^2$$

attains for vectors $y = [y_1, \ldots, y_n]^T$ such that

$$\langle y, y \rangle = \sum_{i=1}^{n} |y_i|^2 = 1.$$

As $\max_k d_k = \lambda_1$ and $\min_k d_k = \lambda_n$, we have

$$\langle Dy, y \rangle = \sum_{i=1}^{n} d_i |y_i|^2 \le \sum_{i=1}^{n} \lambda_1 |y_i|^2 = \lambda_1 \sum_{i=1}^{n} |y_i|^2 = \lambda_1, \quad (2.10)$$

$$\langle Dy, y \rangle = \sum_{i=1}^{n} d_i |y_i|^2 \ge \sum_{i=1}^{n} \lambda_n |y_i|^2 = \lambda_n \sum_{i=1}^{n} |y_i|^2 - \lambda_n. \quad (2.11)$$

Note that the inequality (2.10) turns into equality if and only if $y_i = 0$ for any i such that $d_i < \lambda_1$, i.e., if and only if $Dy = \lambda_1 y$. Returning to the original matrix A, the vector x thus satisfies the equality

$$DU^*x = \lambda_1 U^*x,$$

or,

$$U^*DU^*x = \lambda_1 x,$$

that is,

$$Ax = \lambda_1 x, \qquad \langle x, x \rangle = 1.$$

Conversely, if $\hat{x}$ is any normalized eigenvector associated with λ_1, then $\langle A\hat{x}, \hat{x} \rangle = \lambda_1$.

Finally, we thus have $\langle Ax, x \rangle \le \lambda_1$ for $\langle x, x \rangle = 1$, and equality takes place if and only if x is a normalized eigenvector associated with λ_1. The second statement of the theorem follows from (2.11) in a similar way. $\qquad\square$

The theorem allows to estimate or possibly also find the maximal and minimal eigenvalues of a Hermitian or symmetric matrix. Substituting the vector $e_k = [0,\ldots,0,1,0,\ldots,0]^T$ with 1 in the kth position for y, we have $\langle y,y\rangle = 1$ and $\langle Ay,y\rangle = a_{kk}$. Hence

$$\lambda_n \leq a_{kk} \leq \lambda_1 \quad \text{for} \quad k = 1,\ldots,n.$$

This implies that

$$\lambda_1 \geq \max_k a_{kk}, \tag{2.12}$$

$$\lambda_1 \leq \min_k a_{kk}. \tag{2.13}$$

We shall be concerned with the bounds for eigenvalues in more detail in Chapter 9.

It is possible to express also further eigenvalues. We present here the expression for the next largest eigenvalue λ_2 without proof (see Exercise 2.10): If y is some of the eigenvectors associated with the largest eigenvalue of the matrix A, then

$$\lambda_2 = \max\{\langle Ax,x\rangle; \langle x,x\rangle = 1, \langle x,y\rangle = 0\} \tag{2.14}$$

(hence $\langle Ax,x\rangle$ is maximized over all such normalized vectors x that are orthogonal to y).

Another useful generalization of Theorem 2.7.3 is the following Courant-Fischer theorem.

Theorem 2.7.4 *Let A be a Hermitian (or symmetric) matrix of order n with eigenvalues $\lambda_1 \geq \lambda_2 \geq \cdots \geq \lambda_n$. Let $\mathscr{L}_m$ denote the set of all m-dimensional linear subspaces of the complex (or real) n-dimensional space of column vectors C. Then*

$$\lambda_m = \max_{L\in\mathscr{L}_m} \min_{0\neq x\in L} \frac{\langle Ax,x\rangle}{\langle x,x\rangle},$$

$$\lambda_{n-m+1} = \min_{L\in\mathscr{L}_m} \max_{0\neq x\in L} \frac{\langle Ax,x\rangle}{\langle x,x\rangle}, \quad m = 1,\ldots,n.$$

This theorem implies the following theorem on separation.

Theorem 2.7.5 *Let A be a Hermitian matrix of order n with eigenvalues $\lambda_1 \geq \lambda_2 \geq \cdots \geq \lambda_n$. Let $A^{(k)}$ be any of its principal submatrices of order k.*
If $\lambda_1^{(k)} \geq \cdots \geq \lambda_k^{(k)}$ are eigenvalues of $A^{(k)}$, then

$$\lambda_j \geq \lambda_j^{(k)} \geq \lambda_{n-k+j}, \quad j = 1, \ldots, k.$$

Exercises

2.1.* Show that an orthonormal set of vectors is linearly independent.

2.2.* Show that real symmetric matrices of order n form a linear space. Find its dimension.

2.3.* Prove Theorem 2.4.4.

2.4.* Prove Theorem 2.4.7.

2.5.* Prove Theorem 2.5.2 in the way suggested in the text.

2.6. Prove that the rank of a symmetric matrix equals the number of its nonzero eigenvalues.

2.7.* Prove that a positive semidefinite matrix of rank 1 has the form uu^*, where $u \neq 0$ is a column vector.

2.8. Prove that a positive semidefinite matrix of rank r is the sum of r positive semidefinite matrices of rank 1.

2.9. Let A be a positive semidefinite matrix of rank 1. If $A = B + C$, where B and C are also positive semidefinite, prove that both the matrices B and C are nonnegative scalar multiples of A.

2.10. Prove the formula (2.14).

2.11.* If A is a Hermitian or real symmetric positive definite (positive semidefinite) matrix prove that there exists a positive definite matrix (a positive semidefinite matrix) W such that $W^2 = A$.

2.12.* Let $A_0, A_1, \ldots, A_m$ be symmetric or Hermitian matrices. If A_0 is positive definite prove that there exists $\varepsilon > 0$ such that the matrix $B(x) = A_0 + xA_1 + x^2A_2 + \cdots + x^mA_m$ is also positive definite for any $x \in [-\varepsilon, \varepsilon]$.

2.13. A complex matrix A is said to be *normal* if $AA^* = A^*A$. Show that any Hermitian matrix as well as any unitary matrix are normal.

2.14. Prove that an upper triangular matrix is normal if and only if it is diagonal.

2.15. Prove that a complex square matrix A is normal if and only if there exist a unitary matrix U and a complex diagonal matrix D such that $A = UDU^*$.

2.16.* Prove that for any complex matrix A, the matrices AA^* and A^*A have the same nonzero eigenvalues.

2.17.* Let $\mathscr{C}$ be a class of matrices of the same dimensions containing the zero matrix. The problem of the *quasidirect decomposition* in $\mathscr{C}$ is defined as follows:

1. To find $\mathscr{C}$-indecomposable matrices in $\mathscr{C}$, i.e. matrices $A \in \mathscr{C}$ for which $A = P + Q, P \in \mathscr{C}, Q \in \mathscr{C}, P + Q$ quasidirect (cf. Exercise 1.10), implies $P = 0$ or $Q = 0$.

2. To express any matrix in $\mathscr{C}$ as a quasidirect sum of $\mathscr{C}$-indecomposable matrices in $\mathscr{C}$ and, in addition, to decide about the uniqueness.

Solve this problem if $\mathscr{C}$ is the class of all (a) symmetric, (b) Hermitian matrices.

2.18. If the sum $A + B$ of Hermitian matrices A and B is quasidirect prove that $\text{In}(A + B) = \text{In } A + \text{In } B$.

3

Graphs and Matrices

In this chapter we study the relationship between graphs and matrices. Here, the graph is viewed as the basic (topological) concept of the graph theory (cf. [33]). We restrict ourselves to finite graphs.

3.1 Digraphs

The *directed graph* or *digraph* $\vec{G} = (V, H)$ is an ordered pair of two finite sets V and H, where the set H consists of some ordered pairs of elements of V, i.e. $H \subset V \times V$. The elements of the set V are called *vertices* (sometimes also *nodes*), the elements of the set H are called *arcs*.

If the number of vertices of a digraph is small, we can represent the digraph by means of a *diagram* in such a way that the vertices of the digraph are represented by points in the plane and the arcs of the digraph are represented by arcs (or segments) with an arrow; these arcs or segments go from the point where "the arc begins" to the point where "the arc ends". An arc may both begin and end at the same point. It is then called a *loop*.

Example 3.1 The digraph $\vec{G} = (V, H)$, where
   $V = \{1, 2, 3, 4, 5\}$ and
   $H = \{(1, 2), (1, 5), (2, 3), (3, 5), (4, 3), (4, 4), (5, 1)\}$,
is represented by the diagram in Fig. 1.

Representing a graph by a diagram, we have to distinguish the points which correspond to the vertices of the graph (and denote them by small circles or bold dots) from possible intersections of

arcs that may occur in diagrams and do not correspond to the
vertices of the graph.

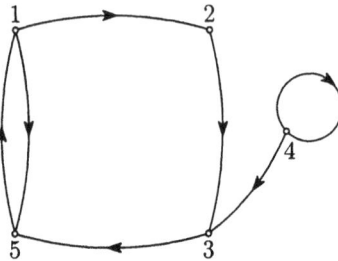

Fig. 1

This representation has also given rise to some terms of the
graph theory. The sequence of vertices $(u_1, u_2, \ldots, u_k)$ such that
$(u_1, u_2), (u_2, u_3), \ldots, (u_{k-1}, u_k)$ are arcs, is called *a directed walk*.
If all the vertices are distinct, the walk is called a *directed path*.
E.g. in $\vec{G}$ of Example 3.1, $(1, 2, 3, 5, 1, 5)$ is a directed walk (from 1
to 5) and $(1, 2, 3, 5)$ is a directed path (from 1 to 5). The number
of arcs in the directed walk or path is then called the *length of
walk* or *path*. We can thus consider the *shortest walk* from the
vertex u to the vertex v. It is such a walk from u to v whose
length is minimal. The shortest walk is clearly a path and hence
we have the following theorem.

Theorem 3.1.1 *If there is a directed walk from a vertex u to a
vertex v in a digraph, then there exists a directed path from u to
v in this digraph as well.*

An important concept of the theory of digraphs is the cycle. A
cycle C in the digraph $\vec{G}$ is, roughly speaking, a path whose first
and last point coincide: $C = (u_1, u_2, \ldots, u_m, u_1)$, $(u_i, u_{i+1}) \in H$,
$i = 1, \ldots, m - 1$, $(u_m, u_1) \in H$. E.g. in the digraph in Fig. 1,
$(1, 2, 3, 5, 1)$ and $(1, 5, 1)$ are cycles. The number m is called the
length of the cycle $(u_1, u_2, \ldots, u_m, u_1)$. Here, it is supposed that
all the vertices $u_1, \ldots, u_m$ are mutually distinct. A loop is considered to be also a cycle, say (u, u), of length 1. In Fig. 1, $(4, 4)$
is a loop.

Let $\vec{G}_1 = (V_1, H_1)$ and $\vec{G}_2 = (V_2, H_2)$ be digraphs. $\vec{G}_1$ is
called a *subdigraph* of $\vec{G}_2$ if $V_1 \subset V_2$ and $H_1 \subset H_2$. $\vec{G}_1$ is called

an *induced subdigraph* of $\vec{G}_2$ if $H_1 = H_2 \cap (V_1 \times V_1)$, i.e., if each arc of $\vec{G}_2$ having both its vertices in V_1 is also an arc of $\vec{G}_1$. Clearly, the induced subdigraph $\vec{G}'$ of a given digraph $\vec{G} = (V, H)$ is uniquely determined by its set of vertices; we thus also speak about a subdigraph $\vec{G}'$ of the digraph $\vec{G}$ *induced* by the set V' of vertices.

Formally, the *empty digraph*, whose set of vertices as well as set of arcs are empty, is considered to be also a digraph. Any other digraph is said to be *non-empty*.

In the study of the structure of digraphs, acyclic digraphs and strongly connected digraphs are important tools.

A digraph is said to be *acyclic* if it contains no cycle.

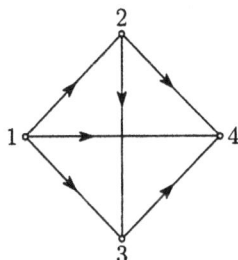

Fig. 2

Example 3.2 The digraph in Fig. 2 is acyclic.

Before proving the fundamental theorem on acyclic digraphs, let us introduce a useful concept. A vertex u in the digraph $\vec{G}$ is called a *source* if there exists no arc in $\vec{G}$ ending in u.

Theorem 3.1.2 *Let* $\vec{G} = (V, H)$ *be a digraph,* $|V| = n$. *Then the following properties are equivalent:*

1. $\vec{G}$ *is acyclic.*

2. *In each non-empty subdigraph* $\vec{G}'$ *of* $\vec{G}$ *there exists a source.*

3. *There exists a labelling of the set* V *of vertices by numbers* $1, 2, \ldots, n$ *such that each arc of* $\vec{G}$ *goes from a vertex labelled by a smaller number to a vertex labelled by a greater number. (Such a labelling is shown in Fig. 2).*

4. $\vec{G}$ *contains no loop and, for each pair of distinct vertices u and v, either there exists no directed path from u to v in $\vec{G}$, or there exists no directed path from v to u in $\vec{G}$.*

Proof. We shall prove the implications 1. → 2., 2. → 3., 3. → 4. and 4. → 1. Thus let the digraph $\vec{G}$ be acyclic and assume that 2. is not true. Then there exists a non-empty subdigraph $\vec{G}'$ of $\vec{G}$ such that each vertex in $\vec{G}'$ is end-vertex of some arc in $\vec{G}'$. Let us begin at some point u_0 of the digraph $\vec{G}'$ (such a point exists). There is an arc ending in u_0, say (u_1, u_0), an arc (u_2, u_1) ending in u_1, an arc (u_3, u_2) ending in u_2, etc. As $\vec{G}'$ has a finite number of vertices the case of $u_p = u_q$, $p > q \geq 0$, has to occur with the smallest p. But then $(u_p, u_{p-1}, \ldots, u_p)$ is a cycle in $\vec{G}'$ and thus in $\vec{G}$ as well, which is a contradiction. Therefore, 1. → 2.

Let the digraph $\vec{G}$ have the property 2. Since $\vec{G}$ itself is a subdigraph of $\vec{G}$ there exists a source in $\vec{G}$; we label it by number 1. Let $\vec{G}_1$ be the subdigraph of $\vec{G}$ obtained by removing the vertex 1 and the arcs beginning at this vertex. We label the source in $\vec{G}_1$, existing by 2., by number 2 and denote by $\vec{G}_2$ the digraph obtained from $\vec{G}_1$ by removing the point 2 and the arcs beginning at this vertex. We again label the source in $\vec{G}_2$ by number 3. Proceeding in this way until all the vertices are labelled, the resulting labelling fulfils 3. since if there were an arc going from i to k, $i > k$, in $\vec{G}$ the vertex k would not have the property that no arc ends in $\vec{G}_{k-1}$ since i is a vertex of $\vec{G}_{k-1}$.

Suppose now that $\vec{G}$ satisfies 3. Then $\vec{G}$ has no loop. Further, if u and v are distinct vertices, we can assume without loss of generality that u is labelled by a smaller number than v. But then there is no directed path from v to u: otherwise the vertex of this path with the maximal number would not be the last one and there would exist an arc of this path beginning at this vertex, which contradicts 3. Thus the implication 3. → 4. holds.

It remains to prove that a digraph satisfying 4. is acyclic. Suppose thus that the digraph $\vec{G}$ fulfils 4. but contains a cycle. This cycle is not a loop and it thus contains two distinct vertices, u and v. For them, however, there exist both a directed path from u to v and a directed path from v to u. This contradiction proves the last implication.   □

A digraph $\vec{G}$ is said to be *strongly connected* if there exists a directed path from each vertex of $\vec{G}$ to any other vertex of $\vec{G}$.

Example 3.3 The digraph in Fig. 1 is not strongly connected as there is no directed path from the vertex 3 to the vertex 4. The subdigraph of this digraph induced by the set of vertices $\{1, 2, 3, 5\}$ is strongly connected. The subdigraph induced by the set of vertices $\{4\}$ is also strongly connected and the subdigraphs induced by the sets $\{1, 5\}, \{1\}, \{2\}, \{3\}$, and $\{5\}$ as well. No other subsets induce strongly connected subdigraphs in $\vec{G}$.

Example 3.4 A cycle is a strongly connected digraph. In fact, we used this assertion, proving that the statements 1., 2., 3. and 4. of Theorem 3.1.2 are equivalent. Equivalence of several statements means that each statement implies any other. However, it suffices to prove only a part of all the implications $i \rightarrow k$, namely such that the digraph with all the vertices k and those arcs (i, k), for which we prove the implication $i \rightarrow k$, is strongly connected.

Let us define the *strong component* of a digraph $\vec{G}$ as any maximal strongly connected induced subdigraph of $\vec{G}$, i.e. such an induced subdigraph that is strongly connected but is contained in no other strongly connected subdigraph of $\vec{G}$.

It follows from Example 3.3 that there are two strong components in the digraph $\vec{G}$ from Fig. 1: the subdigraph induced by the set $\{1, 2, 3, 5\}$ and that induced by the set $\{4\}$. Each point of this digraph is thus contained in one and only one strong component. This is not by chance as we shall see in Theorem 3.1.3. Before we formulate this principal theorem we introduce the concept of the condensation of a digraph. Let $\vec{G} = (V, H)$ be a digraph, let $R = [V_1, \ldots, V_r]$ be a partition of the set V of vertices (i.e. $V_1 \cup \cdots \cup V_r = V$, $V_i \cap V_j = \emptyset$ for $i \neq j$; $i, j = 1, \ldots, r$; $V_i \neq \emptyset$ for all $i = 1, \ldots, r$). Then the *condensation* $\vec{G}_R$ of the digraph $\vec{G}$ with respect to the partition R is the digraph having r vertices $V_1, \ldots, V_r$; (V_i, V_j) is an arc in $\vec{G}_R$ if $i \neq j$ and if there exist vertices $u \in V_i$ and $v \in V_j$ such that $(u, v) \in H$.

Theorem 3.1.3 *The sets of vertices $V_1, \ldots, V_r$ of all the strong components $\vec{G}_1, \ldots, \vec{G}_r$ of the digraph $\vec{G} = (V, H)$ form a partition of the set of vertices V. The condensation $\vec{G}_R$ of $\vec{G}$ with respect to this partition $R = [V_1, \ldots, V_r]$ is an acyclic digraph.*

Two distinct vertices u and v belong to the same set V_k if and only if there exist both a directed path from u to v and a directed path from v to u in $\vec{G}$.

Proof. We first note that if there is a directed path in $\vec{G}$ from the vertex u to the vertex v and from v to the vertex w, then there exists a directed path from u to w in $\vec{G}$ by Theorem 3.1.1. This immediately implies that if $\vec{G}_1$ and $\vec{G}_2$ are strongly connected subdigraphs of $\vec{G}$ having a common vertex, then $\vec{G}_1 \cup \vec{G}_2$ is also a strongly connected subdigraph of $\vec{G}$. As a consequence of this fact we find that the sets of vertices of two distinct strong components are disjoint. Moreover, each vertex is contained in a strongly connected subdigraph of $\vec{G}$ (e.g. in the subdigraph represented by this single vertex), i.e. in one and only one strong component. The sets of vertices $V_1, \ldots, V_r$ of all the strong components of $\vec{G}$ thus form a partition $R = [V_1, \ldots, V_r]$ of the set of vertices V.

In order to prove the second statement of the theorem, let us suppose that there exists a cycle $(V_{k_1}, V_{k_2}, \ldots, V_{k_s}, V_{k_1})$ in the digraph $\vec{G}_R$ obtained as a condensation with respect to the partition R. We have $s > 1$ since there are no loops in $\vec{G}_R$ by definition. Thus $V_{k_1} \neq V_{k_2}$. There is a vertex, say x, in V_{k_1} and there is a vertex, say y, in V_{k_2} as well. We shall show that there exists a directed path from x to y in $\vec{G}$. As (V_{k_1}, V_{k_2}) is an arc of $\vec{G}_R$, there exist vertices $u_{k_1} \in V_{k_1}$, and $v_{k_2} \in V_{k_2}$ such that (u_{k_1}, v_{k_2}) is an arc of $\vec{G}$; similarly, there exist vertices $u_{k_2} \in V_{k_2}$ and $v_{k_3} \in V_{k_3}$ such that $(u_{k_2}, v_{k_3}) \in H$, etc., finally $(u_{k_{s-1}}, v_{k_s}) \in H$. Hence there exists a directed walk $(x, \ldots, u_{k_1}, v_{k_2}, \ldots, u_{k_2}, v_{k_3}, \ldots, u_{k_{s-1}}, v_{k_s}, \ldots, y)$ in $\vec{G}$ since there exists a directed path from x to u_{k_1} in the strong component $\vec{G}[V_{k_1}]$ induced by V_{k_1}, a directed path from v_{k_2} to u_{k_2} in $\vec{G}[V_{k_2}]$ etc., and, finally, a directed path from v_{k_s} to y in $\vec{G}[V_{k_s}]$. By Theorem 3.1.1, there exists a directed path from x to y in $\vec{G}$. Similarly, there exists a directed path from y to x in $\vec{G}$ since there exists an arc (u_{k_s}, v_{k_1}) in $\vec{G}$, $u_{k_s} \in V_{k_s}$, $v_{k_1} \in V_{k_1}$, and a directed path from y to u_{k_s} in $\vec{G}[V_{k_s}]$. Finally, there exists a directed path from v_{k_1} to x in $\vec{G}_i[V_{k_1}]$. But this means that the union of the directed paths from x to y and from y to x is such a strongly connected subdigraph $\vec{S}$ of $\vec{G}$ that has a common vertex with the component V_{k_1}. Therefore $\vec{S} \subset \vec{G}[V_{k_1}]$ so that $y \in V_{k_1}$.

But $y \in V_{k_s} \neq V_{k_1}$, which is a contradiction. The graph $\vec{G}_R$ is thus acyclic.

Finally, if u and v are distinct vertices in the same set V_k, then there exist both a directed path from u to v and a directed path from v to u in $\vec{G}[V_k]$ and, all the more, in $\vec{G}$. Assume that u and v belong to distinct sets V_k and V_j, $k \neq j$. Then both the directed paths, i.e. from u to v and from v to u, cannot be in $\vec{G}$ since it would imply the existence of a directed path both from V_k to V_j and from V_j to V_k in $\vec{G}_R$, which would contradict the acyclicity by 4. of Theorem 3.1.2. □

Corollary 3 *If a digraph $\vec{G}$ has r strong components, then they can be labelled by integers $1, 2, \ldots, r$ in such a way that an arc of $\vec{G}$ never goes from a component with a greater number to a component with a smaller number.*

Proof. Follows from Theorems 3.1.1 and 3.1.3. □

Example 3.5 There are only two strong components in the simple digraph of Fig. 1. One of them contains the vertex 4 and the other contains the vertices $1, 2, 3, 5$.

3.2 Digraph of a matrix

We now turn to the relationship between digraphs and square matrices. This will be useful in the study of some numerical methods in Chapters 12 and 14.

Let $A = [a_{ik}]$ be an $n \times n$ square matrix. Denote by N the set of integers $1, 2, \ldots, n$. With the matrix A we associate a digraph $\vec{G}(A)$ with n vertices in the following way: $\vec{G}(A) = (N, H)$, where H is the set of such pairs (i, k), $i \in N$, $k \in N$, for which $a_{ik} \neq 0$.

Example 3.6 The digraph $\vec{G}(A)$ of the matrix

$$A = \begin{bmatrix} 0 & 2 & 0 & 0 & 1 \\ 0 & 0 & -4 & 0 & 0 \\ 0 & 0 & 0 & 0 & 5 \\ 0 & 0 & -2 & 1 & 0 \\ 3 & 0 & 0 & 0 & 0 \end{bmatrix}$$

is the digraph in Fig. 1.

We may assume that the vertices of $\vec{G}(A)$ are labelled (in the sense of the property 3. of Theorem 3.1.2) since one and only one positive integer of N corresponds to each vertex of $\vec{G}(A)$. We clearly have the next theorem.

Theorem 3.2.1 *Performing a simultaneous permutation of the rows and columns of the matrix A we change only the numbering of vertices in the digraph $\vec{G}$ (the numbering is permuted).*

It means that if we study such properties and algorithms that do not change when simultaneous permutations of rows and columns are performed and that depend only on the fact whether the individual entries are zero or nonzero, then the matrix is described by its digraph regardless of the labelling of the vertices of the digraph. If we wish to express also the values of the entries of the matrix, we can use an *arc-valued digraph*, in which each arc (i, k) is assigned a value. This value is usually called *weight* of the arc and we speak about weighted digraphs. In this case we choose the number a_{ik} for this value. Arcs are not assigned zero values, such arcs are simply omitted.

Conversely, if we wish to associate a matrix to a digraph $\vec{G}$ whose vertices are numbered, the *adjacency matrix* $U(\vec{G}) = [u_{ik}]$ is usually chosen for this matrix. It is defined as an $n \times n$ matrix with entry $u_{ik} = 1$ if $i \neq k$ and there is an arc from the vertex labelled by i to the vertex labelled by k. Otherwise, $u_{ik} = 0$.

We can speak, for example, about *eigenvalues of a digraph, the determinant of a digraph* etc. (These quantities clearly do not change when a simultaneous permutation of rows and columns of the matrix is performed and are thus independent of the labelling of points of the digraph.)

As we shall see especially in Chapter 4 on nonnegative matrices the concept of reducibility or irreducibility of a square matrix is important.

A square matrix A is said to be *reducible* if it has the form

$$\begin{bmatrix} A_1 & B \\ 0 & A_2 \end{bmatrix} \tag{3.1}$$

where A_1 and A_2 are square matrices of order at least 1 or if A can be transformed into the form (3.1) by a simultaneous permutation of rows and columns.

A square matrix is said to be *irreducible* if it is not reducible.

Example 3.7 The matrix of Example 3.6 is reducible, which can be found after interchanging the first and fourth row and the first and fourth column.
The matrix

$$\begin{bmatrix} 0 & 0 & 1 & 0 \\ 0 & 0 & 0 & 1 \\ 0 & 1 & 0 & 0 \\ 1 & 0 & 0 & 0 \end{bmatrix} \tag{3.2}$$

is irreducible as can be shown by performing all the remaining $4! - 1 = 23$ simultaneous permutations of rows and columns. We shall present a better method for extablishing reducibility in a short while.

We have the following important theorem.

Theorem 3.2.2 a) *A square matrix is irreducible if and only if its digraph is strongly connected.*

b) *By a simultaneous permutation of rows and columns (that corresponds to a permutation matrix P), any square matrix can be transformed into the upper block triangular form whose diagonal blocks are no longer reducible,*

$$PAP^T = \begin{bmatrix} A_{11} & A_{12} & A_{13} & \ldots & A_{1r} \\ 0 & A_{22} & A_{23} & \ldots & A_{2r} \\ 0 & 0 & A_{33} & \ldots & A_{3r} \\ \vdots & \vdots & \vdots & \ddots & \vdots \\ 0 & 0 & 0 & \ldots & A_{rr} \end{bmatrix} . \tag{3.3}$$

The diagonal blocks are – apart from permutations within the blocks – determined uniquely and there is a one-to-one correspondence between the blocks and the strong components of the digraph of the given matrix. The ordering of the blocks on the block diagonal corresponds to the labelling of the strong components from Corollary 3. (If there are several such possible labellings, there are several possible upper block triangular forms into which the matrix can be transformed.)

Proof. Let $A = [a_{ik}]$ be a square matrix of order n. Set $N = \{1, 2, \ldots, n\}$ and let $\vec{G}(A) = (N, H)$ be the digraph of this matrix. Suppose first that the matrix A is reducible, i.e., that some simultaneous permutation of rows and columns transforms A into the form

$$\begin{bmatrix} A_1 & B \\ 0 & A_2 \end{bmatrix} \qquad (3.4)$$

with square matrices A_1 and A_2 of order at least 1. Denote by N_1 and N_2 the sets of the vertices of the digraph $\vec{G}(A)$ that correspond to the rows of A_1 and A_2, respectively. The form (3.4) of the matrix implies that in $\vec{G}(A)$ there is no arc going from a vertex of N_2 to a vertex of N_1. It means that the digraph $\vec{G}(A)$ is not strongly connected since N_2 is non-empty and there is no directed walk from any vertex in N_2 to a fixed vertex N_1: If there were such a walk, then an arc would go from the last vertex of this walk that is still in N_2 to the next vertex that is already in N_1. We have thus proved that if the matrix A is reducible, then the digraph $\vec{G}(A)$ is not strongly connected, i.e., if $\vec{G}(A)$ is strongly connected, then A is irreducible.

Conversely, suppose that $\vec{G}(A)$ is not a strongly connected digraph and let $N_1, \ldots, N_r$ (hence $r > 1$) be the sets of vertices of all the strong components of the digraph $\vec{G}(A)$ ordered as in Theorem 3.1.3, i.e., in such a way that in $\vec{G}(A)$ there are no arcs from N_j to N_i for $j > i$. Let us perform the following permutation of rows and columns of A. We first start with the rows (and columns) corresponding to the elements of N_1, continue with the rows (and columns) in N_2 etc. Finally, we end with N_r. The matrix obtained is really of the form (3.3) with irreducible matrices A_{ii} as no arc goes from a vertex in N_j to a vertex in N_i for $j > i$. The matrices A_{ii} are irreducible since the corresponding strong components are strongly connected.

The statement b) is thus proved if we take into account that, conversely, in the digraph of a block triangular matrix with irreducible diagonal blocks, strong components correspond to diagonal blocks. But the rest of the statement a) is also proved since, indeed, the matrix A is reducible if the digraph $\vec{G}(A)$ is not strongly connected.   □

Example 3.8 By the theorem, the matrix (3.2) of Example 3.7 is irreducible as its digraph is the cycle $(1, 3, 2, 4, 1)$.

Theorem 3.2.2, in particular its statement b), can be applied in numerical mathematics, especially to the solution of equations and the computation of matrix eigenvalues. If we are able to transform a matrix into the form (3.3), the solution of a system with this matrix, the computation of an inverse of such a matrix and, in particular, the computation of eigenvalues, which are represented by a collection of the eigenvalues of the matrices $A_{11}, \ldots, A_{rr}$, is substantially easier. The form (3.3) is called the *normal form of a reducible matrix* and will be employed also in Chapter 5.

The digraph $\vec{G} = (V, H)$ may possess the property that with each arc $(i, k) \in H$, the arc (k, i) belongs also to H. Such a digraph will be said to be *symmetric*. The digraph of a symmetric or Hermitian matrix is clearly always symmetric.

3.3 Undirected graphs. Trees

A symmetric digraph can be simply treated as an undirected graph. In graph theory, the *finite undirected graph* (or briefly *graph*) $G = (V, H)$ is introduced as an ordered pair of two finite sets (V, H), where V is the set of vertices and H is the set of some *unordered* pairs of the elements of V, which are called *edges*. A finite undirected graph can also be represented by means of a plane diagram in such a way that vertices of the graph are represented by points in the plane and edges of the graph by segments or arcs joining the corresponding two (possibly also identical) points in the plane. In contrast to the representation of digraphs, the arcs are not equipped with arrows.

For example, Fig. 3 displays a diagram of a graph with eight vertices and nine edges.

It is usually required that an undirected graph contains neither *loops* (i.e. the edges (u, u) where $u \in V$) nor more edges joining the same pair of vertices (the so-called *multiple edges*).

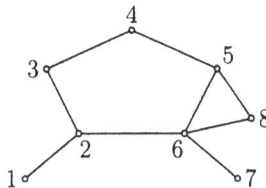

Fig. 3

If (u, v) is an edge of a graph, we say that this edge is *incident* with the vertices u and v or that the vertices u and v are incident with this edge. In a graph containing no loops, a vertex is said to have *degree* k if it is incident exactly with k edges. The vertices of degree 0 are called *isolated*, the vertices of degree 1 are called *end-vertices*. The edge incident with an end-vertex is called *pending edge*.

We have introduced the concepts of a directed walk and a directed path in digraphs. Analogous concepts in undirected graphs are a walk and a path. A *walk* in a graph G is a sequence of vertices (not necessarily distinct), say $(u_1, u_2, \ldots, u_s)$, such that each two consecutive vertices u_k and u_{k+1} $(k = 1, \ldots, s - 1)$ are joined by an edge in G. A *path* in a graph G is then such a walk in which all the vertices are distinct. *A polygon* in G is a walk whose first and last points are identical and, if the last vertex is removed, all the remaining points are distinct. At the same time, this first (and also last) vertex of the walk representing a polygon is not considered distinguished in the polygon. In Fig. 3, the walk $(2, 6, 5, 4, 3, 2)$ is a polygon. Taking into account the number of vertices if the polygon, we speak about a *triangle, a quadrangle*, etc. The polygon with points $2, 6, 5, 4, 3$ in Fig. 3 is a pentagon, the polygon with points $5, 8, 6$ is a triangle.

The terms analogous to those introduced for digraphs can be defined here, too. We speak about a *subgraph* of a given graph and about a *union* of graphs. A *connected graph* is defined as a graph in which there exists a path between any two distinct vertices. An analogue of Theorem 3.1.1 also holds.

Theorem 3.3.1 *If there exists a walk between the vertices u and v in a graph, then there exists a path between u and v in this graph as well.*

Remark. As the set of vertices and edges of the walk represent also a graph, there exists even such a path between u and v in the original graph that is contained in the walk of the theorem.

If the graph G is not connected, we introduce the notion of a *component* of G as such a subgraph of G which is connected but is not contained in any other connected subgraph of G.

The graph in Fig. 4 has three components.

The first component has the vertices $1, 2$ and 3, the second the vertices $4, 5, 6$, and 7, and the third has only the vertex 8.

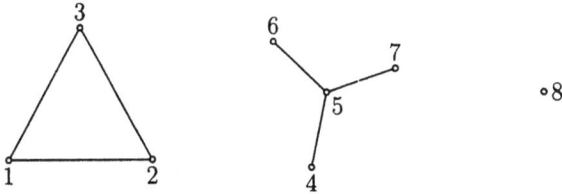

Fig. 4

In contrast to the study of strong components of a digraph, the situation is substantially simpler here: The condensation of a graph G with respect to the partition of the set of vertices into subsets of vertices corresponding to the components is a graph consisting of isolated vertices.

With connected graphs, it is important to study the question how connectivity changes when some edge is removed (the set of vertices remaining the same), or when some vertex as well as all the edges incident with it are removed. An edge of a graph is called a *bridge* if it is not a pending edge and if the graph has more components after removing this edge. A vertex of a connected graph such that the graph has again more components after removing this vertex (together with the incident edge) is called a *cut-vertex*.

The following theorems are useful for the study of cut-vertices and connectivity in general.

Theorem 3.3.2 *If the longest path in the graph G joins the vertices u and v, then neither u, nor v are cut-vertices.*

We leave the proof to the reader as an exercise (see Exercise 3.2).

Theorem 3.3.3 *A connected graph with n vertices, without loops and multiple edges, has at least $n - 1$ edges. If it has more than $n - 1$ edges, it contains a polygon as a subgraph.*

Proof. We shall use induction on n. The theorem is true for $n = 0$ and $n = 1$. Suppose that the number of vertices of a connected graph G is $n > 1$ and that the theorem is true for connected graphs with $n - 1$ vertices. Let the longest path in G join the vertices u and v. By Theorem 3.3.2, the graph G_1 obtained from G by removing the vertices u and the edges incident with u is connected. By the induction hypothesis, the graph G_1 (with $n - 1$ vertices) has at least $n - 2$ edges and, moreover, if G_1 has more that $n - 2$ edges, then it contains a polygon. As u is not an isolated vertex in G the number of edges of G is at least by 1 greater than that of G_1, i.e. at least $n - 1$. Further, if this number is greater than $n - 1$, then either the number of edges of G_1 is already greater than $n - 2$ and then there exists a polygon in G_1, or u is incident with at least two edges (u, w_1) and (u, w_2), $w_1 \neq w_2$. In the graph G_1, however, there exists a path between w_1 and w_2. Completing it with the edges (u, w_1) and (u, w_2), we obtain a polygon in the graph G.   □

We now present a theorem on an important type of connected graphs.

Theorem 3.3.4 *Let G be a connected graph, without loops and multiple edges, with n vertices. Then the following conditions are equivalent:*

1. *The graph G has exactly $n - 1$ edges.*

2. *Each edge of G is either a pending edge, or a bridge.*

3. *There exists one and only one path between any two distinct vertices of G.*

4. *The graph G contains no polygon as a subgraph, but adding any new edge (and no new vertex) to G, we always obtain a polygon.*

5. *The graph G contains no polygon.*

Proof. We shall prove the implications 1. → 2., 2. → 3., 3. → 4., 4. → 5., and 5. → 1.

1. → 2. Removing an arbitrary edge (but no vertex) from the graph G with n vertices and $n - 1$ edges, we obtain a graph with $n - 2$ edges which is not connected by Theorem 3.3.3.

2. → 3. Let 2. hold and let there exist two different paths between two distinct vertices p and q in the graph G ($p = w_0$, $w_1, \ldots, w_s = q$) and ($p = z_0, z_1, \ldots, z_t = q$). Let $w_1 = z_1, \ldots$, $w_k = z_k$, but $w_{k+1} \neq z_{k+1}$ (where $k = 0$ may occur). Then the edge (w_k, w_{k+1}) is neither pending (w_k is also incident with the edge (w_{k-1}, w_k) for $k \geq 1$ or with (w_0, z_1) for $k = 0$, w_{k+1} is also incident with the edge (w_s, z_{t-1}) for $k + 1 = s$), nor a bridge as in G there exists a walk $(w_k, z_{k+1}, z_{k+2}, \ldots, z_t = w_s, w_{s-1}, \ldots, w_{k+1})$ from w_k to w_{k+1} which does not contain the edge (w_k, w_{k+1}).

3. → 4. Let 3. hold. If the graph G contained a polygon, this polygon would have at least three vertices. Between two of them, therefore, there would exist two different paths in G. The graph G thus contains no polygon. Let there be two distinct vertices p and q but not the edge (p, q) in G. Adding the edge (p, q) to G, we obtain a polygon as there was a path from p to q in G.

4. → 5. Clear.

5. → 1. Follows immediately from Theorem 3.3.3. □

A connected graph satisfying one (and then all) of the conditions 1. to 5. of Theorem 3.3.4 is called a *tree*. E.g. the graph in Fig. 5 is a tree and any path is a tree, too. Another example of a tree is a *star*, i.e. a graph with n vertices, $n - 1$ of which are end-vertices and the last one is joined with all these end-vertices.

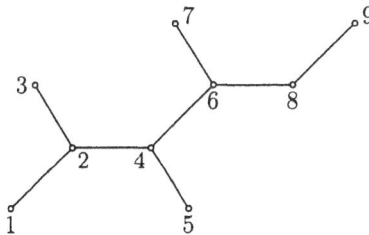

Fig. 5

A subgraph of a connected graph G which has the same vertices as G and which is a tree is called a *spanning tree* of G.

Theorem 3.3.5 *There always exists a spanning tree of a connected graph. Moreover, choosing an arbitrary subgraph S of a connected graph G that contains no polygon, we can find a spanning tree of G that contains S as a subgraph.*

Proof. Remove successively the edges that do not belong to S from the graph G. Do it in such a way that the connectivity is preserved. This process is finished after a finite number of steps. Its result is a connected graph K having the same vertices as the graph G and containing all the edges of S. If K contained a polygon this polygon would not be a subgraph of the graph S by the property 5. of Theorem 3.3.4. Some edge of this polygon that does not belong to S could therefore be removed without affecting connectivity, which contradicts the fact that K is the final result of the process of removing the edges. K is thus the spanning tree of the graph G containing S. □

We now turn back to digraphs for a while. We can also speak about (directed) spanning trees here.

If $\vec{G}$ is a digraph then its subdigraph $\vec{S}$ is called a *directed spanning tree* of $\vec{G}$ if $\vec{S}$ has the same set of vertices as $\vec{G}$ and is "strongly acyclic", i.e., $\vec{S}$ contains no cycle even when the direction of some arcs is changed. Such a directed spanning tree contains, of course, $n - 1$ arcs if n is the number of vertices of $\vec{G}$. By the property 2. of Theorem 3.1.2, there is at least one source in $\vec{S}$.

The following auxiliary theorem holds.

Theorem 3.3.6 *If a directed tree $\vec{S}$ has a unique source p, then there exists a unique directed path from the vertex p to any other vertex of $\vec{S}$.*

Proof. Denote by S the symmetrized graph obtained from $\vec{S} = (V, H)$ in such a way that we add the arc (s, r) to each arc (r, s) of $\vec{S}$ and then consider the graph as undirected. By the property 3. of Theorem 3.3.4, for each vertex $q \neq p$ there exists a unique path C between p and q in S, namely the path $C = (p = r_0, r_1, \ldots, r_s = q)$. We shall show that $(r_0, r_1, \ldots, r_s)$ is a directed path from P to q in $\vec{S}$. Suppose that $(r_k, r_{k+1}) \in H$ does not hold for some $k = 0, \ldots, s - 1$, i.e., (r_k, r_{k+1}) is not an arc of $\vec{S}$. Hence $(r_{k+1}, r_k) \in H$. Since $r_{k+1} \neq p$, r_{k+1} is not a source and there exists a vertex q_1 of $\vec{S}$ such that $(q_1, r_{k+1}) \in H$. Similarly, we have $q_1 \neq p$ as otherwise S would contain the polygon $(p, r_1, \ldots, r_k, r_{k+1}, p)$. Thus there exists a vertex q_2 such that $(q_2, q_1) \in H$, a vertex $q_3 \neq p$ such that $(q_3, q_2) \in H$, etc. Since $\vec{S}$

has a finite number of vertices it has to happen that for some j, q_j is equal to some previous point q_i (or r_{k+1}), which leads to a polygon in S, i.e. a contradiction. Each path from p to q in S is thus directed as a directed path from p to q in $\vec{S}$. □

The following theorem is of importance.

Theorem 3.3.7 *Let a square matrix $A = [a_{ik}]$ of order n have all its column sums equal to zero. Then, of course, $\det A = 0$, and one of the solutions of the system*

$$Ax = 0, \qquad x = [x_1, \ldots, x_n]^T, \tag{3.5}$$

is

$$x_i = \sum_{\vec{K}_i} \pi(\vec{K}_i), \qquad i = 1, \ldots, n, \tag{3.6}$$

where the summation is taken over all such directed spanning trees $\vec{K}_i$ of the digraph $\vec{G}(A)$ of A that have their only source at the vertex i, and $\pi(\vec{K}_i)$ denotes the product of the numbers a_{pq} over all the arcs (p, q) of $\vec{K}_i$.

Proof. By assumption,

$$a_{ii} = -\sum_{k \neq i} a_{ki}, \qquad i = 1, \ldots, n. \tag{3.7}$$

Let us prove that

$$\sum_{i=1}^{n} a_{1i} x_i = 0$$

holds for the numbers x_i from (3.6). This can be rewritten with regard to (3.7) as

$$\sum_{i=2}^{n} a_{1i} x_i = \left(\sum_{k=2}^{n} a_{k1} \right) x_1,$$

or

$$\sum_{i=2}^{n} a_{1i} \sum_{\vec{K}_i} \pi(\vec{K}_i) = \sum_{k=2}^{n} a_{k1} \sum_{\vec{K}_1} \pi(\vec{K}_1). \tag{3.8}$$

On the right-hand side of (3.8), there is a sum of the form $\sum_{\vec{G}_1} \pi(\vec{G}_1)$ ($\pi(\vec{P})$ is the product of the numbers a_{pq} over all the arcs (p, q) of a digraph $\vec{P}$) over all such subdigraphs $\vec{G}_1$ of $\vec{G}$ with n arcs that have no source and the only cycle containing the vertex 1. This can be seen from the fact that the digraphs $\vec{G}_1$ are obtained by the addition of the arc $(k, 1)$ to the directed spanning trees $\vec{K}_1$. Each of them is obtained once and only once since, by Theorem 3.3.6, there is one and only one directed path from 1 to k in $\vec{K}_1$ and, conversely, the spanning tree $\vec{K}_1$ corresponding to each digraph $\vec{G}_1$ is such a digraph that is obtained by removing from $\vec{G}_1$ the arc going to the vertex 1 (in the cycle containing the vertex 1).

The left-hand side of (3.8), however, is equal to the same sum $\sum_{\vec{G}_1} \pi(\vec{G}_1)$. The reason is that the addition of the arc $(1, i)$ of $\vec{G}(A)$ to the spanning tree $\vec{K}_i$, $i \neq 1$, (thus if $a_{1i} \neq 0$) gives one and only one digraph $\vec{G}_1$ since by Theorem 3.3.6 one and only one directed path goes from i to 1 in $\vec{K}_1$. Conversely, such a digraph $\vec{G}_1$ is obtained from one and only one directed spanning tree $\vec{K}_j$, $j \neq 1$, namely from that constructed from $\vec{G}_1$ by removing the (only) arc $(j, 1)$ from $\vec{G}_1$ (in the cycle containing the vertex 1). The proof has thus been completed since the verification of all the other equations in (3.5) is analogous. $\Box$

Example 3.9 Let

$$A = \begin{bmatrix} -(a_{21} + a_{31}) & a_{12} & a_{13} \\ a_{21} & -(a_{12} + a_{32}) & a_{23} \\ a_{31} & a_{32} & -(a_{13} + a_{23}) \end{bmatrix}.$$

Then

$$x_1 = a_{12}a_{13} + a_{12}a_{23} + a_{13}a_{32},$$
$$x_2 = a_{21}a_{13} + a_{21}a_{23} + a_{23}a_{31},$$
$$x_3 = a_{31}a_{12} + a_{32}a_{21} + a_{31}a_{32}.$$

Here the relation $Ax = 0$ for $x = [x_1. x_2, x_3]^T$ can, of course, be verified directly.

Remark. If the rank of the matrix A is $n - 1$, it is even true that not all the unknowns in the solution $x_1, \ldots, x_n$ given in (3.6) are zero. In this case, one can show that the number x_i, $i = 1, \ldots, n$, is equal to the cofactor of each entry a_{ij} of the matrix A premultiplied by $(-1)^{n-1}$ for $j = 1, \ldots, n$. In this way we can also calculate the determinant of a square matrix C of order m or solve the linear system

$$Cy = d \tag{3.9}$$

with a nonsingular matrix C. Indeed, if e^T is an m-component row vector, $e^T = [1, \ldots, 1]$, then the matrix

$$A = \begin{bmatrix} C & -d \\ -e^T C & e^T d \end{bmatrix} \tag{3.10}$$

fulfils the assumptions of Theorem 3.3.7. If $x_1, \ldots, x_{m+1}$ are the numbers from (3.6) then $x_{m+1} = (-1)^m \det C \neq 0$ and $y = [y_1, \ldots, y_m]^T$ with $y_k = x_k / x_{m+1}$, $k = 1, \ldots, m$.

This procedure for solving the system (3.9) can, however, be recommended only in the case when there is only a small number of arcs in the digraph $\vec{G}(A)$ of the matrix (3.10) (i.e., a small number of nonzero off-diagonal entries in A), or when this digraph and its spanning trees can easily be analysed. (Cf. Exercise 3.6.)

3.4 Bigraphs

Let us turn to bipartite graphs. An undirected graph G is said to be *bipartite* if each polygon in G has an even number of edges. We present the basic theorem on bipartite graphs.

Theorem 3.4.1 *Let $G = (V, H)$ be an undirected graph, without loops and multiple lines, having at least two vertices. Then the following conditions are equivalent:*

1. *G is a bipartite graph.*

2. *Whenever u and v are distinct vertices of G then either the length of any two paths joining u and v is an even number, or the length of any two paths joining u and v is an odd number.*

3. *There exists a partition V_1, V_2 of the set of vertices V (i.e., $V = V_1 \cup V_2$, $V_1 \cap V_2 = \emptyset$, $V_1 \neq \emptyset$, $V_2 \neq \emptyset$) such that if the*

edge (u, v) belongs to H, then either $u \in V_1$ and $v \in V_2$, or $u \in V_2$ and $v \in V_1$.

4. If $U(G)$ is the adjacency matrix of G (viewed as a symmetric digraph) for some chosen numbering of vertices, then there exists a permutation matrix P such that

$$PU(G)P^T = \begin{bmatrix} 0 & A_{12} \\ A_{12}^T & 0 \end{bmatrix} \qquad (3.11)$$

where the diagonal blocks are square matrices of order at least 1.

Proof. We shall prove the implications $1. \to 2., 2. \to 3., 3. \to 1.$ and the equivalence $3. \leftrightarrow 4.$

$1. \to 2.$ Let 1. hold. Let u and v be two distinct vertices of the graph G. Further let $C_1 = (u = w_0, w_1, \ldots, w_s = v)$ and $C_2 = (u = z_0, z_1, \ldots, z_t = v)$ be paths joining u and v. By induction on the sum $s + t$ of the lengths of the two paths we shall prove that $s + t$ is even and, therefore, s and t are either both odd, or both even. If $s + t = 2$, i.e. $s = t = 1$, the implication is true. Let thus $s + t > 2$ for the paths C_1 and C_2 and suppose that the implication is true for any two paths between any two vertices of G whose sum of lengths is less than $s + t$. If C_1 and C_2 have no common vertices except for u and v, then C_1 and C_2 form a polygon of length $s + t$ and thus $s + t$ is even. Therefore the implication holds. If they have some common vertex distinct both from u and v, e.g. $w_j = z_k$, $j > 0$, $k > 0$, then let j be the least such subscript for which this holds. Then $(w_0, w_1, \ldots, w_j, z_{k-1}, z_{k-2}, \ldots, z_0)$ is a polygon of length $j + k$ and thus $j + k$ is even. The sum of the lengths of the remaining parts of the paths C_1 and C_2 between w_j and v (of length $s - j$) and between z_k and v (of length $t - k$) is less than $s + t$ and thus the sum $s + t - j - k$ is even by the induction hypothesis. But since $j + k$ is also even, $s + t$ is even.

$2. \to 3.$ Let $G_1, \ldots, G_r$ be all the components of the graph G. If each component contains only one vertex, the partition in 3. does exist. If some component, say G_1, contains more than one vertex, we choose a fixed vertex u_i in each component G_i. We then introduce V_1 as the set of all the vertices in V whose distance from some of the vertices u_i (i.e., the length of the shortest path to u_i) is even and V_2 as the set of all the vertices in V whose distance from some of the vertices u_i is odd. Clearly V_1 and V_2 are well defined (for each vertex, the distance from one and only one vertex u_i is defined), $V = V_1 \cup V_2$, $V_1 \cap V_2 = \emptyset$, $V_1 \neq \emptyset$, $V_2 \neq \emptyset$. We shall show that the partition $[V_1, V_2]$ fulfils the requirement of 3. Let $(u, v) \in H$, i.e., $u \neq v$ and both the vertices u and v belong to the same component, say G_1. Let $C = (u = w_0, w_1, \ldots, w_s = u_1)$ be the shortest path (or one of them) between u and u_1. We shall distinguish two cases:

a) The length s of C is even and thus $u \in V_1$. If v is one of the vertices $w_1, \ldots, w_s$, then $v = w_j$ for j odd as (u, v) is a path between u and v of length 1. Therefore $v \in V_2$ since the length of the path $(v = w_j, w_{j+1}, \ldots, w_s = u_1)$ is $s - j$, and thus odd, and even the length of the shortest path between v and u_1 is odd by 2. If the vertex v is not contained in C, then $(v, u = w_0, w_1, \ldots, w_s = u_1)$ is a path between v and u_1. Its length is odd, and therefore $v \in V_2$ again.

b) s is odd and thus $u \in V_2$. We can prove that $v \in V_1$ in the way analogous to that used in a).

3. $\rightarrow$ 1. Let 3. hold and let $K = (u_0, u_1, \ldots, u_s = u_0)$ be a polygon of length s in the graph G. If $u_0 \in V_1$, then $u_1 \in V_2$, $u_2 \in V_1$, etc., i.e. s is even. If $u_0 \in V_2$, then $u_1 \in V_1$, $u_2 \in V_2$, etc., thus s is again even.

To show the equivalence of 3. and 4., assume that 3. holds. Choose a labelling of the vertices of G such that the numbers assigned to the vertices in V_1 are less than those assigned to the vertices in V_2. If the adjacency matrix corresponding to the initial labelling was $U(G)$, then it is now of the form

$$PU(G)P^T = \begin{bmatrix} 0 & A_{12} \\ A_{12}^T & 0 \end{bmatrix},$$

where V_1 corresponds to the subscripts of the first block row (and column) and V_2 to the subscripts of the second block row (and column). Conversely, if the adjacency matrix is of the form (3.11) for some labelling, then we denote by V_1 the set of vertices corresponding to the first block row (and column) of the matrix on the right-hand side of (3.11) and by V_2 the set of vertices corresponding to the second block row (and column). Then the partition $[V_1, V_2]$ fulfils the conditions of 3. $\qquad\square$

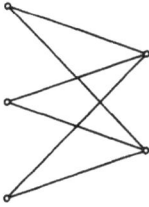 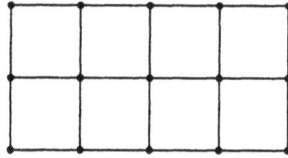

Fig. 6   Fig. 7

Example 3.10. The graph in Fig. 6 is bipartite. (It is called the *complete bipartite graph* $K_{3,2}$.)

Any tree is also a bipartite graph. The *rectangular lattice* M_{pq} ($M_{3,5}$ is in Fig. 7) is also a bipartite graph.

We have seen that the digraph $\vec{G}(A)$ of a matrix A does not change (except for the numbering of vertices) when the rows and columns of A are simultaneously permuted. In the conclusion of this chapter we show another way of associating a digraph with a (not necessarily square) matrix.

A *bigraph* $\vec{B} = (V_1, V_2, H)$ is a digraph with the set of vertices $V_1 \cup V_2$ partitioned into two non-empty disjoint subsets V_1 and V_2; V_1 is the first subset, V_2 the second. The set of arcs H of the bigraph contains only arcs which go from a vertex in V_1 to a vertex in V_2: $H \subset V_1 \times V_2$. As a rule, we shall assume – especially when a matrix is represented by a bigraph – that the subset V_1 is labelled by the numbers $1, 2, \ldots, |V_1|$ and the subset V_2 by the numbers $1, 2, \ldots, |V_2|$.

With a given $m \times n$ matrix $A = [a_{ik}]$ we associate a bigraph $\vec{B}(A) = (V_1, V_2, H)$ where the first set of vertices $V_1 = \{1, 2, \ldots, m\}$ has m vertices, the second set $V_2 = \{1, 2, \ldots, n\}$ has n vertices, and H is represented by those pairs (i, k), $i \in V_1, k \in V_2$, for which $a_{ik} \neq 0$.

Example 3.11 The bigraph of the matrix

$$A = \begin{bmatrix} 1 & -1 & 0 \\ 0 & 3 & -2 \end{bmatrix}$$

is shown in Fig. 8.

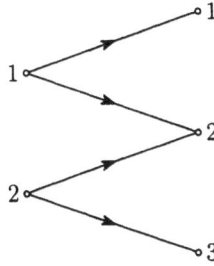

Fig. 8

As compared with the digraph of a matrix, the advantage of the bigraph $\vec{B}(A)$ of a matrix A is that any independent permutation of the rows of A and the columns of A affects only the numbering in the sets V_1 and V_2 of the bigraph. On the other hand, the digraph $\vec{G}(A)$ of a square matrix A of order n has only n vertices while the bigraph $\vec{B}(A)$ has $2n$ vertices.

We shall return to the application of bigraphs in Chapter 14. Here we only present some additional facts. The bigraphs $\vec{B}_1 = (V_1, V_2, H^{(1)})$ and $\vec{B}_2 = (W_1, W_2, H^{(2)})$ with labelled sets of vertices can be added if $V_1 = W_1$ and $V_2 = W_2$ and multiplied if $V_2 = W_1$. Under these conditions, the *sum* of the bigraphs $\vec{B}_1$ and $\vec{B}_2$ is such a bigraph $\vec{B}_3 = (V_1, V_2, H^{(3)})$ for which $(i, k) \in H^{(3)}$, where $i \in V_1 = W_1$ and $k \in V_2 = W_2$, if and only if at least one of the statements $(i, k) \in H^{(1)}$, $(i, k) \in H^{(2)}$ is true. The *product* of the bigraphs $\vec{B}_1$ and $\vec{B}_2$ is such a bigraph $\vec{B}_4 = (V_1, W_2, H^{(4)})$ for which $(i, k) \in H^{(4)}$, where $i \in V_1$ and $k \in W_2$, if and only if there exists at least one vertex $j \in V_2 = W_1$ for which $(i, j) \in H^{(1)}$ and $(j, k) \in H^{(2)}$ as well.

The bigraph $\vec{B}^T = (V_2, V_1, H^T)$ is called the *transpose* of $\vec{B} = (V_1, V_2, H)$ if $(k, i) \in H^T \leftrightarrow (i, k) \in H$ holds for $i \in V_1$ and $k \in V_2$. If $V_1 = W_1$ and $V_2 = W_2$, we can speak about the intersection of the bigraphs $\vec{B}_1$ and $\vec{B}_2$, their union (which is equal to their sum), inclusion etc. If $V_1 \subset W_1, V_2 \subset W_2$, and $H^{(1)} \subset H^{(2)}$, then the bigraph $\vec{B}_1$ is contained in the bigraph $\vec{B}_2$; it is called a subbigraph. $\vec{B}_1$ is an *induced subbigraph* if $H^{(1)} = H^{(2)} \cap (V_1 \times V_2)$, i.e., if $(i, k) \in H^{(1)}$ if and only if $i \in V_1$, $k \in V_2$, and $(i, k) \in H^{(2)}$. The addition, multiplication, and further operations with bigraphs are thus in a one-to-one correspondence with the addition, multiplication, and further operations with Boolean matrices (cf. Chapter 4).

Theorem 3.4.2 *If A_1 and A_2 are two $m \times n$ matrices, then*

$$\vec{B}(A_1 + A_2) \subset \vec{B}(A_1) + \vec{B}(A_2).$$

If the product $A_1 A_2$ is defined, then

$$\vec{B}(A_1 A_2) \subset \vec{B}(A_1)\vec{B}(A_2).$$

Finally, $\vec{B}(A^T) = (\vec{B}(A))^T$.

In the conclusion, we shall prove a theorem analogous to the statement a) of Theorem 3.2.2.

A square matrix A is called *partly decomposable* if there exists a permutation matrix P such that PA is reducible. If a square matrix is not partly decomposable, then it is said to be *fully indecomposable*. This statement can be formulated also in other words.

Theorem 3.4.3 *A square matrix A is partly decomposable if and only if there exist permutation matrices P and Q such that*

$$PAQ = \begin{bmatrix} A_1 & B \\ 0 & A_2 \end{bmatrix} \qquad (3.12)$$

where A_1 and A_2 are square matrices of order at least 1. The matrix A is fully indecomposable if there is no pair of permutation matrices P and Q satisfying the relation (3.12).

We shall now introduce the concept of the stable subset of the set V_1 in a *square bigraph* $\vec{B} = (V_1, V_2, H)$, i.e. a bigraph with $|V_1| = |V_2|$. A subset $W_1 \subset V_1$ is said to be *stable* in $\vec{B}$ if the inequality

$$|W_2| \leq |W_1| \qquad (3.13)$$

holds for the set of vertices

$$W_2 = \{k \in V_2; \text{ there exists } i \in W_1 \text{ such that } (i, k) \in H\}.$$

Clearly, both the empty set and the whole set V_1 are stable; they will be called *improper stable subsets*. The *proper stable subset* is a stable subset W_1 such that $\emptyset \neq W_1 \neq V_1$. The following theorem now holds.

Theorem 3.4.4 *A square matrix A is partly decomposable if and only if its bigraph $\vec{B}(A)$ contains a proper stable subset. In other words, a square matrix A is fully indecomposable if and only if its bigraph contains no proper stable subset.*

Proof. Let the matrix A be partly decomposable, i.e., let (3.12) hold for some permutation matrices P and Q. Thus there exists a proper subset $W_1 \subset V_1$ (corresponding to the indices in the second block row) in the bigraph $\vec{B}(PAQ) = (V_1, V_2, H)$ such that the set W_2 (corresponding to the indices in the second block column) satisfies the inequality $|W_2| \le |W_1|$. W_1 is thus a stable subset in $\vec{B}(PAQ)$ and in $\vec{B}(A)$ as well since $\vec{B}(A)$ is obtained from $\vec{B}(PAQ)$ only by a different labelling of V_1 and V_2.

Conversely, let W_1 be an arbitrary proper stable subset of the bigraph $\vec{B}(A) = (V_1, V_2, H)$. Label V_1 and V_2 in such a way that all the elements not contained in W_1 are labelled first and the elements of W_1 are labelled afterwards. The way of labelling W_2 is analogous (and, if necessary, arbitrary elements are added to W_2 in order that $|W_2| = |W_1|$). After the corresponding permutations of rows and columns, the matrix A will be of the form (3.12). The rest of the proof is clear. □

Exercises

3.1.* Prove that a digraph without loops is acyclic if and only if each its strong component has just one vertex.

3.2. Prove Theorem 3.3.2.

3.3. Prove that if G is a tree, then each its vertex is either an end-vertex or a cut-vertex.

3.4. Give an example of a connected graph whose each vertex is either an end-vertex or a cut-vertex and which is not a tree.

3.5.* Let S be a tree with n vertices, let u_1 be its fixed vertex. Show that there exists a labelling $\varphi(u)$ of the vertices of the graph S by mutually different numbers $1. 2, \dots, n$ such

that the following statement holds: If u_2 and u_3 are distinct
vertices such that a path from u_1 to u_3 contains u_2, then
$\varphi(u_2) < \varphi(u_3)$.

3.6. Employing the remark following Theorem 3.3.7, calculate
the determinant of the $n \times n$ matrix

$$\begin{bmatrix} 2 & -1 & 0 & \dots & 0 & 0 \\ -1 & 2 & -1 & \dots & 0 & 0 \\ \vdots & \vdots & \vdots & \ddots & \vdots & \vdots \\ 0 & 0 & 0 & \dots & 2 & -1 \\ 0 & 0 & 0 & \dots & -1 & 2 \end{bmatrix}.$$

3.7. Find all proper stable subsets of the bigraph of the matrix
of Example 3.6.

3.8. Prove that a square matrix A is partly decomposable (fully
indecomposable) if and only if its transpose A^T is partly
decomposable (fully indecomposable).

3.9. Prove the following Frobenius-König theorem [40]: Let A
be an $n \times n$ matrix. Then all terms of $\det A$ are equal to
zero if and only if there exists an $s \times t$ zero submatrix of A
with $s + t = n + 1$.

3.10. Prove that a matrix A is fully indecomposable if and only
if there exists a permutation matrix P such that PA is
irreducible and has all diagonal entries nonzero.

4

Nonnegative Matrices. Stochastic and Doubly Stochastic Matrices

In this chapter, we are concerned with matrices whose entries are nonnegative real numbers possessing possibly some additional properties. Such matrices occur in engineering (reactor problems), in the theory of stochastic processes, and they are also useful in numerical analysis, e.g. in estimates, as will be seen later.

4.1 Nonnegative matrices

A matrix A is said to be *nonnegative* if all of its entries are nonnegative. In this case, we write $A \geq 0$. If all of the entries are positive, A is said to be *positive* and we write $A > 0$. Similarly, if A and B are matrices of equal dimensions, $A \geq B$ or $A > B$ means that $A - B \geq 0$ or $A - B > 0$, respectively.

Nonnegative matrices have some properties that are evident.

Theorem 4.1.1 *The sum of nonnegative matrices of the same dimensions is again a nonnegative matrix. If the product of two nonnegative matrices, A and B, is defined, then AB is a nonnegative matrix. The product AB of a positive matrix A and a nonnegative matrix $B \neq 0$ is a nonnegative matrix different from the zero matrix.*

Corollary 4 *If* x *and* y *are* n-*component column vectors such that*

$$x \geq y$$

and if A *is an* m × n *nonnegative matrix, then*

$$Ax \geq Ay.$$

If $x \geq y, x \neq y$, *and if* $A > 0$, *then*

$$Ax > Ay.$$

We now introduce the concept of the *zero-nonzero structure* of a matrix. Two matrices of equal dimensions, $P = [p_{ik}]$ and $Q = [q_{ik}]$, are said to be of the same zero-nonzero structure if for all i, k we have

$$p_{ik} \neq 0 \qquad \text{if and only if} \qquad q_{ik} \neq 0.$$

Clearly, this is an equivalence relation; the zero-nonzero structure is then defined to be a class of matrices equivalent under this relation.

The most natural approach to the study of zero-nonzero structures uses the so-called *Boolean matrices*. These are matrices whose entries are 0 and 1. Operations of addition, scalar multiplication, and matrix multiplication can be defined for the Boolean matrices. All these operations are defined like with ordinary matrices but the addition and multiplication of the entries 0 and 1 is performed in the Boolean manner, that is

$$
\begin{aligned}
&0 + 0 = 0,\\
&0 + 1 = 1 + 0 = 1,\\
&1 + 1 = 1;\\
&0 \cdot 0 = 0 \cdot 1 = 1 \cdot 0 = 0,\\
&1 \cdot 1 = 1.
\end{aligned}
\tag{4.1}
$$

With every matrix $A = [a_{ik}]$ whose entries are (real or complex) numbers we can then associate a Boolean matrix A_B defined as $A_B = [\alpha_{ik}]$ where $\alpha_{ik} = 1$ if $a_{ik} \neq 0$, and $\alpha_{ik} = 0$ if $a_{ik} = 0$. The matrix A_B is called the *Boolean representation* of A.

The proof of the following theorem on nonnegative matrices is immediate and is left to the reader (Exercise 4.1).

Theorem 4.1.2 *The zero-nonzero structure of the sum or product of two non-negative matrices is uniquely determined by zero-nonzero structures of both the summands or factors. Moreover, we have (with $P \geq 0$ and $Q \geq 0$)*

$$(P + Q)_B = P_B + Q_B, \tag{4.2}$$

$$(PQ)_B = P_B Q_B, \tag{4.3}$$

where, of course, the operations of addition and multiplication on the right-hand sides are Boolean.

The zero-nonzero structure of a matrix may be also characterized by the bigraph or (for square matrices) by the digraph of the matrix.

We shall now investigate the connection between the zero-nonzero structure of a power of a nonnegative square matrix A and walks in the digraph $\vec{G}(A)$ of this matrix.

Theorem 4.1.3 *Given a nonnegative square matrix A and a positive integer k, then the (i, j)-entry of A^k is nonzero if and only if there is a walk of length k from the point i to j in $\vec{G}(A)$.*

Proof. We use induction on k. The theorem holds for $k = 1$. Supposing that the theorem is true for k we shall prove that it holds for $k+1$ as well. Set $A^k = B = [b_{pq}]$, $A = [a_{pq}]$, $A^{k+1} = [c_{pq}]$. Then

$$c_{ij} = \sum_p b_{ip} a_{pj}. \tag{4.4}$$

First, suppose that there is a walk of length $k + 1$ from i to j in the digraph $\vec{G}(A)$, say $(i, p_1, \ldots, p_k, j)$. By the induction hypothesis, $b_{ip_k} > 0$. Since also $a_{p_k j} > 0$ and all terms in (4.4) are nonnegative, we have $c_{ij} > 0$. Hence the (i, j)-entry of A^{k+1} is nonzero. Conversely, suppose that the (i, j)-entry of A^{k+1} is different from zero (and thus positive). Then the sum in (4.4) contains a positive term, say $b_{it} a_{tj} > 0$. This means $b_{it} > 0$ and $a_{tj} > 0$; by induction hypothesis (as the (i, t)-entry of A^k is nonzero), there is a walk of length k from i to t in $\vec{G}(A)$. Therefore, since (t, j) is an arc in $\vec{G}(A)$, there exists a walk of length $k + 1$ from i to j. $\qquad\square$

Theorem 4.1.4 *Let A be an irreducible nonnegative square matrix of order n, and let $k_0, k_1, \ldots, k_{n-1}$ be positive numbers. Then the matrix $k_0 I + k_1 A + k_2 A^2 + \cdots + k_{n-1} A^{n-1}$ is positive. In particular, $(I + A)^{n-1} > 0$.*

Proof. Let A be nonnegative and irreducible. Since the matrix $\sum_{i=0}^{n-1} k_i A^i$ is a sum of nonnegative matrices, it is sufficient to show that for arbitrary fixed i and j the (i, j)-entry of a certain power A^p, $0 \le p \le n - 1$, is positive. This is true for $i = j$ as the diagonal entries of I are positive. Suppose now $i \ne j$. Then there is a path from i to j in the digraph $\vec{G}(A)$ since $\vec{G}(A)$ is strongly connected by Theorem 3.2.2. The length d of the path is at most $n - 1$. By Theorem 4.1.3, however, the (i, j)-entry of A^d is then positive, and the first statement of the theorem is proved. The second statement is a particular case of the first, obtained by the binomial theorem. □

In the sequel, we shall need two lemmas giving bounds for the spectral radius $\rho(A)$ of a matrix A. Before proving them, however, we introduce one notation. Given an (in general complex) matrix $A = [a_{ik}]$, we denote by $m(A)$ the matrix $[|a_{ik}|]$ and call it the *modulus* of A.

Lemma 4.1.5 *If A and B are square matrices, and if $m(A) \le B$, then $\rho(A) \le \rho(B)$. In particular, $\rho(A) \le \rho(m(A))$.*

Proof. Suppose that $A \le B$ but $\rho(A) > \rho(B)$. Denote by s a number satisfying $\rho(A) > s > \rho(B)$ and set $P = (1/s)A$, $Q = (1/s)B$. Then we have

$$\rho(P) = s^{-1}\rho(A) > 1, \quad \rho(Q) = s^{-1}\rho(B) < 1. \qquad (4.5)$$

By Theorem 1.7.13, this means $\lim_{k \to \infty} Q^k = 0$. But $m(P^k) \le (m(P))^k \le Q^k$. It follows that also $\lim_{k \to \infty} P^k = 0$, i.e. $\rho(P) < 1$, which contradicts (4.5). The rest of the proof is clear. □

Lemma 4.1.6 *Let A be a nonnegative matrix and let z be a nonnegative vector. If ξ is a real number such that*

$$Az > \xi z,$$

then

$$\rho(A) > \xi.$$

Proof. We may assume $\xi \geq 0$. Clearly $z \neq 0$. Since $Az > \xi z$ there exists $\varepsilon > 0$ such that

$$Az \geq (\xi + \varepsilon)z.$$

It means that the matrix $B = (\xi + \varepsilon)^{-1}A$ satisfies

$$Bz \geq z,$$

from which

$$B^k z \geq B^{k-1} z \geq \cdots \geq z$$

for any integer $k > 0$. Hence, the matrix B^k does not converge to the zero matrix,

$$\rho(B) \geq 1,$$

and finally

$$\rho(A) \geq \xi + \varepsilon > \xi.$$

$\square$

4.2   The Perron-Frobenius theorem

Now we are able to prove the *Perron-Frobenius theorem* [26] on nonnegative matrices.

Theorem 4.2.1 *Let $A \geq 0$ be an irreducible square matrix of order n, $n > 1$. Then $\rho(A)$ is a simple positive eigenvalue of A, and there is a positive eigenvector associated with the eigenvalue $\rho(A)$. No nonnegative eigenvector is associated with any other eigenvalue of A.*

Remark. The eigenvalue $\rho(A)$ is usually called the *Perron root* of A, and the associated eigenvector is the *Perron vector* of A.
Proof. First we prove a lemma (cf. [52]).

Lemma (Perron). *If $A > 0$, then $\rho(A)$ is a positive eigenvalue of A and there is only one linearly independent eigenvector belonging to the eigenvalue $\rho(A)$. Moreover, this eigenvector may be chosen positive.*

Proof. The lemma is true for $n = 1$. Let $n > 1$ and $A > 0$. There exists an eigenvalue λ of A such that

$$\rho(A) = |\lambda|. \tag{4.6}$$

Denote by u an eigenvector belonging to λ,

$$Au = \lambda u, \qquad u \neq 0. \tag{4.7}$$

If v and w are complex numbers, and α, β are positive real numbers, it is clear that

$$|\alpha v + \beta w| \leq \alpha |v| + \beta |w|,$$

where equality is valid if and only if v and w belong to the same half-line with initial point 0 or, in other words, if and only if there is a complex unit (a number of modulus 1) η such that $\eta v \geq 0$ and $\eta w \geq 0$ hold simultaneously. The inequality may easily be generalized to the case

$$\left| \sum_{i=1}^{n} \alpha_i v_i \right| \leq \sum_{i=1}^{n} \alpha_i |v_i|$$

for arbitrary complex numbers $v_1, \ldots, v_n$ and positive real numbers $\alpha_1, \ldots, \alpha_n$. Here, equality takes place if and only if there exists a complex unit η such that

$$\eta v_i \geq 0, \qquad i = 1, \ldots, n.$$

We now use this result to show that the vector u of (4.7) has the property that there is a complex unit η such that

$$\eta u_i \geq 0 \qquad \text{for} \quad i = 1, \ldots, n. \tag{4.8}$$

To prove this, assume that (4.8) does not hold. Then we have

$$|\lambda| |u_k| = \left| \sum_{j=1}^{n} a_{kj} u_j \right| < \sum_{j=1}^{n} a_{kj} |u_j|$$

in the kth equation of (4.7) by the above statement. This is true for every $k = 1, \ldots, n$, and thus

$$Am(u) > |\lambda| m(u).$$

But by Lemma 4.1.6,

$$|\lambda| < \rho(A),$$

which contradicts (4.6).

The inequalities (4.8) imply that the vector

$$v = \eta u, \qquad v \neq 0,$$

is nonnegative and, by (4.7), it satisfies

$$Av = \lambda v. \tag{4.9}$$

If $v_k \neq 0$, and thus $v_k > 0$, then the kth equation of (4.9) gives

$$\lambda > 0.$$

Hence $\lambda = \rho(A)$ and using (4.9) again, we finally obtain $v > 0$.

In particular, we have proved the following implication: *If λ is an eigenvalue such that*

$$|\lambda| = \rho(A)$$

and if u is an associated eigenvector, then

$$m(u) > 0.$$

Now suppose that there are two linearly independent eigenvectors, say $v = [v_i]$, and $w = [w_i]$, associated with the eigenvalue λ. Since $v \neq 0$, there exists an integer k such that $v_k \neq 0$. The vector $z = w - (w_k v_k^{-1})v$ is also an eigenvector of A belonging to λ since $z \neq 0$. But $z_k = 0$, which contradicts the result proved above which states that $m(z) > 0$. This completes the proof of the lemma. $\qquad\square$

We now return to the proof of Theorem 4.2.1. The matrix A is irreducible. By Theorem 4.1.4, the matrix $(I + A)^{n-1}$ is positive and thus the matrix $(I + A^T)^{n-1} = ((I + A)^{n-1})^T$ is positive as well. By Perron's lemma, there exists a vector $y > 0$ such that

$$y^T(I + A)^{n-1} = \rho((I + A)^{n-1})y^T. \tag{4.10}$$

Denote by λ a (complex) eigenvalue of A such that

$$|\lambda| = \rho(A). \tag{4.11}$$

Let x be an eigenvector (hence $x \neq 0$) associated with λ, i.e.

$$Ax = \lambda x.$$

Then we have

$$|\lambda| m(x) \leq Am(x),$$

or

$$\rho(A) m(x) \leq Am(x).$$

Further

$$\rho^2(A) m(x) \leq \rho(A) Am(x) = A\rho(A) m(x) \leq A^2 m(x),$$

and, in general,

$$\rho^k(A) m(x) \leq A^k m(x), \quad k = 1, 2, \ldots. \tag{4.12}$$

Hence

$$(1 + \rho(A))^{n-1} m(x) \leq (I + A)^{n-1} m(x) \tag{4.13}$$

and, left-multiplying by the positive vector y^T,

$$(1 + \rho(A))^{n-1} (y^T m(x)) \leq y^T (I + A)^{n-1} m(x);$$

the right-hand side equals

$$\rho((I + A)^{n-1})(y^T m(x))$$

according to (4.10). Since $y^T m(x)$ is a positive number, we have

$$(1 + \rho(A))^{n-1} \leq \rho((I + A)^{n-1}). \tag{4.14}$$

By Theorem 1.7.12, the eigenvalues of $(I + A)^{n-1}$ are of the form $(1 + \alpha)^{n-1}$, where α runs through the eigenvalues of A. This means that there is an eigenvalue μ of A such that

$$|(1 + \mu)^{n-1}| = \rho((I + A)^{n-1}). \tag{4.15}$$

On the other hand, we have $|\mu| \leq \rho(A)$. Substituting into (4.14) we therefore have

$$(1 + \rho(A))^{n-1} \leq |(1 + \mu)^{n-1}|,$$

and further

$$1 + \rho(A)) \leq |1 + \mu| \leq 1 + |\mu| \leq 1 + \rho(A).$$

This implies

$$\mu \geq 0,$$

and thus

$$\mu = \rho(A).$$

Equality is valid also in all the inequalities that we have added, i.e., in (4.12). In particular, for $k = 1$ we have

$$Am(x) = \rho(A)m(x),$$

or

$$Am(x) = \mu m(x)$$

and, by (4.13) and (4.15),

$$(I + A)^{n-1}m(x) = (1 + \mu)^{n-1}m(x) = \rho((I + A)^{n-1})m(x).$$

Using Perron's lemma we now get

$$m(x) > 0.$$

It follows that there exists only one linearly independent eigenvector associated with the eigenvalue μ by the same argument as that used in the proof of Perron's lemma. Moreover, $\rho(A) > 0$ since A is a non-zero matrix ($n > 1$!).

To complete the proof, we must show, first of all, that $\rho(A)$ is a simple eigenvalue of A. This follows, however, from Schur's lemma (Theorem 1.7.15). Indeed, only one linearly independent eigenvector of A, say u, is associated with the eigenvalue $\rho(A)$. Moreover, $u > 0$. The respective eigenvector v of the matrix A^T, which is also irreducible by Theorem 3.2.2, can be chosen positive as well: $v > 0$. Therefore, $v^T u > 0$ and Schur's lemma implies that $\rho(A)$ is a simple eigenvalue.

Finally, let us prove that no nonnegative eigenvector is associated with any other eigenvalue. Suppose that $Az = \xi z$, $z \geq 0$, and $\xi \neq \rho(A)$. We have shown that A^T has a positive eigenvector, say $w > 0$:

$$A^T w = \rho(A)w.$$

But then
$$w^T A z = w^T \xi z = \xi(w^T z),$$
and, on the other hand,
$$w^T A z = \rho(A)(w^T z),$$
which is a contradiction since $\rho(A) - \xi \neq 0$ and $w^T z > 0$. $\quad\quad\square$

4.3   Cyclic matrices

Among other facts, the Perron-Frobenius theorem states that the eigenvalues of an irreducible nonnegative matrix A are arranged in the complex plane in such a way that the intersection of the positive real axis with the boundary of the smallest disk, with center at the origin, which contains all the eigenvalues of A (its radius is equal to $\rho(A)$), always is an eigenvalue. This does not mean that no other eigenvalues can have their moduli equal to $\rho(A)$. We now present a very interesting theorem which relates the number of the eigenvalues of A lying on the boundary of the disk to the combinatorial properties of the zero-nonzero structure of A and to the form of the characteristic polynomial of A ([54, 27, 59]).

Theorem 4.3.1 *Let A be an irreducible nonnegative square matrix of order n. Let h be a positive integer. The following properties of A and h are equivalent:*

1. *There exist exactly h distinct eigenvalues of A whose moduli are equal to $\rho(A)$.*

2. *There exists a permutation matrix P such that PAP^T has the form*

$$PAP^T = \begin{bmatrix} 0 & A_{12} & 0 & \ldots & 0 \\ 0 & 0 & A_{23} & \ldots & 0 \\ \vdots & \vdots & \vdots & \ddots & \vdots \\ 0 & 0 & 0 & \ldots & A_{h-1,h} \\ A_{h1} & 0 & 0 & \ldots & 0 \end{bmatrix},$$

where the diagonal blocks are square matrices, and there is no permutation matrix which transforms A into an analogous form having more than h block rows.

3. The greatest common divisor of the lengths of all the cycles in the digraph $\vec{G}(A)$ of the matrix A is h.

4. If $(-1)^n(\lambda^n + k_{n_1}\lambda^{n_1} + k_{n_2}\lambda^{n_2} + \cdots + k_{n_s}\lambda^{n_s})$ is the characteristic polynomial of the matrix A, and

$$k_{n_1} \neq 0, \; k_{n_2} \neq 0, \; \ldots, \; k_{n_s} \neq 0, \; n > n_1 > \cdots > n_s \geq 0,$$

then the greatest common divisor of the numbers

$$n - n_1, \; n_1 - n_2, \; \ldots, \; n_{s-1} - n_s$$

is h.

5. If $S(Z)$ denotes the spectrum of the matrix Z, then h is the maximal positive integer k such that

$$S([\exp(2\pi i/k)]A) = S(A).$$

Proof. Denote by h_t, $t = 1, \ldots, 5$, the number h in the property t. To prove the theorem it is sufficient to show that $h_1 \leq h_2$, $h_2 \leq h_3$, $h_3 \leq h_4$, $h_4 \leq h_5$, $h_5 \leq h_1$.

The basis for the proof of the inequality $h_1 \leq h_2$ is the following lemma.

Lemma. *Let A be an irreducible nonnegative square matrix. If $\sigma\rho(A)$ is an eigenvalue of A, and $|\sigma| = 1$, then there exists a diagonal matrix D such that $m(D) = I$, and*

$$AD = \sigma DA. \tag{4.16}$$

Conversely, if (4.16) holds for some σ, $|\sigma| = 1$, then $\sigma\rho(A)$ is an eigenvalue of A, and Du, where $u > 0$ is the Perron vector of A, is an associated eigenvector of A.

To prove the lemma, suppose

$$Az = \sigma\rho(A)z. \quad z \neq 0. \quad |\sigma| = 1. \tag{4.17}$$

Then

$$\rho(A)m(z) = m(\sigma\rho(A)z) = m(Az) \leq Am(z). \qquad (4.18)$$

By the Perron-Frobenius Theorem 4.2.1, there exists a vector $y > 0$ such that $A^T y = \rho(A)y$.

Therefore

$$\rho(A)y^T m(z) \leq y^T Am(z) = \rho(A)y^T m(z).$$

Since $y^T m(z) > 0$ and $\rho(A) > 0$, equality takes place in (4.18):

$$Am(z) = \rho(A)m(z).$$

Hence $m(z)$ is a (positive) Perron vector. Therefore, there exists a unique diagonal matrix D such that

$$z = Dm(z),$$

that is

$$m(D) = I.$$

From (4.17) we then have

$$ADm(z) = \sigma\rho(A)DM(z),$$

and thus

$$(\sigma^{-1}D^{-1}AD)m(z) = \rho(A)m(z) = Am(z).$$

Now, since the matrix $Z = \sigma^{-1}D^{-1}AD$ satisfies

$$m(Z) = |\sigma^{-1}|m(D^{-1})m(A)m(D) = m(A) = A,$$

the above relation may be written as

$$(m(Z) - Z)m(z) = 0.$$

In view of $m(z) > 0$ we thus have

$$m(Z) = Z,$$

or

$$A = \sigma^{-1}D^{-1}AD,$$

which establishes (4.16).

Conversely, if (4.16) holds for some σ, $|\sigma| = 1$, and if $u > 0$ is the Perron vector of A, we have

$$A(Du) = \sigma DAu = \sigma D\rho(A)u = \sigma\rho(A)Du.$$

Hence, Du is an eigenvector associated with the eigenvalue $\sigma\rho(A)$ of the matrix A. This completes the proof of the lemma.

We now turn to the proof of the inequality $h_1 \leq h_2$. Denote by $\lambda_1, \ldots, \lambda_{h_1}$ all the distinct eigenvalues of A whose moduli are equal to $\rho(A)$. Therefore,

$$\lambda_k = \rho(A)\exp(\omega_k i), \quad 0 \leq \omega_k < 2\pi.$$

We may suppose that the eigenvalues are ordered as

$$0 = \omega_1 < \omega_2 < \omega_3 < \cdots < \omega_{h_i} < 2\pi.$$

Setting

$$\sigma = \exp(\omega_2 i)$$

we find that $\sigma\rho(A)$ is an eigenvalue of A with $|\sigma| = 1$. Therefore, we can apply the lemma and conclude that there exists a diagonal matrix D such that $m(D) = I$ and

$$AD = \sigma DA.$$

However, this means that

$$AD^2 = (AD)D = \sigma(DA)D = \sigma D(AD) = \sigma^2 D^2 A$$

and, by induction

$$AD^k = \sigma^k D^k A, \quad k = 1, 2, \ldots. \tag{4.19}$$

Moreover, (4.16) implies (D is nonsingular)

$$AD^{-1} = \sigma^{-1}D^{-1}A,$$

and thus

$$AD^{-k} = \sigma^{-1}D^{-k}A.$$

Hence, (4.19) holds for any integer k. Since $m(D^k) = I$ for any integer k, we deduce from the lemma that $\rho(A)\sigma^k$ is an eigenvalue of A for every integer k. Denote by σ^t the first member of the sequence $1, \sigma, \sigma^2, \ldots$ which is equal to some of the preceding members, say σ^s, $0 \le s < t$. If $h = t - s$, then

$$\sigma^h = 1, \tag{4.20}$$

the numbers $1, \sigma, \sigma^2, \ldots, \sigma^{h-1}$ being distinct. Let us prove that

$$h = h_1$$

and

$$\lambda_k = \rho(A)\sigma^{k-1}, \quad k = 1, \ldots, h. \tag{4.21}$$

Formula (4.21) is obviously valid for $k = 1$ and $k = 2$. But we have shown that the numbers $\rho(A)\sigma^{k-1}$, $k = 1, \ldots, h$, are distinct eigenvalues of A whose moduli are equal to $\rho(A)$. Therefore, $h \le h_1$. Suppose that $h < h_1$. Then one of the eigenvalues, e.g. $\lambda_j = \rho(A)\exp(\omega_j i)$, is not included among the numbers $\rho(A)\sigma^{k-1}$, $k = 1, \ldots, h$. This means that ω_j is not of the form $(2\pi/h)k$, where k is an integer. Let

$$\frac{2\pi}{h}(s - 1) < \omega_j < \frac{2\pi}{h}s, \quad s \text{ an integer,} \quad 1 \le s \le h. \tag{4.22}$$

If we set $\exp(\omega_j i) = \tau$, then τ satisfies the hypothesis of the lemma. Hence, there exists a diagonal matrix $\tilde{D}$ such that $m(\tilde{D}) = I$ and

$$A\tilde{D} = \tau\tilde{D}A.$$

From (4.19) we have

$$AD^{1-s} = \sigma^{1-s}D^{1-s}A.$$

Therefore,

$$AD^{1-s}\tilde{D} = \sigma^{1-s}D^{1-s}A\tilde{D} = \sigma^{1-s}\tau D^{1-s}\tilde{D}A.$$

Applying the lemma (since $m(D^{1-s}\tilde{D}) = m(D^{1-s})m(\tilde{D}) = I$) we conclude that $\lambda = \sigma^{1-s}\tau\rho(A)$ is an eigenvalue of A. But using (4.20) and (4.22) we find that $\lambda = \rho(A)\exp(\omega i)$, where

$$\omega = \omega_j - \frac{2\pi}{h}(s - 1)$$

satisfies

$$0 < \omega < \frac{2\pi}{h}.$$

This contradicts the fact that λ is one of the numbers $\lambda_1, \lambda_2, \ldots,$ λ_{h_1} since $\lambda_2 = \sigma\rho(A)$ has been taken to have the least positive argument $\omega_2 = 2\pi/h$ (in view of $\sigma^h = 1$) of all the eigenvalues $\lambda_2, \ldots, \lambda_{h_1}$. Thus, $h = h_1$ and we have also found that (4.21) holds with

$$\sigma = \exp\left(2\pi i/h_1\right). \tag{4.23}$$

Defining σ by (4.23) and using (4.19) we have (we write h instead of h_1)

$$AD^h = \sigma^h D^h A,$$

so that $D^h u$ is an eigenvector belonging to $\sigma^h \rho(A) = \rho(A)$ by the lemma. Here, $u > 0$ is the Perron vector of the matrix A. By the Perron-Frobenius theorem, u is the only linearly independent eigenvector belonging to $\rho(A)$. Hence, $D^h u$ is a scalar multiple of u and since $m(D) = I$, we have

$$D^h u = \alpha u, \qquad |\alpha| = 1,$$

or

$$D^h = \alpha I, \qquad |\alpha| = 1.$$

Thus, the diagonal entries of the matrix D are of the form $\xi\sigma^j$, $j = 0, 1, \ldots, h-1$, $|\xi| = 1$, and we can find a permutation matrix P such that

$$PDP^T = \xi \operatorname{diag}\{I_1, \sigma I_2, \sigma^2 I_3, \ldots, \sigma^{h-1} I_h\}. \tag{4.24}$$

Here, $I_1, \ldots, I_h$ are identity matrices and, for the present, we do not exclude the possibility that some of them are missing (are of "order zero"). With the permutation matrix P, we form the matrix PAP^T and partition it correspondingly. We thus have

$$PAP^T = \begin{bmatrix} A_{11} & A_{12} & \ldots & A_{1h} \\ A_{21} & A_{22} & \ldots & A_{2h} \\ \vdots & \vdots & \ddots & \vdots \\ A_{h1} & A_{h2} & \ldots & A_{hh} \end{bmatrix}, \tag{4.25}$$

where the square matrix A_{kk} is of the same order as I_k, $k = 1, \ldots, h$.

The equality $AD = \sigma DA$ implies

$$(PAP^T)(PDP^T) = \sigma(PDP^T)(PAP^T).$$

Substituting from (4.24) and (4.25) we find that the block A_{ik} satisfies

$$A_{ik}\sigma^{k-1}I_k = \sigma\sigma^{i-1}I_iA_{ik},$$

and thus

$$(1 - \sigma^{i-k+1})A_{ik} = 0, \qquad i, k = 1, \ldots, h.$$

Since $\sigma^k = 1$ if and only if k is divisible by h, it follows that

$$A_{ik} = 0 \qquad \text{for all} \quad i, k = 1, \ldots, h,$$

except for the cases when $i - k + 1$ is an integral multiple of h. This occurs only with $k = i + 1$, $i = 1, \ldots, h - 1$, and $i = h$, $k = 1$. Hence, PAP^T simplifies to

$$PAP^T = \begin{bmatrix} 0 & A_{12} & 0 & \ldots & 0 \\ 0 & 0 & A_{23} & \ldots & 0 \\ \vdots & \vdots & \vdots & \ddots & \vdots \\ 0 & 0 & 0 & \ldots & A_{h-1,h} \\ A_{h1} & 0 & 0 & \ldots & 0 \end{bmatrix}. \qquad (4.26)$$

If the order of some of the matrices I_k were zero, then the kth row and column would be absent from the representation (4.26). Clearly, this would mean that the $(k - 1)$th (hth for $k = 1$) block row of PAP^T is zero (if the order of I_{k-1} is at least one; otherwise, the row in question is the row $k - 2$, etc.), contrary to the assumption that A, and thus PAP^T as well, is irreducible. Therefore, we have shown that the representation (4.26) has $h = h_1$ block rows. The representation described in 2. has h_2 block rows, which is the maximal possible number, and thus

$$h_2 \geq h_1$$

as was to be proved.

To prove $h_2 \leq h_3$, denote by $N_1, N_2, \ldots, N_{h_2}$ the sets of vertices in the digraph $\vec{G}(PAP^T)$ that correspond to the block rows of PAP^T. Now let $C = (u_1, u_2, \ldots, u_s, u_1)$ be a cycle in $\vec{G}(PAP^T)$. If $u_1 \in N_k$, then $u_2 \in N_{k+1}$ (or N_{k+1-h_2}), $u_3 \in N_{k+2}, \ldots$, up to $u_s \in N_{k+s-1}$ (or $N_{k+s-1-ih_2}$, where i is an integer), and eventually again $u_1 \in N_{k+s-jh_2}$ for an integer j. But since $u_1 \in N_k$, we have $s = jh_2$, i.e., h_2 is a divisor of the length s of the cycle C. This holds for any cycle in $\vec{G}(PAP^T)$ and thus also for any cycle in $\vec{G}(A)$ (cf. Theorem 3.2.1), so that h_2 is a divisor of the greatest common divisor h_3 of the lengths of all the cycles as well. Hence, $h_2 \leq h_3$.

To prove $h_3 \leq h_4$, we first show that if s is a positive integer not divisible by h_3 and if S is a subset of $N = \{1, 2, \ldots, n\}$ such that $|S| = s$, then the principal minor of A formed from the rows and columns whose indices are elements of S is zero:

$$\det A(S) = 0.$$

We return to the definition (1.5). If we express $\det A(S)$ by (1.5), we find that each term contains at least one cyclic product $a_{j_1 j_2} a_{j_2 j_3} \ldots a_{j_k j_1}$ whose length k is not divisible by h_3. By 3., such a product is zero so that $\det A(S) = 0$.

Let

$$(-\lambda)^n + c_1(-\lambda)^{n-1} + \cdots + c_n$$

be the characteristic polynomial of A. It follows that each coefficient c_j with j not divisible by h_3 is zero (note that c_k is the sum of all the principal minors of A of order k by Theorem 1.6.2). Thus, we conclude that the differences of the exponents n_j of the powers $(-\lambda)^{n_j}$ with nonzero coefficients are divisible by h_3. The greatest common divisor h_4 of these differences thus satisfies $h_3 \leq h_4$ as required.

We now prove $h_4 \leq h_5$. Suppose that the characteristic polynomial $\varphi(\lambda)$ of the matrix A has the form described in 4. It follows that $\varphi(\lambda)$ can be expressed in the form $\lambda^{n_s} \psi(\lambda^{h_4})$ where ψ is a polynomial. The transformation $\lambda = \tilde{\lambda} \exp(2\pi/h_4)$ leaves invariant the set of the roots of this polynomial, which is the spectrum of A. By Theorem 1.7.12, this fact can also be stated as $S(A) = S([\exp(2\pi i/h_4)]A)$. As h_5 is the maximal k such that $S(A) = S([\exp(2\pi i/k)]A)$ we have $h_4 \leq h_5$.

It remains to prove $h_5 \leq h_1$. Since, by Theorem 4.2.1, $\rho(A)$ is a positive eigenvalue of A, the property 5. implies that $\rho(A)\exp(2\pi i/h_5)$ is an eigenvalue of A; similarly, we find that

$$\rho(A)\exp(2\pi i/h_5)\exp(2\pi i/h_5) = \rho(A)\exp(2\pi i \cdot 2/h_5),$$

etc. up to $\rho(A)\exp(2\pi i(h_5 - 1)/h_5)$ are eigenvalues of A. These are h_5 distinct eigenvalues, and thus $h_5 \leq h_1$.   □

Corollary 5 *If A is an irreducible nonnegative matrix having exactly h distinct eigenvalues of modulus $\rho(A)$, then these eigenvalues considered as points in the complex plane are the vertices of a regular polygon of h sides, with center at the origin, one of the vertices being $\rho(A)$ (of course, if $h = 1$, we have only the point $\rho(A)$, if $h = 2$, we get two points $\rho(A)$ and $-\rho(A)$). All these h eigenvalues are simple.*

The corollary easily follows from the statements 1. and 5. of the theorem. The eigenvalues are simple since e.g. (4.16) gives

$$D^{-1}AD = \sigma A, \qquad \sigma = \frac{2\pi i}{h},$$

and thus the multiplicity of $\rho(A)$ and $\sigma\rho(A)$ (and of the other eigenvalues $\sigma^k\rho(A)$ as well) is the same, and equals 1 by the Perron-Frobenius theorem.

Remark. An irreducible nonnegative matrix with $h = 1$ is said to be *primitive*. If $h > 1$, then the matrix is *cyclic of index h*.

In the sequel, we consider general (i.e., not necessarily irreducible) nonnegative square matrices.

Theorem 4.3.2 *Let A be a nonnegative square matrix. Then $\rho(A)$ is an eigenvalue of A, and there exists a nonnegative eigenvector of A associated with this eigenvalue.*

Proof. By Theorem 3.2.2, there exists a permutation matrix P such that PAP^T has the form

$$PAP^T = \begin{bmatrix} A_{11} & A_{12} & \dots & A_{1r} \\ 0 & A_{22} & \dots & A_{2r} \\ \vdots & \vdots & \ddots & \vdots \\ 0 & 0 & \dots & A_{rr} \end{bmatrix}.$$

where the blocks A_{kk}, $k = 1, \ldots, r$, are irreducible. Let $r \geq 2$. Clearly, $\rho(A) = \rho(PAP^{-1}) = \rho(PAP^T) = \max_k \rho(A_{kk})$. Hence, there exists an integer j, $1 \leq j \leq r$, such that

$$j = \min\{k; \rho(A) = \rho(A_{kk})\}.$$

Now, PAP^T can be written in the form

$$PAP^T = \begin{bmatrix} B_{11} & B_{12} & B_{13} \\ 0 & A_{jj} & B_{23} \\ 0 & 0 & B_{33} \end{bmatrix}, \qquad (4.27)$$

where both B_{11} and B_{33} are also upper block triangular (and one of them may be missing).

By Theorem 4.2.1, there exists a positive vector y_j such that

$$A_{jj}y_j = \rho(A)y_j.$$

If $j = 1$, then

$$\begin{bmatrix} y_1 \\ 0 \end{bmatrix}$$

is a nonnegative eigenvector of the matrix PAP^T in (4.27). If $j > 1$, let us show that there exists a nonnegative vector z such that

$$\begin{bmatrix} z \\ y_j \\ 0 \end{bmatrix}$$

is an eigenvector of PAP^T. Hence, it is sufficient to find $z \geq 0$ satisfying

$$B_{11}z + B_{12}y_j = \rho(A)z. \qquad (4.28)$$

We have $\rho(B_{11}) < \rho(A)$ by definition of j. The spectral radius of the matrix

$$Z = (\rho(A))^{-1}B_{11}$$

is thus less than 1, so that the series

$$I + Z + Z^2 + \ldots$$

converges to $(I - Z)^{-1}$ by Theorem 1.7.14. Since $Z \geq 0$, the matrix $(I - Z)^{-1}$ is a sum of nonnegative matrices and, therefore,

$$(I - Z)^{-1} \geq 0.$$

Clearly, the vector z given by

$$z = (\rho(A))^{-1}(I - Z)^{-1}B_{12}y_j$$

satisfies (4.28) and, moreover, we have $z \geq 0$ as required.   □

Remark. The eigenvalue $\rho(A)$ of a nonnegative matrix A is also called *Perron root* of A and an associated nonnegative eigenvector is then called *Perron vector* (in general, several linearly independent Perron vectors may exist if A is reducible).

We now establish several simple statements that supplement the Perron-Frobenius theorem and Theorem 4.3.2. The following theorem characterizes the nonnegative matrices whose Perron root is zero.

Theorem 4.3.3 *Let A be a nonnegative square matrix of order $n \geq 2$. Then the following conditions are equivalent:*

1. $A^n = 0$.

2. *There exists a positive integer k such that $A^k = 0$.*

3. *The digraph $\vec{G}(A)$ of the matrix A is acyclic.*

4. *There exists a permutation matrix P such that PAP^T is an upper triangular matrix with zeros on the main diagonal.*

5. $\rho(A) = 0$.

Proof. We prove the implications 1. → 2., 2. → 3., 3. → 4., 4. → 5., and 5. → 1.

1. → 2. Evident.

2. → 3. If there is a cycle in the digraph $\vec{G}(A)$ we cannot have $A^k = 0$ for any k by Theorem 4.1.3.

3. → 4. By the result of Exercise 3.1, each strong component of an acyclic digraph contains just one point; thus, there exists a permutation matrix P such that PAP^T is an upper triangular matrix. All the diagonal entries are zero as otherwise $\vec{G}(A)$ would contain a loop, which is a cycle of length 1.

4. → 5. Since PAP^T is an upper triangular matrix with zeros on the main diagonal, all the eigenvalues of PAP^T are zero. Therefore, A has only zero eigenvalues and $\rho(A) = 0$.

5. $\rightarrow$ 1. It follows from 5. that all the eigenvalues of A are zero. Hence, all the Jordan blocks in the Jordan normal form A_1 of the matrix A have the form

$$\begin{bmatrix} 0 & 1 & 0 & \dots & 0 \\ 0 & 0 & 1 & \dots & 0 \\ \vdots & \vdots & \vdots & \ddots & \vdots \\ 0 & 0 & 0 & \dots & 1 \\ 0 & 0 & 0 & \dots & 0 \end{bmatrix}$$

and their order is at most n. Thus $A^n = 0$. □

Theorem 4.3.4 *Let A be a nonnegative square matrix having a positive eigenvector. Then, this vector is a Perron vector, that is, it is associated with the eigenvalue $\rho(A)$.*

Proof. Denote the vector of the theorem by u, so that $u > 0$ and $Au = cu$. By Theorem 4.3.2, there exists a nonnegative eigenvector v of A^T associated with $\rho(A^T) = \rho(A)$. We now have

$$v^T A u = c v^T u,$$

and, on the other hand,

$$v^T A u = \rho(A) v^T u.$$

Since $v^T u > 0$ we obtain $c = \rho(A)$. □

4.4 Stochastic matrices

An important subclass of the nonnegative matrices is the class of stochastic matrices. A *stochastic matrix* $A = [a_{ik}]$ is a square matrix satisfying

1. $a_{ik} \geq 0$ for all i and k,
2. $\sum_k a_{ik} = 1$ for all i.

Stochastic matrices play an important part in the study of stochastic processes with a finite number of states. The entries in a stochastic matrix are then the probabilities of transitions from one state to another. One usually investigates whether, starting from an initial vector of probabilities, the process will be stabilized and what the limiting vector will be. This leads to the study of the behaviour of $A^k u$ as $k \to \infty$.

Condition 2. of the definition of a stochastic matrix may be also written as

$$Ae = e$$

where $e = [1, 1, \dots, 1]^T$. By Theorem 4.3.4, e is a Perron vector of the nonnegative matrix A. This implies the next theorem immediately.

Theorem 4.4.1 *Every stochastic matrix is a nonnegative matrix which has the eigenvector $e = [1, 1, \dots, 1]^T$ associated with the Perron root 1. Hence, the spectral radius of a stochastic matrix is always 1.*

The spectral properties of stochastic matrices follow from those of nonnegative matrices as may be seen from the next theorem.

Theorem 4.4.2 *Let P be a nonnegative matrix having a positive Perron vector u, and let $\rho(P) > 0$. Then there exists a diagonal matrix D with positive diagonal entries, and a real number $k > 0$ such that the matrix*

$$A = kDPD^{-1}$$

is stochastic. In addition, $k = (\rho(P))^{-1}$.

Proof. We have $Pu = pu$, $u > 0$, $p = \rho(P)$. Hence, there exists a diagonal matrix D with positive diagonal entries such that $Du = e = [1, \dots, 1]^T$. Set $k = (\rho(P))^{-1}$. Then

$$Ae = kDPD^{-1}e = kDPu = kD\rho(P)u = Du = e,$$

and, moreover, $A \geq 0$. The number k is determined uniquely since $1 = \rho(A) = k\rho(DPD^{-1}) = k\rho(P)$. □

Theorem 4.4.3 *The product of stochastic matrices of the same order is again a stochastic matrix.*

Proof. It follows immediately from the fact that the product $A_1 A_2$ of two stochastic matrices, A_1 and A_2, is a nonnegative matrix and $A_1 A_2 e = A_1 e = e$.   □

The following theorem describes the behaviour of A^k in the most important case.

Theorem 4.4.4 *If A is an irreducible primitive stochastic matrix, then the powers A^k converge (as $k \to \infty$) to the matrix ev^T of rank one, where $e = [1, \ldots, 1]^T$ and $v = [v_1, \ldots, v_n]^T$ is the eigenvector of A^T associated with the eigenvalue 1, normalized in such a way that $v^T e = 1$.*

Proof. By Theorem 1.7.6, A is similar to a matrix in Jordan normal form whose diagonal block corresponding to the eigenvalue 1 is of order one and the moduli of other eigenvalues are less than 1. Hence,

$$PAP^{-1} = A_J = \begin{bmatrix} 1 & 0 \\ 0 & Z \end{bmatrix}, \qquad \rho(Z) < 1.$$

Consequently,

$$A_J^k = \begin{bmatrix} 1 & 0 \\ 0 & Z^k \end{bmatrix},$$

and thus $A_J^k \to E$,

$$E = \begin{bmatrix} 1 & 0 \\ 0 & 0 \end{bmatrix} = e_1 e_1^T,$$

where $e_1 = [1, 0, \ldots, 0]^T$. Therefore,

$$A^k = P^{-1} A_J^k P \to P^{-1} E P = \tilde{E},$$

and

$$\tilde{E} = P^{-1} e_1 e_1^T P. \tag{4.29}$$

But we have

$$Ae = e, \quad A^T v = v, \quad v^T e = 1.$$

Thus

$$(PAP^{-1})Pe = Pe, \quad (PAP^{-1})^T (P^T)^{-1} v = (P^T)^{-1} v,$$

and we obtain

$$A_J Pe = Pe, \qquad A_J^T (P^T)^{-1} v = (P^T)^{-1} v.$$

Since $A_J e_1 = e_1$, $A_J^T e_1 = e_1$, and the eigenvectors associated with the eigenvalue 1 are determined uniquely up to a multiplicative factor, we get

$$Pe = k_1 e_1, \qquad (P^T)^{-1} v = k_2 e_1.$$

The condition $v^T e = 1$ implies $1 = v^T e = (v^T P^{-1})(Pe) = k_2 e_1^T k_1 e_1 = k_2 k_1$. Therefore, by (4.29) we have $\tilde{E} = P^{-1} e_1 e_1^T P = k_1^{-1} e k_2^{-1} v^T = e v^T$, which completes the proof. □

4.5 Doubly stochastic matrices

In the conclusion of this chapter we consider another important subclass of the class of nonnegative matrices (and also of that of stochastic matrices), the class of doubly stochastic matrices. A square matrix $A = [a_{ik}]$ is said to be *doubly stochastic* if both A and A^T are stochastic matrices, i.e. if we have (with a matrix of order n)

$$a_{ik} \geq 0, \quad \sum_j a_{ij} = 1, \quad \sum_j a_{jk} = 1, \quad i, k = 1, \ldots, n.$$
(4.30)

This may be also written as

$$A \geq 0, \qquad Ae = e, \qquad A^T e = e, \qquad e = [1, \ldots, 1]^T.$$

We now state two theorems on doubly stochastic matrices; their proofs are evident.

Theorem 4.5.1 *The Perron root of a doubly stochastic matrix equals 1. The vector $e = [1, 1, \ldots, 1]^T$ is a Perron eigenvector of both A and A^T.*

Theorem 4.5.2 *The product of doubly stochastic matrices is again a doubly stochastic matrix.*

Before proceeding further we recall the concept of a convex
set in a Euclidean point space E. A set K of the points in E
is said to be *convex* if it possesses the following property: If
A_1 and A_2 are points in K, and if λ_1 and λ_2 are nonnegative
real numbers satisfying $\lambda_1 + \lambda_2 = 1$, then also $\lambda_1 A_1 + \lambda_2 A_2 \in$
K. Here $\lambda_1 A_1 + \lambda_2 A_2$ is defined to be the point B whose each
(e.g., rectangular, but, generally speaking affine) coordinate b_i
is of the form $\lambda_1 a_i^{(1)} + \lambda_2 a_i^{(2)}$, where $a_i^{(1)}$ and $a_i^{(2)}$ are the ith
coordinates of the points A_1 and A_2, respectively. (For example,
with $\lambda_1 = \frac{1}{2}$ we obtain the midpoint of the pair of points A_1 and
A_2.)

A *convex polyhedron* K in E is the set of all the points of the
form

$$\sum_{i=1}^{N} \lambda_i A_i, \tag{4.31}$$

where $A_1, \ldots, A_N$ are points in E, and $\lambda_1, \ldots, \lambda_N$ are nonnegative real numbers satisfying $\sum_{i=1}^{N} \lambda_i = 1$. It may happen that
some of the points A_i may be deleted from (4.31) without changing the convex polyhedron in question (this is the case when the
point itself may be expressed in the form (4.31)). If $A_1, \ldots, A_N$
contain no such a redundant point, they are called the *vertices*
of K.

Considering now a real square matrix $A = [a_{ik}]$ of order n as
a point with coordinates

$$[a_{11}, a_{12}, \ldots, a_{1n}, a_{21}, a_{22}, \ldots, a_{2n}, \ldots, a_{n1}, \ldots, a_{nn}]$$

in the n^2-dimensional Euclidean space $\tilde{E}$, we may ask what are
the properties of the set formed in $\tilde{E}$ by doubly stochastic matrices.

The proof of the next theorem is left to the reader (Exercise
4.7).

Theorem 4.5.3 *The matrices $A = [a_{ik}]$ satisfying the conditions*

$$\sum_{j=1}^{n} a_{ij} = 1, \qquad \sum_{j=1}^{n} a_{jk} = 1, \qquad i, k = 1, \ldots, n. \tag{4.32}$$

form an $(n-1)^2$-dimensional point subspace E of the space $\tilde{E}$.

Now, we are in a position to state and prove the following famous theorem (cf. [4]).

Theorem 4.5.4 (Birkhoff) *In the space $\tilde{E}$ of Theorem 4.5.3, the set of doubly stochastic matrices of order n is represented by a convex polyhedron in the subspace E. The vertices of the polyhedron are exactly all the points corresponding to permutation matrices (i.e., to the matrices which in each row and each column have some entry one and all other entries zero).*

Proof. Let $P_1, \ldots, P_N$, where $N = n!$, be the $n!$ different permutation matrices of order n. Clearly, P_i, $i = 1, \ldots, N$, are doubly stochastic matrices. Every matrix A of the form

$$A = \sum_{i=1}^{N} \lambda_i P_i, \qquad \lambda_i \geq 0, \qquad \sum_{i=1}^{N} \lambda_i = 1, \qquad (4.33)$$

is doubly stochastic as well since $A \geq 0$ and

$$Ae = \sum_{i=1}^{N} \lambda_i P_i e = \left(\sum_{i=1}^{N} \lambda_i \right) e = e,$$

$$A^T e = \sum_{i=1}^{N} \lambda_i P_i^T e = \left(\sum_{i=1}^{N} \lambda_i \right) e = e.$$

Our task is to prove that any doubly stochastic matrix A may be expressed in the form (4.33). The proof is by induction on the number $\nu(A)$ of the nonzero entries of A. If $\nu(A) = n$ ($\nu(A) < n$ is impossible), the statement is true as A is a permutatioon matrix in this case.

Therefore, let $\nu(A) > n$ for the matrix $A = [a_{ik}]$ and suppose that the statement is true for any doubly stochastic matrix $\tilde{A}$ satisfying $\nu(\tilde{A}) < \nu(A)$. In what follows, we denote by J the open interval $J = (0, 1)$.

Since $\nu(A) > n$, there is a row in the matrix A, say row i_1, with at least two nonzero entries, which belong to J: $a_{i_1 j_1} \in J, a_{i_1 j_2} \in J$ ($j_1 \neq j_2$). As $a_{i_1 j_2} < 1$ there is another nonzero entry located in the j_2th column: $a_{i_2 j_2} \in J$, $i_2 \neq i_1$. Next, we obtain j_3 such that $a_{i_2 j_3} \in J$, further i_3 such that $a_{i_3 j_3} \in J$, etc. As there are n different indices only it must sometimes happen for the first time that we get an index j_t which occurred as a column index before, or an index i_t which occurred as a row index before. However, no matter whether the former or the latter case takes place we arrive at the following situation: For some $s \geq 2$ we have s distinct indices $p_1, \ldots, p_s$ and s distinct indices $q_1, \ldots, q_s$ such that $a_{p_i q_i} \in J$, $i = 1, \ldots, s$, and, moreover, $a_{p_1 q_s} \in J$, $a_{p_i q_{i-1}} \in J$, $i = 2, \ldots, s$. We now set

$$\alpha = \min_{i=1,\ldots,s} a_{p_i q_i}, \tag{4.34}$$

$$\beta = \min(a_{p_1 q_s}, \min_{i=2,\ldots,s} a_{p_i q_{i-1}}). \tag{4.35}$$

Let A_1 be the matrix obtained from A by replacing $a_{p_i q_i}$ by $a_{p_i q_i} - \alpha$ for $i = 1, \ldots, s$, $a_{p_1 q_s}$ by $a_{p_1 q_s} + \alpha$, and $a_{p_i q_{i-1}}$ by $a_{p_i q_{i-1}} + \alpha$ for $i = 2, \ldots, s$. The other entries are left unaltered. Similarly, let A_2 be the matrix obtained from A by replacing $a_{p_i q_i}$ by $a_{p_i q_i} + \beta$ for $i = 1, \ldots, s$, $a_{p_i q_s}$ by $a_{p_i q_s} - \beta$, and $a_{p_i q_{i-1}}$ by $a_{p_i q_{i-1}} - \beta$ for $i = 2, \ldots, s$, whereas the other entries are left unaltered. By (4.34) and (4.35), both A_1 and A_2 are nonnegative matrices having the same row and column sums as A. Thus, A_1 and A_2 are doubly stochastic matrices. However, $\nu(A_1) < \nu(A)$ since A_1 has zero entry in the position corresponding to the entry of A realizing the minimum in (4.34). Similarly, $\nu(A_2) < \nu(A)$. Hence, by the induction hypothesis, both A_1 and A_2 are of the form (4.33),

$$A_1 = \sum_{i=1}^{N} \lambda_i^{(1)} P_i, \qquad \lambda_i^{(1)} \geq 0, \qquad \sum_{i=1}^{N} \lambda_i^{(1)} = 1,$$

$$A_2 = \sum_{i=1}^{N} \lambda_i^{(2)} P_i, \qquad \lambda_i^{(2)} \geq 0, \qquad \sum_{i=1}^{N} \lambda_i^{(2)} = 1.$$

But we clearly have

$$A = \beta(\alpha + \beta)^{-1} A_1 + \alpha(\alpha + \beta)^{-1} A_2,$$

from which

$$A = \sum_{i=1}^{N} \lambda_i P_i,$$

where $\lambda_i \geq 0$ for $i = 1, \ldots, N$, and

$$\sum_{i=1}^{N} \lambda_i = 1.$$

The theorem now follows as it is easy to verify that no permutation matrix P_i, $i = 1, \ldots, N$, is redundant, i.e., expressible in the form (4.33) otherwise than with $\lambda_i = 1$ and $\lambda_j = 0$ for $j \neq i$. □
  A consequence of Birkhoff's theorem is the following statement (Exercise 4.8).

Theorem 4.5.5 *If A is a doubly stochastic matrix of order n, then there exists a permutation $(i_1, \ldots, i_n)$ of the indices $1, \ldots, n$ such that $a_{1i_1} a_{2i_2} \ldots a_{ni_n} > 0$. In other words, the expression for $\det A$ given in* (1.5) *contains a nonzero term.*

We shall see in Chapter 9 that the multiplication of a vector by a doubly stochastic matrix "averages" the components of the vector in a sense (Theorem 9.3.8).

Exercises

4.1. Prove Theorem 4.1.2.

4.2. Let A be a reducible nonnegative matrix. Is it true that there are at least two linearly independent nonnegative eigenvectors of A?

4.3. If a nonnegative matrix A has a nonnegative eigenvector which is not positive, prove that A is reducible.

4.4. Let A be a nonnegative matrix distinct from the zero matrix. If at least one of the eigenvectors of A is positive, prove that $\rho(A) > 0$.

4.5.* Let $U = [u_{ik}]$ be a unitary matrix. Show that the matrix $A = [|u_{ik}|^2]$ is doubly stochastic.

4.6. Let $A = [a_{ik}]$ be a reducible doubly stochastic matrix of order n. Suppose that it is partitioned into the form

$$A = \begin{bmatrix} A_{11} & A_{12} \\ A_{21} & A_{22} \end{bmatrix},$$

where A_{11}, A_{22} are square matrices and $A_{12} = 0$. Show that then $A_{21} = 0$ as well.

4.7. Prove Theorem 4.5.3.

4.8. Prove Theorem 4.5.5.

5

M-Matrices
(Matrices of Classes K and K_0)

In this chapter we investigate matrices of a certain type which occur relatively often in applications.

Example 5.1. Examining input-output relations in closed economy we arrive at a system of linear equations

$$
\begin{array}{ccccccccc}
x_1 & - & \alpha_{12}x_2 & - & \ldots & - & \alpha_{1n}x_n & = & y_1, \\
\alpha_{21}x_1 & + & x_2 & - & \ldots & - & \alpha_{2n}x_n & = & y_2, \\
& & & & & & & & \vdots \\
-\alpha_{n1}x_1 & - & \alpha_{n2}x_2 & - & \ldots & + & x_n & = & y_n.
\end{array}
\tag{5.1}
$$

Here, x_i denotes the gross output of the ith sector, α_{ik} the input of the products of the ith sector (in units of the output of the ith sector) needed to produce one unit of output in the kth sector, y_i net product of the ith sector. If the system under consideration is selfsupporting, then $y_i \geq 0$ for all i, and of course we also have $x_i > 0$ for all i.

Example 5.2. To solve the Dirichlet problem for a plane domain Ω means to seek a function u (having continuous second derivatives) which assumes prescribed values on the boundary of Ω, whereas it satisfies the condition

$$
\frac{\partial^2 u}{\partial x^2} + \frac{\partial^2 u}{\partial y^2} = 0
\tag{5.2}
$$

in the domain Ω. This problem may be solved approximately e.g. using the finite difference method on a square net.

We construct a square plane net (Fig. 9), denoting by Ω_0 the system of those net points contained in Ω. Then, instead of the function u, we consider the function u_0 defined at the points of the net domain Ω_0. Condition (5.2) is replaced by the condition stating that the value of u_0 at a point of Ω_0 equals the arithmetic mean of the four neighbouring values of the function u_0. Some of these neighbouring values (near the boundary) may be considered as known, some values are unknown. The system of equations corresponding to Fig. 9 has the form

$$
\begin{aligned}
u_1 \;-\; \tfrac{1}{4}u_2 \qquad\qquad -\; \tfrac{1}{4}u_6 \qquad\qquad &= \tfrac{1}{4}a_1 + \tfrac{1}{4}b_1, \\
-\tfrac{1}{4}u_1 \;+\; u_2 \;-\; \tfrac{1}{4}u_3 \qquad\qquad -\; \tfrac{1}{4}u_7 &= \tfrac{1}{4}a_2, \\
\text{etc.,} &
\end{aligned}
$$

$$(5.3)$$

where u_i denotes the unknown value of the "discrete" function u_0 at the point with the subscript i, a_k, b_k are the values of the function u (or values obtained by interpolating the prescribed data) on the boundary of the domain Ω_0.

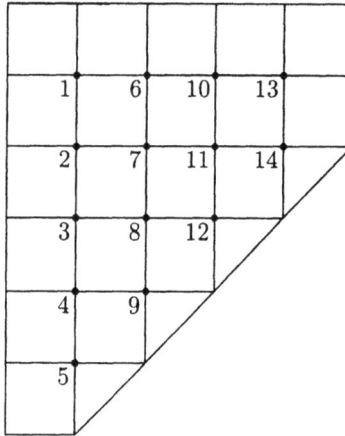

Fig. 9

We can see that the system (5.3) is of the form

$$Au_0 = f_0,$$

where the diagonal entries of the matrix $A = [a_{ik}]$ are equal to 1 and the off-diagonal ones are equal to $-\tfrac{1}{4}$ in the case that i

and k are neighbouring points, or are zero if i and k are not neighbouring points of the net. The components of the vector f_0 corresponding to the "interior" net points (e.g., the components $7, 8, \ldots, 10, 13, 14$) are zero. Since any point $z \in \Omega_0$ has at most four neighbours in Ω_0, it follows that

$$Ae_0 \geq 0,$$

where e_0 is the column vector all components of which are equal to 1. Moreover, the matrix A is symmetric.

Note that in both examples the matrix $A = [a_{ik}]$ had the following two properties: The off-diagonal entries of A were non-positive, i.e.,

$$a_{ik} \leq 0 \qquad \text{for all} \quad i, k, \ i \neq k, \tag{5.4}$$

and for a vector $u > 0$ we had

$$Au \geq 0, \qquad Au \neq 0. \tag{5.5}$$

As we shall see later, for an irreducible matrix A the properties (5.4) and (5.5) mean that A is a so-called M-matrix or a matrix of class K. Such matrices are the main subject of this chapter.

First of all, we introduce a notation. The *class* Z_n $(n \geq 1)$ will be the set of all real square matrices of order n whose off-diagonal entries are all nonpositive:

$$Z_n = \{A = [a_{ik}], i, k = 1, \ldots, n; a_{ik} \leq 0, i \neq k\}. \tag{5.6}$$

The union of all Z_n will be denoted by Z:

$$Z = \bigcup_{n=1,2,\ldots} Z_n. \tag{5.7}$$

5.1 Class K

Before presenting the definition of M-matrices or matrices of class K we prove the following fundamental theorem (cf. [14, 2]).

Theorem 5.1.1 *Let A be a matrix from Z_n. Then the following properties are equivalent:*

1. *There exists a vector* $x \geq 0$ *such that* $Ax > 0$.

2. *There exists a vector* $x > 0$ *such that* $Ax > 0$.

3. *There exists a diagonal matrix* D *with positive diagonal entries, such that the entries of the matrix* $AD = (w_{ik})$ *satisfy the condiotion* $w_{ii} > \sum_{k \neq i} |w_{ik}|$ *for each* i.

4. *Whenever* $B \in Z_n$ *and* $B \geq A$, *then* B *is nonsingular*.

5. *Every real eigenvalue of any principal submatrix of* A *is positive*.

6. *All principal minors of* A *are positive*.

7. *For each* $k = 1, \ldots, n$, *the sum of all the principal minors of order* k *of the matrix* A *is positive*.

8. *Every real eigenvalue of the matrix* A *is positive*.

9. *There exist a matrix* $C \geq 0$ *and a number* $k > \rho(C)$ *such that* $A = kI - C$.

10. *There exists a splitting* $A = P - Q$ *of the matrix* A *such that* $P^{-1} \geq 0$, $Q \geq 0$, *and* $\rho(P^{-1}Q) < 1$.

11. A *is nonsingular and* $A^{-1} \geq 0$.

12. $\det A(N_k) > 0$, $k = 1, \ldots, n$, *where* $N_k = \{1, 2, \ldots, k\}$.

13. *There exist a lower triangular matrix* R *and an upper triangular matrix* S *such that both* R *and* S *have positive diagonal entries and* $A = RS$.

14. *There exist a lower triangular matrix* $R \in Z_n$ *and an upper triangular matrix* $S \in Z_n$ *such that both* R *and* S *have positive diagonal entries and* $A = RS$.

15. *There exists a splitting* $A = P - Q$ *such that* $P^{-1} \geq 0$ *and* $Q \geq 0$, *and for every such splitting,* $\rho(P^{-1}Q) < 1$.

16. *There exists a diagonal matrix* D *with positive diagonal entries, such that the symmetric part* $\frac{1}{2}(B + B^T)$ *of the matrix* $B = DAD^{-1}$ *is positive definite*.

17. *There exists a diagonal matrix H with positive diagonal entries, such that the symmetric part $\frac{1}{2}(C + C^T)$ of the matrix $C = AH$ is positive definite.*

18. *The real part of any eigenvalue of A is positive.*

Proof. Let us prove the following implications: 1. → 2., 2. → 3., 3. → 4., 4. → 5., 5. → 6., 6. → 7., 7. → 8., 8. → 9., 9. → 10., 10. → 11., 11. → 1., 6. → 12., 12. → 13., 13. → 14., 14. → 11., 11. → 15., 15. → 9., 9. → 16., 16. → 17., 17. → 18., 18. → 8. As the digraph with the points $1, 2, \ldots, 18$ and the arcs $(1, 2)$, $(2, 3)$, etc., (see Fig. 10) which correspond to the above implications is (in the sense of Chapter 3) strongly connected, this will prove the equivalence.

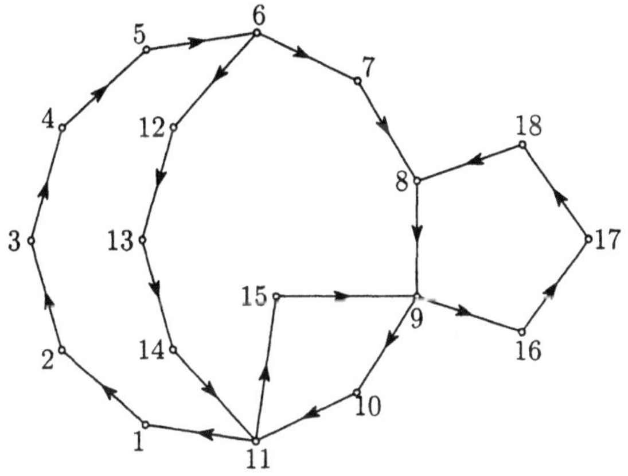

Fig. 10

1. → 2. Let x be the vector from 1., i.e., let $x \geq 0$, $Ax > 0$. Set $e = [1, 1, \ldots, 1]^T$. Since $Ax > 0$, there exists $\varepsilon > 0$ such that $Ax + \varepsilon Ae > 0$. Then, the vector $x + \varepsilon e$ is positive and satisfies 2. since $A(x + \varepsilon e) = Ax + \varepsilon Ae > 0$.

2. → 3. Let $x = [x_1, \ldots, x_n]^T$ be the vector from 2. Choose as D the diagonal matrix $\mathrm{diag}(x_1, \ldots, x_n)$; then $w_{ik} = a_{ik}x_k$. The fact that $Ax > 0$ can be written as

$$a_{ii}x_i > \sum_{k \neq i}(-a_{ik})x_k, \qquad i = 1, \ldots, n,$$

which is

$$w_{ii} > \sum_{k \neq i} |w_{ik}|, \qquad i = 1, \ldots, n.$$

3. $\rightarrow$ 4. Let $D = \mathrm{diag}(d_1, \ldots, d_n)$ be the matrix from 3. and suppose that 4. does not hold. This means that there exists a matrix $B \in Z_n$, $B \geq A$, which is singular. Hence, by Theorem 1.5.3, there exists a vector $y = [y_1, \ldots, y_n] \neq 0$ such that

$$By = 0.$$

Denote by t the subscript for which

$$\left| \frac{y_t}{d_t} \right| = \max_k \left| \frac{y_k}{d_k} \right|.$$

Then

$$b_{tt}y_t = \sum_{k \neq t} (-b_{tk})y_k, \qquad (5.8)$$

and also

$$a_{tt}d_t > \sum_{k \neq t} |a_{tk}d_k|.$$

Now, $B \geq A$, and thus

$$b_{tt} \geq a_{tt},$$
$$b_{tk} \geq a_{tk} \quad \text{for} \quad k \neq t,$$

or

$$|b_{tk}| \leq |a_{tk}| \quad \text{for} \quad k \neq t$$

since $B \in Z_n$ and $A \in Z_n$.

Therefore, (5.8) yields (we have $|y_t/d_t| \neq 0$)

$$|b_{tt}y_t| \leq \sum_{k \neq t} |b_{tk}y_k| \leq \sum_{k \neq t} |a_{tk}|d_k \left| \frac{y_k}{d_k} \right|$$

$$\leq \left[\sum_{k \neq t} |a_{tk}d_k| \right] \left| \frac{y_t}{d_t} \right| < a_{tt}d_t \left| \frac{y_t}{d_t} \right|$$

$$\leq b_{tt}|y_t| \leq |b_{tt}y_t|,$$

which is a contradiction.

4. → 5. Let $\emptyset \neq M \subset N = \{1, 2, \ldots, n\}$ and let λ be a real eigenvalue of the matrix $A(M)$. We prove that 4. implies $\lambda > 0$. Suppose that $\lambda \leq 0$. We define the matrix $B = [b_{ik}]$ by choosing

$$
\begin{aligned}
b_{ii} &= a_{ii} - \lambda && \text{for all } i, \\
b_{ik} &= a_{ik} && \text{for } i \in M, \ k \in M, \ i \neq k, \\
b_{ik} &= 0 && \text{otherwise.}
\end{aligned}
$$

Since $\lambda \leq 0$, we have $B \geq A$. Also, we easily see that $B \in Z_n$. By 4., B is nonsingular. However, $\det B = \prod_{i \notin M} \det B(M) = 0$ as $B(M) = A(M) - \lambda I(M)$ and λ is an eigenvalue of $A(M)$.

5. → 6. Let $\emptyset \neq M \subset N = \{1, 2, \ldots, n\}$. By 5., all the real eigenvalues of $A(M)$ are positive. Since $A(M)$ is a real matrix, the non-real eigenvalues of $A(M)$ (if any) occur as pairs of conjugate complex numbers; therefore, the product of these non-real eigenvalues is positive. Hence, the product of all the eigenvalues of $A(M)$ is positive. By (1.39) this product is the determinant of the matrix $A(M)$ and we thus have

$$
\det A(M) > 0.
$$

6. → 7. Obvious.

7. → 8. We use the formula

$$
\det(A + (-x)I) = (-x)^n + c_1(-x)^{n-1} + \cdots + c_n \tag{5.9}
$$

of Theorem 1.6.2, where c_k is the sum of all principal minors of order k of the matrix A. Suppose that 7. is true, i.e., $c_k > 0$, $k = 1, \ldots, n$. Since (5.9) cannot have a nonpositive real root, 8. holds.

8. → 9. Set $k = \max_i a_{ii}$. Then, the matrix $C = kI - A$ is nonnegative. By Theorem 4.3.2, $\rho(C)$ is an eigenvalue of C. Since $A = kI - C$ is a polynomial in C we conclude from Theorem 1.7.12 that $\alpha = k - \rho(C)$ is an eigenvalue of A. By 8., the (real) eigenvalue $\alpha = k - \rho(C)$ is positive and thus

$$
k > \rho(C).
$$

9. → 10. It suffices to choose $P = kI$, $Q = C$ with k and C defined in 9. Indeed, this leads to

$$
\rho(P^{-1}Q) = \rho\left(\frac{1}{k}C\right) = \frac{1}{k}\rho(C) < 1.
$$

10. → 11. Supposing that 10. holds, we may write $A = P(I - Z)$, where $Z = P^{-1}Q$. Now, since $\rho(Z) < 1$, it follows by Theorem 1.7.14 that $A^{-1} = (I + Z + Z^2 + \ldots)P^{-1}$. However, $Z \geq 0$ and thus $A^{-1} \geq 0$.

11. → 1. Choose $x = A^{-1}e$, where $e = [1, 1, \ldots, 1]^T$. By 11., $A^{-1} \geq 0$. The vector x is thus nonnegative and, moreover, $Ax = e > 0$.

6. → 12. Obvious.

12. → 13. By 12., A is strongly nonsingular and, by Theorem 1.4.3, there exist a lower triangular matrix $R = B$ whose diagonal entries are 1's and an upper triangular matrix $S = DC$ such that $A = RS$. By Theorem 1.4.3 and 12., the diagonal entries of S are positive.

13. → 14. Let $R = [r_{ik}]$ and $S = [s_{ik}]$, where $r_{ii} = 1$, $s_{ii} > 0$, $i = 1, \ldots, n$. It is sufficient to show that $R \in Z_n$ and $S \in Z_n$, i.e., that

$$r_{ij} \leq 0, \quad s_{ij} \leq 0 \qquad \text{for } i \neq j. \tag{5.10}$$

Let us prove this by induction on $i + j$. Since $a_{12} = r_{11}s_{12} = s_{12}$ and $a_{21} = r_{21}s_{11}$, we have $r_{21} \leq 0$, $s_{12} \leq 0$. Hence, (5.10) is valid for $i + j = 3$. Now, let $i + j > 3$ and suppose that the inequalities (5.10) are true for all the pairs i', j', $i' \neq j'$, which satisfy $i' + j' < i + j$.

The relation $A = RS$ can be written as

$$a_{ij} = \sum_{k=1}^{t} r_{ik}s_{kj}, \tag{5.11}$$

where $t = \min(i, j)$. We distinguish two cases. First, let $i > j$. Then from (5.11) we get

$$r_{ij}s_{jj} = a_{ij} - \sum_{k=1}^{j-1} r_{ik}s_{kj}.$$

We have $a_{ij} \leq 0$ on the right-hand side. Furthermore, $\sum_{k=1}^{j-1} r_{ik}s_{kj} \geq 0$ since all the r_{ik}'s and s_{kj}'s in this sum satisfy the induction hypothesis and are thus nonpositive. Therefore, $r_{ij}s_{jj} \leq 0$, and finally $r_{ij} \leq 0$.

Now, suppose $i < j$. Then we use (5.11) to calculate ($r_{ii} = 1$)

$$s_{ij} = a_{ij} \sum_{k=1}^{i-1} r_{ik} s_{kj}$$

and by a similar argument we obtain $s_{ij} \leq 0$.

14. $\rightarrow$ 11. By the remark following Theorem 1.3.8 we have: If a lower triangular matrix R with positive diagonal entries belongs to Z, then $R^{-1} \geq 0$. Similarly, for any upper triangular matrix $S \in Z$ with positive diagonal entries we have $S^{-1} \geq 0$. Hence, if 14. holds, then $R^{-1} \geq 0$, $S^{-1} \geq 0$ and $A^{-1} = S^{-1} R^{-1} \geq 0$.

11. $\rightarrow$ 15. The existence of such splitting is clear by choosing $P = A$, $Q = 0$. Hence the matrix $P^{-1}Q$ is nonnegative. By Theorem 4.3.2, there is a vector $z \geq 0$, $z \neq 0$, such that $P^{-1}Qz = \sigma z$, where $\sigma = \rho(P^{-1}Q)$. Therefore, we have $Qz = \sigma Pz$ and supposing, contrary to 15., $\rho(P^{-1}Q) \geq 1$, we obtain $Az = (P - Q)z = (\sigma^{-1} - 1)Qz \leq 0$. Multiplying the inequality $Az \leq 0$ by A^{-1} (which is nonnegative by 11.) we get $z \leq 0$, whence $z = 0$, which is a contradiction. Therefore, $\rho(P^{-1}Q) < 1$.

15. $\rightarrow$ 9. By 15., the matrix A cannot be nonpositive. In such case, the (existing) splitting $A = P - Q$, $P^{-1} \geq 0$, $Q \geq 0$ would imply $P \leq Q$. Left-multiplying by ρ^{-1}, $I \leq P^{-1}Q$ which contradicts $\rho(P^{-1}Q) < 0$. Thus, $k = \max_i a_{ii}$ is positive. Set $C = kI - A$. The matrices $P = kI$ and $Q = C$ satisfy the hypotheses of 15. Hence, $\rho(P^{-1}Q) < 1$ or $\rho(k^{-1}C) < 1$. Therefore, $k > \rho(C)$ and 9. holds.

9. $\rightarrow$ 16. Assuming 9., we have a matrix $C \geq 0$ and a number $k > \gamma = \rho(C)$ such that $A = kI - C$. First, suppose that C is irreducible. Then according to the Perron-Frobenius Theorem 4.2.1, there is a positive vector $u = [u_i]$ such that $Cu = \gamma u$. The matrix C^T is irreducible and nonnegative, too, so that again there is a positive vector $v = [v_i]$ such that $C^T v = \gamma v$. Now, consider the diagonal matrix $D = \text{diag}(d_1, \ldots, d_n)$ and the (positive) vector $w = [w_i]$ given by the relations $d_i = (v_i/u_i)^{1/2}$, $w_i = (u_i v_i)^{1/2}$, $i = 1, \ldots, n$. Since $D^{-1}w = u$ and $Dw = v$, it follows that $(DCD^{-1})w = \gamma w$, $(DCD^{-1})^T w = \gamma w$. Hence, if we define $B = DAD^{-1}$, then the matrix $B \in Z$ satisfies

$$Bw = \beta w, \qquad B^T w = \beta w$$

with $\beta = k - \gamma > 0$. Therefore,

$$\frac{1}{2}(B + B^T)w = \beta w, \quad w > 0.$$

Since we have proved in fact that the properties 2. and 6. are equivalent, it follows from the above relation that all the principal minors of the matrix $\frac{1}{2}(B + B^T)$ are positive and using Theorem 2.5.1 we see that this matrix is positive definite.

Now, suppose that the matrix C is reducible. By Theorem 3.2.2, there exists a permutation matrix P such that

$$PCP^T = \begin{bmatrix} C_{11} & C_{12} & \cdots & C_{1r} \\ 0 & C_{22} & \cdots & C_{2r} \\ \vdots & \vdots & \ddots & \vdots \\ 0 & 0 & \cdots & C_{rr} \end{bmatrix},$$

where the square matrices C_{ii} are no longer reducible. Denote by G the matrix PAP^T. Then $G = [G_{ik}]$ is block triangular as well and we have $G_{ik} = 0$ for $i > k$, $G_{ik} = -C_{ik}$ for $i \neq k$, and $G_{ii} = kI_i - C_{ii}$. By the preceding part of the proof regarding the irreducible case, there exist diagonal matrices D_i, $i = 1, \ldots, r$, all having positive diagonal entries, such that the symmetric parts $\frac{1}{2}(B_i + B_i^T)$ of the matrices $B_i = D_i G_{ii} D_i^{-1}$ are positive definite. Furthermore, for $\varepsilon > 0$ we introduce the block diagonal matrix $L(\varepsilon)$ by

$$L(\varepsilon) = \mathrm{diag}(D_1, \varepsilon^{-1}D_2, \varepsilon^{-2}D_3, \ldots, \varepsilon^{1-r}D_r).$$

Then

$$L(\varepsilon)G(L(\varepsilon))^{-1} = \begin{bmatrix} B_1 & B_{12} & B_{13} & \cdots & B_{1r} \\ 0 & B_2 & B_{23} & \cdots & B_{2r} \\ \vdots & \vdots & \vdots & \ddots & \vdots \\ 0 & 0 & 0 & \cdots & B_r \end{bmatrix},$$

where $B_{ik} = \varepsilon^{k-i}D_i G_{ik} D_k^{-1}$ for $i < k$.

The symmetric part of the block diagonal matrix $\mathrm{diag}(B_1, \ldots, B_r)$ is positive definite. By the result of Exercise 2.12, there exists a number $\varepsilon_0 > 0$ such that also the symmetric part of the matrix $L(\varepsilon_0)G(L(\varepsilon_0))^{-1}$ is positive definite. Denote by D the diagonal matrix $P^T L(\varepsilon_0)P$. Then the symmetric part of $B = DAD^{-1}$ is positive definite as well since

$$B = P^T L(\varepsilon_0)PAP^T(L(\varepsilon_0))^{-1}P = P^T L(\varepsilon_0)G(L(\varepsilon_0))^{-1}P$$

and the symmetric part of the matrix B can be obtained by permuting the rows and columns of the symmetric part of $L(\varepsilon_0) \times G(L(\varepsilon_0))^{-1}$.

16. $\to$ 17. Let $H = D^{-2}$, where D is the diagonal matrix of 16. Since the symmetric part of the matrix DAD^{-1} is positive definite, Theorem 2.5.1 shows that also the symmetric part of AH is positive definite as

$$AH + HA^T = D^{-1}(DAD^{-1} + D^{-1}A^T D)D^{-1}.$$

17. $\to$ 18. Assuming 17., we see that the matrix $Z = \frac{1}{2}(AH + HA^T)$ is positive definite. Then every nonzero complex vector x satisfies

$$\langle (AH + HA^T)x, x \rangle > 0,$$

thus

$$\langle AHx, x \rangle + \langle x, AHx \rangle > 0,$$

or

$$\mathrm{Re}\langle AHx, x \rangle > 0.$$

Let λ be an eigenvalue of A and $y \neq 0$ an associated eigenvector. Define the vector x by $Hx = y$. Thus, $x \neq 0$ and, furthermore, $\mathrm{Re}\langle Ay, H^{-1}y \rangle = \mathrm{Re}(\lambda\langle y, H^{-1}y \rangle) > 0$. Since H^{-1} is positive definite, $\langle y, H^{-1}y \rangle = \langle H^{-1}y, y \rangle > 0$, and therefore $\mathrm{Re}\,\lambda > 0$.

18. $\to$ 8. Obvious. $\square$

The theorem just established is the basis for defining the *M-matrices* [49] or the *matrices of class K* [14] as those matrices in Z which satisfy (for the corresponding n) one of the conditions of Theorem 5.1.1, hence all of them.

We remark that Theorem 5.1.1 collects the most important properties of the matrices of class K. However, it is evident that some of the assertions 1. to 18. are "weaker" than others (for instance, 8. is weaker than 5., 12. is weaker than 6., etc.). This fact is of apparent importance. If we want to prove that a matrix from Z belongs to K we verify some of the weaker properties (e.g. 12.). On the other hand, if we already know that $A \in K$ we rather use the stronger properties to develop further consequences.

The reader may have noticed that some of the statements of Theorem 5.1.1 are identical with those of Theorem 2.5.1. This fact leads e.g. to the following statement, which has been used, after all, in the proof of Theorem 5.1.1.

Theorem 5.1.2 *Let $A \in Z$ be a symmetric matrix. Then $A \in K$ if and only if A is positive definite.*

5.2 Class K_0

We may ask whether it is possible to find an analogy of positive semidefinite matrices in Z. The answer is affirmative. However, we first prove an important theorem.

Theorem 5.2.1 *Let $A \in Z_n$. Then the following properties are equivalent:*

1. *$A + \varepsilon I \in K$ for any $\varepsilon > 0$.*

2. *Every real eigenvalue of any principal submatrix of A is nonnegative.*

3. *All principal minors of A are nonnegative.*

4. *For each $k = 1, 2, \ldots, n$, the sum of all the principal minors of order k of the matrix A is nonnegative.*

5. *Every real eigenvalue of A is nonnegative.*

6. *There exist a matrix $C \geq 0$ and a number $k \geq \rho(C)$ such that $A = kI - C$.*

7. *Every eigenvalue of the matrix A has a nonnegative real part.*

Proof. We shall prove the theorem by verifying the implications 1. $\to$ 2., 2. $\to$ 3., 3. $\to$ 4., 4. $\to$ 5., 5. $\to$ 6., 6. $\to$ 1., 1. $\to$ 7., 7. $\to$ 1.

1. $\rightarrow$ 2. Let $\emptyset \neq M \subset N = \{1, 2, \ldots, n\}$ and let λ be a real eigenvalue of the principal submatrix $A(M)$. Suppose that $\lambda < 0$ and choose $\varepsilon = -\lambda$, so that $\varepsilon > 0$. The matrix $B = A + \varepsilon I$ belongs to K according to 1. Property 5. of Theorem 5.1.1 states that all the real eigenvalues of $B(M)$ are positive. But $\lambda + \varepsilon = 0$ is an eigenvalue of $B(M) = A(M) + \varepsilon I(M)$, which is a contradiction.

2. $\rightarrow$ 3. This follows from (1.39) in the way analogous to the proof of the implication 5. $\rightarrow$ 6. in Theorem 5.1.1. By 2., the real eigenvalues of each principal submatrix of A are nonnegative and the non-real ones occur as pairs of conjugate complex numbers; their product is therefore nonnegative.

3. $\rightarrow$ 4. Obvious.

4. $\rightarrow$ 5. As in the proof of the implication 7. $\rightarrow$ 8. in Theorem 5.1.1, we employ the formula (1.36), i.e. (5.9). The assumption 4. means that $c_k \geq 0$ for $k = 1, \ldots, n$. Suppose that 5. does not hold. Then there is a real negative eigenvalue λ_0 of the matrix A, i.e., $\lambda_0 < 0$ and $\det(A - \lambda_0 I) = 0$. After the substitution of λ_0 into (5.9), the left-hand side vanishes while the right-hand side consists of a sum of nonnegative numbers, the first of which, $(-\lambda_0)^n$, is positive. This is a contradiction.

5. $\rightarrow$ 6. This can be proved in the way analogous to the proof of the implication 8. $\rightarrow$ 9. in Theorem 5.1.1.

6. $\rightarrow$ 1. Let $\varepsilon > 0$. Then assuming 6., we have the following property of the matrix $B = A + \varepsilon I$: There exist a matrix $C \geq 0$ and a number $k' = k + \varepsilon > \rho(C)$ such that $B = k'I - C$. By 9. of Theorem 5.1.1, we have $B \in K$, i.e. 1. holds.

The equivalence of 1. and 7. follows from condition 18. of Theorem 5.1.1. $\qquad\square$

The class of all the matrices in Z that (for the corresponding n) satisfy one of the conditions 1. to 7. of Theorem 5.2.1 (and thus all of them) will be denoted by K_0. The *matrices of the class* K_0 are usually also called M-matrices; if one wishes to emphasize that the matrix belongs to K, one speaks about a nonsingular M-matrix. This is justified by the next theorem.

Theorem 5.2.2 *We have $K \subset K_0$. A matrix from K_0 belongs to K if and only if it is nonsingular.*

Proof. Let $A \in K$. Then $A \in Z$ and A possesses the property 6. of Theorem 5.1.1. But it then has the property 3. of Theorem 5.2.1 as well, i.e. $A \in K_0$.

Let now $A \in K_0$. If A is singular, the relation $A \in K$ is not true by property 11. of Theorem 5.1.1. If A is nonsingular, then 0 is not its eigenvalue. By 5. of Theorem 5.2.1, any real eigenvalue of A is nonnegative and thus positive. Therefore A satisfies 8. of Theorem 5.1.1 and $A \in K$. $\qquad\Box$

We add several remarks to Theorem 5.2.1. Some of the assertions 1. to 7. of this theorem are modifications of the assertions of Theorem 5.1.1. On the other hand, some assertions of Theorem 5.1.1 find no analogy here. This is – as a rule – caused by the fact that such a modification cannot be carried out (see Exercises 5.2, 5.3, and 5.4).

An easy consequence of a comparison of Theorem 5.2.1 with Theorem 2.5.2 on positive semidefinite matrices is the next theorem.

Theorem 5.2.3 *Let the matrix $A \in Z$ be symmetric. Then $A \in K_0$ if and only if A is positive semidefinite.*

The further theorems are concerned with certain properties of matrices in K and K_0.

Theorem 5.2.4 *If $A \in K$ then $A^T \in K$ as well. If $A \in K_0$ then $A^T \in K_0$ as well.*

Proof. We obtain the assertion of the theorem employing twice the property 6. of Theorem 5.1.1 and the property 3. of Theorem 5.2.1. $\qquad\Box$

Theorem 5.2.5 *If $A \in K$, $B \in Z$, and $B \geq A$, then $B \in K$. If $A \in K_0$, $B \in Z$, and $B \geq A$, then $B \in K_0$.*

Proof. The first assertion follows by employing twice the property 1. of Theorem 5.1.1: Since $A \in K$, then by 1. there exists a vector $x \geq 0$ such that $Ax > 0$. As $B \geq A$, $Bx > 0$ as well, and thus $B \in K$ by 1. To prove the second part we employ twice the property 1. of Theorem 5.2.1. $\qquad\Box$

Theorem 5.2.6 *If $A \in K_0$ is irreducible and singular, then there exists a positive vector u such that $Au = 0$. Furthermore, the rank of A is by 1 less than its order.*

Proof. If the order of A is 1, i.e. $A = [0]$, the theorem holds. Let thus $n \geq 2$. By 6. of Theorem 5.2.1, there exists a matrix $C \geq 0$ and a number $k \geq \rho(C)$ such that $A = kI - C$. We note that C is irreducible and its order is at least 2. By the Perron-Frobenius Theorem 4.2.1, we have $\rho(C) > 0$, this eigenvalue is simple and there exists a vector $u > 0$ such that $Cu = \rho(C)u$. Since A is singular, 0 is its eigenvalue, and thus k is an eigenvalue of C by Theorem 1.7.12. Moreover, $|k| \leq \rho(C)$ as $\rho(C)$ is the spectral radius of the matrix C. Since $|k| \leq \rho(C) \leq k$, we have $\rho(C) = k$, $Cu = ku$, and thus $Au = 0$. Since $\rho(C)$ is a simple eigenvalue of the matrix C, the rank of the matrix $C - \rho(C)I = -A$ is $n - 1$, and the rank of A is the same. □

Theorem 5.2.7 *If $A \in K_0$ is irreducible, then there exists a vector $x > 0$ such that $Ax \geq 0$.*

Proof. If A is nonsingular, then $A \in K$ by Theorem 5.2.2 and the assertion follows from 1. of Theorem 5.1.1. If A is singular, the assertion is a consequence of Theorem 5.2.6. □

Theorem 5.2.8 *If $A \in K_0$, then there exists a vector $x \geq 0$, $x \neq 0$, such that $Ax \geq 0$.*

Proof. By property 6. of Theorem 5.2.1, there exist a nonnegative matrix C and a number $k \geq \rho(C)$ such that $A = kI - C$. Denoting by x the nonnegative vector associated with $\rho(C)$ according to Theorem 4.3.2, we have $Ax = (k - \rho(C))x \geq 0$ and, as $x \neq 0$, this vector fulfils the assertion of the theorem. □

Theorem 5.2.9 *If $A \in Z$ and if there exists a positive vector x such that $Ax \geq 0$, then $A \in K_0$.*

Proof. Choose an arbitrary positive number ε. By hypothesis then $(A + \varepsilon I)x > 0$. Finally, $A \in K_0$ by 2. of Theorem 5.1.1 and 1. of Theorem 5.2.1. □

Theorem 5.2.10 *Let $A \in Z$ be an irreducible matrix. Then the following statements are equivalent:*

1. *There exists a vector $x > 0$ such that $Ax > 0$.*

2. *There exists a vector $x > 0$ such that $Ax \geq 0$, $Ax \neq 0$.*

3. *$A \in K$.*

4. *$A^{-1} > 0$.*

Proof. Let us prove the implications 1. $\rightarrow$ 2., 2. $\rightarrow$ 3., 3. $\rightarrow$ 4., 4. $\rightarrow$ 1.

   1. $\rightarrow$ 2. Obvious.

   2. $\rightarrow$ 3. We have $A \in K_0$ by Theorem 5.2.9. By Theorem 5.2.2, it is then sufficient to show that A is nonsingular. Thus, suppose that A is singular. Then A^T is singular and irreducible as well and $A^T \in K_0$. By Theorem 5.2.6, there exists a positive vector u such that $A^T u = 0$. Then $u^T Ax = (A^T u)^T x = 0$ and this is also equal to $u^T (Ax)$. Here, u^T is a positive vector, Ax a nonnegative one. This implies $Ax = 0$, which contradicts the hypothesis. Therefore $A \in K$.

   3. $\rightarrow$ 4. By property 9. of Theorem 5.1.1, there exist a matrix $C \geq 0$ and a number $k > \rho(C)$ such that $A = kI - C$. We thus have

$$k^{-1}A = I - C,$$

where

$$Z = k^{-1}C$$

is a nonnegative irreducible matrix (since C is irreducible) with the spectral radius

$$\rho(Z) = k^{-1}\rho(C) < 1.$$

By Theorem 1.7.14, we can write

$$(k^{-1}A)^{-1} = I + Z + Z^2 + \ldots;$$

the series on the right-hand side converges and its terms are nonnegative. By Theorem 4.1.4, however, the sum of the first n terms of this series is positive. Thus $(k^{-1}A)^{-1} > 0$ and $A^{-1} > 0$ as well.

4. → 1. Suppose $A^{-1} > 0$. For the vector $x = A^{-1}e$, where $e = [1, 1, \ldots, 1]^T$, we then obtain that $x > 0$ and $Ax > 0$. □

We now apply some of the above theorems to the matrix A of the system (5.3). Since $A \in Z$ and $Ae_0 \geq 0$, $Ae_0 \neq 0$, where e_0 is the positive vector of all ones, the condition 2. of Theorem 5.2.10 is satisfied as soon as the matrix A is irreducible. But this situation takes place for a connected domain if the net is sufficiently fine. Theorem 5.2.10 then implies that A belongs to K and hence A is nonsingular. Moreover, $A^{-1} > 0$ and the solution u_0 thus possesses the following property: If the right-hand side f_0 increases, the solution at each point increases as well, i.e., for $u_0 = A^{-1}f_0$ and $f_0 \leq f_1$ we have $u_1 = A^{-1}f_1 \geq A^{-1}f_0 = u_0$.

A theorem analogous to Theorem 2.5.5 also holds.

Theorem 5.2.11 *If $A \in K$, then any principal submatrix of A belongs also to K. If A_{11} is a principal submatrix of A, then the Schur complement $[A/A_{11}]$ of the matrix A_{11} in A belongs again to K.*

Proof. Applying twice the property 6. of Theorem 5.1.1, we obtain the first part of the theorem. To prove its second part, we write (after a permutation, if necessary)

$$A = \begin{bmatrix} A_{11} & A_{12} \\ A_{21} & A_{22} \end{bmatrix},$$

so that

$$[A/A_{11}] = A_{22} - A_{21}A_{11}^{-1}A_{12}.$$

We have $[A/A_{11}] \in Z$ as $A_{22} \in Z$, $A_{21} \leq 0$, $A_{21} \leq 0$, and $A_{11}^{-1} \geq 0$ by the first part of the theorem. Considering Theorem 1.4.2, we observe that $[A/A_{11}]^{-1} = B_{22}$, where B_{22} is the corresponding principal submatrix of A^{-1}. Since $A^{-1} \geq 0$ by property 11. of Theorem 5.1.1, $B_{22} \geq 0$ as well and hence $[A/A_{11}] \in Z$ fulfils 11. of Theorem 5.1.1. We thus have $[A/A_{11}] \in K$. □

5.3 Diagonally dominant matrices

We now introduce a further concept. A square (real or complex) matrix $A = [a_{ik}]$ of order n is said to be *diagonally dominant* if there exist positive numbers $d_1, d_2, \ldots, d_n$ such that

$$|a_{ii}|d_i > \sum_{k \neq i} |a_{ik}|d_k \qquad \text{for } i = 1, \ldots, n. \qquad (5.12)$$

In the related literature the definition is, as a rule, presented in a different form. A matrix $A = [a_{ik}]$ is said to be diagonally dominant if

$$|a_{ii}| > \sum_{k \neq i} |a_{ik}| \qquad \text{for } i = 1, \ldots, n. \qquad (5.13)$$

In our considerations, such a matrix A will be said to be *uniformly diagonally dominant*.

Given a square (real or complex) matrix $A = [a_{ik}]$, we further denote by $H(A)$ the matrix

$$H(A) = [h_{ik}], \qquad h_{ii} = |a_{ii}|, \qquad h_{ik} = -|a_{ik}|, \quad i \neq k. \qquad (5.14)$$

We now have the following theorem.

Theorem 5.3.1 *Let $A = [a_{ik}]$ be a square complex matrix. Then the following are equivalent:*
1. *A is diagonally dominant.*
2. *$H(A) \in K$.*

Proof. 1. $\to$ 2. The property that there are positive numbers $d_1, \ldots, d_n$ such that (5.12) holds can be also stated as follows:

$$H(A)d > 0,$$

where $d = [d_1, \ldots, d_n]^T > 0$. Since $H(A) \in Z$, we have $H(A) \in K$ by 2. of Theorem 5.1.1.

2. $\to$ 1. If $H(A) \in K$, then 2. of Theorem 5.1.1 implies the existence of a vector $x > 0$ such that $H(A)x > 0$. Denoting by $d_1, \ldots, d_n$ the components of the vector x, we obtain (5.12). $\square$

Theorem 5.3.2 *If A is diagonally dominant, A^T is diagonally dominant as well.*

Proof. The theorem follows from property 2. of Theorem 5.3.1 and from Theorem 5.2.4. $\square$

Remark.   An analogous theorem for a uniformly diagonally dominant matrix (of order greater than 1) is not true.

Theorem 5.3.3 *If a matrix A is diagonally dominant, then any its principal submatrix is also diagonally dominant. The coefficients of the diagonal dominance of A that correspond to the index set of the submatrix can be chosen for the coefficients of the diagonal dominance of the submatrix. If A is uniformly diagonally dominant, then any its principal submatrix is also uniformly diagonally dominant.*

Proof. Let n be the order of A, $N = \{1, 2, \ldots, n\}$. Let $A(M)$ be such a principal submatrix of the matrix A that corresponds to the index set $M \subset N$. Since (5.12) holds, we also have

$$|a_{ii}|d_i > \sum_{k \neq i} |a_{ik}|d_k \qquad \text{for all } i \in M.$$

Thus $A(M)$ is diagonally dominant with the coefficients d_i for $i \in M$. The rest follows immediately by choosing $d_i = 1$ in the preceding argument. □

The next theorem shows the fundamental importance of diagonally dominant matrices.

Theorem 5.3.4 *Let A be a diagonally dominant matrix. Then A is nonsingular and the modulus $m(A^{-1})$ of its inverse satisfies*

$$m(A^{-1})H(A) \leq I, \qquad (5.15)$$

$$H(A)m(A^{-1}) \leq I, \qquad (5.16)$$

as well as

$$m(A^{-1}) \leq (H(A))^{-1}. \qquad (5.17)$$

Proof. Suppose first that A fulfils (5.12) and set $D = \operatorname{diag}(d_1, \ldots, d_n)$. The matrix $B = DAD^{-1}$ is then uniformly diagonally dominant and it is easy to see that it is sufficient to prove the nonsingularity for a uniformly diagonally dominant matrix.

Suppose that a matrix A satisfying (5.13) is singular. Hence, there exists a vector $x = [x_1, \ldots, x_n]^T \neq 0$ such that $Ax = 0$. Let k be such that $|x_k| = \max_j |x_j|$. Then

$$a_{kk}x_k = -\sum_{j \neq k} a_{kj}x_j,$$

from which

$$|a_{kk}||x_k| = \left|\sum_{j\neq k} a_{kj}x_j\right| \leq \sum_{j\neq k} |a_{kj}||x_j| \leq$$

$$\leq \left(\sum_{j\neq k} |a_{kj}|\right)|x_k| < |a_{kk}||x_k|$$

as $x_k \neq 0$. This is a contradiction.
If $A^{-1} = [\alpha_{ik}]$, we have

$$a_{ii}\alpha_{ii} = 1 - \sum_{j\neq i} a_{ij}\alpha_{ji},$$

from which

$$|a_{ii}||\alpha_{ii}| \leq 1 + \sum_{j\neq i} |a_{ij}||\alpha_{ji}|.$$

For $i \neq k$, we obtain

$$a_{ii}\alpha_{ik} = -\sum_{j\neq i} a_{ij}\alpha_{jk},$$

from which

$$|a_{ii}||\alpha_{ik}| \leq \sum_{j\neq i} |a_{ij}||\alpha_{jk}|.$$

This proves (5.15). Clearly, (5.16) can be proved in a similar way. The inequality (5.17) follows from (5.15) if we multiply both its sides by $(H(A))^{-1}$, which is a nonnegative matrix by 2. of Theorem 5.3.1 and 11. of Theorem 5.1.1.   □

The next theorem is important for the application of the Gaussian elimination to the linear algebraic system whose matrix is diagonally dominant.

Theorem 5.3.5 *If A is diagonally dominant and A_{11} is its principal submatrix, then the Schur complement $[A/A_{11}]$ of the submatrix A_{11} in A is again a diagonally dominant matrix. Moreover, if the coefficients of the diagonal dominance of A (of order n) are $d_1, \ldots, d_n$, then one can choose as coefficients of the diagonal dominance of $[A/A_{11}]$ those d_j whose indices do not belong to the index set of the submatrix A_{11}. In particular, if A is uniformly diagonally dominant $(d_i = 1, i = 1, \ldots, n)$, then $[A/A_{11}]$ is a uniformly diagonally dominant matrix as well.*

Proof. We prove the theorem by induction on the order m of the matrix A_{11}. Let $m = 1$ and $A_{11} = [a_{11}]$ so that the entries of $[A/[a_{11}]]$ are $a_{ik} - a_{i1}a_{11}^{-1}a_{1k}$, $i, k = 2, \ldots, n$. Assume (5.12). For $k = 2, \ldots, n$ we then have

$$
\begin{aligned}
|a_{kk}|d_k &> \sum_{j \neq k} |a_{kj}|d_j \\
&= \sum_{1 < j \neq k} |a_{kj}|d_j + |a_{k1}|d_1 \\
&= \sum_{1 < j \neq k} |a_{kj}|d_j + \frac{|a_{k1}|}{|a_{11}|}|a_{11}|d_1 \\
&\geq \sum_{1 < j \neq k} |a_{kj}|d_j + \frac{|a_{k1}|}{|a_{11}|}\sum_{1 < j} |a_{1j}|d_j \\
&\geq \sum_{1 < j \neq k} |a_{kj} - a_{k1}a_{11}^{-1}a_{1j}|d_j + |a_{k1}||a_{11}^{-1}||a_{1k}|d_k.
\end{aligned}
$$

This implies

$$
\begin{aligned}
|a_{kk} - a_{k1}a_{11}^{-1}a_{1k}|d_k &\geq (|a_{kk}| - |a_{k1}||a_{11}^{-1}||a_{1k}|)d_k \\
&> \sum_{1 < j \neq k} |a_{kj} - a_{k1}a_{11}^{-1}a_{1j}|d_j,
\end{aligned}
$$

which means that the matrix $[A/[a_{11}]]$ is diagonally dominant with the coefficients $d_2, \ldots, d_n$.

If now $m > 1$, we employ the formula (1.31) to perform the induction step. Considering the matrix

$$
A = \begin{bmatrix} A_{11} & A_{12} \\ A_{21} & A_{22} \end{bmatrix}
$$

and setting $\tilde{A}_{00} = [a_{11}]$, we write

$$
A_{11} = \begin{bmatrix} \tilde{A}_{00} & \tilde{A}_{01} \\ \tilde{A}_{10} & \tilde{A}_{11} \end{bmatrix}, \quad A_{12} = \begin{bmatrix} \tilde{A}_{02} \\ \tilde{A}_{12} \end{bmatrix},
$$

$$
A_{21} = [\tilde{A}_{20}, \tilde{A}_{21}], \quad A_{22} = \tilde{A}_{22},
$$

so that A_{11} is of order $m - 1$. The matrix $[A/[a_{11}]]$ is thus $[A/\tilde{A}_{00}]$ and we denote it by

$$
\begin{bmatrix} \tilde{C}_{11} & \tilde{C}_{12} \\ \tilde{C}_{21} & \tilde{C}_{22} \end{bmatrix} = C,
$$

where $\tilde{C}_{11}$ is a nonsingular matrix of order $m - 1$. By the part of the assertion that we have already proved, C is a diagonally dominant matrix and the numbers $d_2, \ldots, d_n$ can be chosen for the coefficients. By the induction hypothesis ($\tilde{C}_{11}$ is of order $m - 1$), the matrix $[C/\tilde{C}_{11}]$ is again diagonally dominant, with the coefficients $d_{m+1}, \ldots, d_n$. But the formula (1.31) gives

$$[C/\tilde{C}_{11}] = [[A/\tilde{A}_{00}]/[A_{11}/\tilde{A}_{00}]] = [A/A_{11}].$$

This proves the theorem since the last assertion follows from the preceding ones for $d_i = 1, i = 1, \ldots, n$.   □

5.4 Monotone matrices

The matrices of class K represent a special case of the so-called monotone matrices. The definition of monotone matrices is based on the following theorem.

Theorem 5.4.1 *Let A be a real matrix. Then the following two properties are equivalent:*

1. *$A^{-1} \geq 0$.*
2. *Whenever $Ax \geq Ay$ holds for vectors x and y, then $x \geq y$.*

Proof. The implication 1. → 2. is clear (it is sufficient to left-multiply the relation $Ax - Ay \geq 0$ by A^{-1}, which gives $x - y \geq 0$).

We now prove that, conversely, 2. implies 1. The property 2. can be reformulated as follows: Whenever $Az \geq 0$, then $z \geq 0$. This property, however, means in the first place that A is a nonsingular matrix. Indeed, suppose that A is singular. By Theorem 1.5.3, there exists a vector $y \neq 0$ such that $Ay = 0$. The property 2. gives $y \geq 0$. But at the same time we have $A(-y) = 0$, which means that $-y \geq 0$. Hence $y = 0$, which is a contradiction. Now, denote by $e_i = [0, \ldots, 0, 1, 0, \ldots, 0]^T$ the n-component column vector with 1 as the ith component, $i = 1, \ldots, n$ (the order of A being n). Then there exist vectors z_i, $i = 1, \ldots, n$, such that $Az_i = e_i$. Since $e_i \geq 0$, the second property implies $z_i \geq 0$, $i = 1, \ldots, n$. But $z_i = A^{-1}e_i$ are the columns of the matrix A^{-1}, and hence $A^{-1} \geq 0$.   □

A matrix is said to be *monotone* if it possesses one of the properties of Theorem 5.4.1 (and, therefore, both of them).

Monotone matrices have many further properties. An easy consequence of the property 1. is the next theorem.

Theorem 5.4.2 *The product of monotone matrices is again a monotone matrix. Every permutation matrix is monotone.*

We now return to Example 5.1. Here the relation between the vector x of the gross output and the vector y of the net product is expressed by the equation $Ax = y$. We have $A \in Z$ and, supposing that the system is not supplied from outside and that the resulting vector of the net product is positive, we have $x > 0$ and $y > 0$. Hence, by property 2. of Theorem 5.1.1, $A \in K$ and by 11., $A^{-1} \geq 0$. The matrix A (the so-called *Leontiev matrix*) is thus monotone. It means that it is necessary to increase the gross output in order to increase the net product. But, conversely, it is not true that the increase of each individual gross output necessarily leads to the increase of the net product (as $x_1 \geq x_2$ does not imply $Ax_1 \geq Ax_2$ in general). We can, however, easily prove the following theorem.

Theorem 5.4.3 *If $A \in K$ and z is a positive eigenvector of the matrix A, then an increase of the vector x in the relation $Ax = y$ by a positive multiple of the vector z implies the increase of the vector y (by a positive multiple of the vector z, too).*

5.5 Class P

In the conclusion of this chapter, let us investigate a more general class P of matrices, which is also important for applications and comprises e.g. positive definite matrices and matrices of class K. The *class P* consists of all the real square matrices whose all principal minors are positive. The following theorem (cf. [14, 17, 2]) describes the basic properties of the class P.

Theorem 5.5.1 *Let A be a real square matrix. Then the following properties are equivalent:*

1. $A \in P$, *i.e., all the principal minors of A are positive.*

2. *If D is a nonnegative diagonal matrix of the same order as A, then all the principal minors of A + D are different from zero.*

3. *For every column vector* $x = [x_i], x \neq 0$, *there exists an integer* k *such that* $x_k y_k > 0$ *where* $y = [y_i] = Ax$.

4. *For every column vector* $x \neq 0$ *there exists a diagonal matrix* D_x *with positive diagonal entries, such that* $\langle Ax, D_x x \rangle > 0$.

5. *For every column vector* $x \neq 0$ *there exists a nonnegative diagonal matrix* H_x, *such that* $\langle Ax, H_x x \rangle > 0$.

6. *Every real eigenvalue of any principal submatrix of A is positive.*

7. *For every column vector* $x \neq 0$, *there exists an integer* k *such that* $x_k z_k > 0$, *where* $z = [z_i] = A^T x$.

8. *The implication*

$$z \geq 0, \quad SA^T Sz \leq 0 \to z = 0$$

holds for every diagonal matrix S whose diagonal entries are ± 1.

9. *For every diagonal matrix S whose diagonal entries are* ± 1, *there exists a vector* $x \geq 0$, *such that* $SASx > 0$.

Proof. We shall first prove the equivalence of the properties 1. to 6.

1. → 2. Follows from (1.12).

2. → 3. Suppose that 3. does not hold. Then there is a vector $x = [x_i] \neq 0$ such that $x_k y_k \leq 0$ for all k, where $y = Ax = [y_i]$. Set $M = \{i; x_i \neq 0\}$; hence $M \neq \emptyset$. The matrix $A(M)$ and the vectors $x(M)$ and $y(M)$ then satisfy

$$y(M) = A(M)x(M).$$

Further, there exists a nonnegative diagonal matrix $D(M)$ such that

$$y(M) = -D(M)x(M).$$

We then have
$$(A(M) + D(M))x(M) = 0,$$
from which we can see that 2. does not hold (after adding zeros to $D(M)$).

3. → 4. Suppose that 3. holds. Let $x = [x_i] \neq 0$. By 3., there exists an integer k such that $x_k y_k > 0$ for $y = Ax = [y_i]$. Therefore, there exists $\varepsilon > 0$ such that

$$x_k y_k + \varepsilon \sum_{j \neq k} x_j y_j > 0.$$

Choosing $D_x = \text{diag}(d_i)$, $d_k = 1$, $d_j = \varepsilon$ for $j \neq k$, we have $\langle Ax, D_x x \rangle > 0$.

4. → 5. Obvious.

5. → 6. Let M be a nonempty index set and let λ be a real eigenvalue of the matrix $A(M)$ and $x(M)$ an associated eigenvector. We add zeros to $x(M)$ to obtain a vector x satisfying 5. Consider now the matrix H_x of 5. Then $\langle Ax, H_x x \rangle > 0$. However,

$$\langle Ax, H_x x \rangle = \langle A(M)x(M), H_x(M)x(M) \rangle$$
$$= \lambda \langle x(M), H_x(M)x(M) \rangle.$$

As $\langle x(M), H_x(M)x(M) \rangle \geq 0$, we have $\lambda > 0$.

6. → 1. Follows from (1.39) since the product of non-real eigenvalues of any real matrix is positive.

The equivalence of 7. with the properties 1. to 6. follows from the fact that 1., and thus also 3., do not change when A is replaced by its transpose.

To prove the equivalence of 7. to 8., suppose that 8. is not true. Then there exist a diagonal matrix S whose diagonal entries are ± 1 and a vector $u \geq 0$, $u \neq 0$, such that $SA^T Su \leq 0$. Set $y = Su$ so that $y = [y_i] \neq 0$. Since $z = A^T y = [z_i]$ satisfies $y_k z_k \leq 0$ for all k, it follows that 7. does not hold.

Conversely, suppose that 7. is not true. Then there exists a vector $y = [y_i] \neq 0$ such that $z = A^T y = [z_i]$ satisfies $y_k z_k \leq 0$ for all k. Clearly, there exists a diagonal matrix S whose diagonal entries are ± 1 such that $u = Sy \geq 0$ and $Sz \leq 0$. Since $SA^T Su \leq 0$ and $u \neq 0$, we conclude that 8. does not hold.

The equivalence of 8. and 9. is a consequence of the following general theorem on linear inequalities. □

Theorem 5.5.2 *Let C be a real $m \times n$ matrix. Then the follow-ing statements are equivalent:*

1. *There exists a vector $x \geq 0$ such that $Ax > 0$.*
2. *If $y \geq 0$ and $A^T y \leq 0$, then $y = 0$.*

Proof. 1. $\rightarrow$ 2. Let 1. hold and suppose that a vector $y \neq 0$ satisfies $y \geq 0$, $A^T y \leq 0$. Then $y^T Ax = y^T (Ax) > 0$. On the other hand, $y^T Ax = (A^T y)^T x \leq 0$, which is a contradiction.

2. $\rightarrow$ 1. Let 2. hold and denote by C_1 and C_2 the convex cones of nonnegative m- and n-component column vectors, respectively. The set C_3 of all vectors of the form $-A^T x$, where $x \in C_1$, is again a convex cone, which has only the zero vector in common with C_2 by 2. Hence, in the space of n-component vectors there exists a hyperplane described by the equation $u^T x = 0$, such that all vectors $x \in C_2$, $x \neq 0$, satisfy $u^T x > 0$, whereas $u^T x < 0$ holds for all vectors $x \in C_3$, $x \neq 0$. The first condition implies $u > 0$ (it is sufficient to take $x = [1, 0, \ldots, 0]^T$, and so forth); the second condition gives $u^T A^T z > 0$ for all vectors $z \geq 0$ such that $A^T z \neq 0$, i.e., for all $z \geq 0$, $z \neq 0$, by 2. This, in turn, yields $u^T A^T > 0$, or $Au > 0$. Hence, 1. is true. $\square$

The next two theorems follow from Theorem 5.5.1 if we set $D_x = I$ in 4. and $S = I$ in 9.

Theorem 5.5.3 *Given a real square matrix A such that $A + A^T$ is positive definite, then $A \in P$.*

Theorem 5.5.4 *If $A \in P$, then there exists a vector $x \geq 0$ such that $Ax > 0$.*

A generalization of the class P is the *class P_0* of all the real square matrices whose all principal minors are nonnegative. We present the fundamental theorem (for the proof see [17]).

Theorem 5.5.5 *Let A be a real square matrix. Then the follow-ing properties are equivalent:*

1. $A \in P_0$.

2. *If D is a diagonal matrix of the same order as A and if the diagonal entries of D are positive, then $A + D \in P$.*

3. *For every vector* $x = [x_i]$, $x \neq 0$, *there exists an integer* k *such that* $x_k \neq 0$ *and* $x_k y_k \geq 0$, *where* $y = [y_i] = Ax$.

4. *For every vector* $x \neq 0$ *there exists a nonnegative diagonal matrix* D_x *such that* $\langle x, D_x x \rangle > 0$ *and* $\langle Ax, D_x x \rangle \geq 0$.

5. *Every real eigenvalue of any principal submatrix of* A *is non-negative.*

6. $A + \varepsilon I \in P$ *for any* $\varepsilon > 0$.

7. *For every diagonal matrix* S *whose diagonal entries are* $0, 1$, *or* -1, *there exists a vector* $y \neq 0$ *such that* $Sy \geq 0$, $S^2 y = y$, *and* $SAy \geq 0$.

The following analogies of Theorems 5.5.3 and 5.5.4 are easily obtained.

Theorem 5.5.6 *Given a real square matrix* A *such that* $A + A^T$ *is positive semidefinite, then* $A \in P_0$.

Theorem 5.5.7 *If* $A \in P_0$, *then there exists a vector* $x \geq 0$, $x \neq 0$, *such that* $Ax \geq 0$.

The class P finds an important application in the so-called *linear complementarity problem*: Given a real square matrix M and a real vector q, find vectors

$$w \geq 0, \qquad z \geq 0 \tag{5.18}$$

satisfying

$$w = Mz + q, \tag{5.19}$$

$$w^T z = 0. \tag{5.20}$$

(The condition (5.20) clearly means that of any two corresponding components of the vectors w and z at least one is zero.) As Exercise 5.8 shows, the basic problem of linear programming can be considered as a linear complementarity problem. Several methods for the numerical solution of the linear complementarity problem (e.g. Cottle and Dantzig [7]) can be considered as equivalent alternatives to the well known simplex method.

The following existence theorem [2] is presented without proof.

Theorem 5.5.8 *A necessary and sufficient condition for the problem* (5.18), (5.19), (5.20) *with a given matrix M to have a unique solution for any vector q is that* $M \in P$.

Exercises

5.1. Prove that K_0 is the set of the limits of all convergent sequences of the matrices of class K.

5.2. Find whether the following converse of Theorem 5.2.8 is valid: If $A \in Z$ and if there is a vector $x \geq 0$, $x \neq 0$, such that $Ax \geq 0$, then $A \in K_0$.

5.3. Find whether the following converse of Theorem 5.2.9 is valid: If $A \in K_0$, then there exists a vector $x > 0$ such that $Ax \geq 0$.

5.4. Show that there exists a matrix $A \in Z$ (with $n \geq 2$) such that $\det A(N_k) \geq 0$, $k = 1, \ldots, n$, but $A \notin K_0$. (Here, $N_k = \{1, 2, \ldots, k\}$.)

5.5. Consider the matrices of the class K_0. Do they form a convex cone (as it was with positive semidefinite matrices in Chapter 2)?

5.6.* Let $A = [a_{ik}] \in K_0$. Show that there exists a nonnegative eigenvalue $q(A)$ such that all eigenvalues λ of A satisfy the inequality $|\lambda - s| \leq |q(A) - s|$, where $s = \max_k a_{kk}$. Further show that there exists a nonnegative (positive, if A is irreducible) eigenvector of A associated with the eigenvalue $q(A)$.

5.7. Prove that P_0 is the set of the limits of all convergent sequences of the matrices of class P.

5.8. Let A be an $m \times n$ real matrix, let b be an m-component and c an n-component column vectors. Then the solution of the basic problem of linear programming, which consists in minimizing $c^T x$ over all n-component vectors x satisfying

$x \geq 0$ and $Ax \geq b$, can be obtained as a solution of the problem (5.18) to (5.20) with

$$M = \begin{bmatrix} 0 & A \\ -A^T & 0 \end{bmatrix}, \qquad q = \begin{bmatrix} -b \\ c \end{bmatrix}.$$

5.9. Show that the matrix M from Exercise 5.8 belongs to P_0.

5.10. Let A be a real matrix. Prove that exactly one of the following two cases occurs:

1. There exists a vector $x \geq 0$ such that $Ax > 0$.

2. There exists a non-zero vector $y \geq 0$, such that $A^T y \leq 0$.

6

Tensor Product of Matrices. Compound Matrices

In this chapter, we study some types of matrices constructed in a special way. These matrices are of theoretical importance and, on the other hand, they sometimes occur in practical problems.

Example 6.1 Solve the Dirichlet problem (see Example 5.2) for a rectangle. Using the finite difference method, suppose that the interior net points constitute a rectangle with the longer side of length $5h$ and the shorter side of length $3h$, where h is the mesh size of the net (Fig. 11). Labelling the net points by numbers as shown in the figure, we obtain the matrix of the system in the form

$$
M = \frac{1}{4}
\begin{bmatrix}
4 & -1 & & & & -1 & & & & & & & & & \\
-1 & 4 & -1 & & & & -1 & & & & & & & & \\
 & -1 & 4 & -1 & & & & -1 & & & & & & & \\
 & & -1 & 4 & -1 & & & & -1 & & & & & & \\
 & & & -1 & 4 & & & & & -1 & & & & & \\
-1 & & & & & 4 & -1 & & & & -1 & & & & \\
 & -1 & & & & -1 & 4 & -1 & & & & -1 & & & \\
 & & -1 & & & & -1 & 4 & -1 & & & & -1 & & \\
 & & & -1 & & & & -1 & 4 & -1 & & & & -1 & \\
 & & & & -1 & & & & -1 & 4 & & & & & -1 \\
 & & & & & -1 & & & & & 4 & -1 & & & \\
 & & & & & & -1 & & & & -1 & 4 & -1 & & \\
 & & & & & & & -1 & & & & -1 & 4 & -1 & \\
 & & & & & & & & -1 & & & & -1 & 4 & -1 \\
 & & & & & & & & & -1 & & & & -1 & 4 \\
\end{bmatrix}
$$

We shall see in a while how to express this matrix (or a general matrix corresponding to the rectangle with the sides of length rh and sh) in a simple form or, more specifically, in a form better displaying the structure of such a matrix.

157

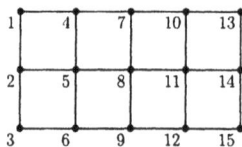

Fig. 11

6.1   Tensor product

Let $A = [a_{ik}]$ be an $m \times n$ matrix and $B = [b_{pq}]$ and $r \times s$ matrix. The $mr \times ns$ matrix

$$A \otimes B = \begin{bmatrix} Ab_{11} & Ab_{12} & \cdots & Ab_{1s} \\ Ab_{21} & Ab_{22} & \cdots & Ab_{2s} \\ \vdots & \vdots & \ddots & \vdots \\ Ab_{r1} & Ab_{r2} & \cdots & Ab_{rs} \end{bmatrix}$$

is then called the *tensor product* of A and B. It is a partitioned matrix and each its block is an $m \times n$ matrix.

Remark. An alternative is to define the tensor product $A \otimes B$ as

$$\begin{bmatrix} a_{11}B & a_{12}B & \cdots & a_{1n}B \\ a_{21}B & a_{22}B & \cdots & a_{2n}B \\ \vdots & \vdots & \ddots & \vdots \\ a_{m1}B & a_{m2}B & \cdots & a_{mn}B \end{bmatrix}.$$

Though both definitions yield analogous results we consider our approach more convenient.

We shall turn back to Example 6.1. We can readily verify that the matrix M can be expressed as

$$M = (I_5 \otimes I_3) - \frac{1}{4}(I_5 \otimes Z_3) - \frac{1}{4}(Z_5 \otimes I_3), \qquad (6.1)$$

where I_5 and I_3 are the identity matrices of order 5 and 3, and

$$Z_3 = \begin{bmatrix} 0 & 1 & 0 \\ 1 & 0 & 1 \\ 0 & 1 & 0 \end{bmatrix}, \qquad Z_5 = \begin{bmatrix} 0 & 1 & & & \\ 1 & 0 & 1 & & \\ & 1 & 0 & 1 & \\ & & 1 & 0 & 1 \\ & & & 1 & 0 \end{bmatrix}.$$

We now establish the basic properties of the tensor product of matrices.

Theorem 6.1.1 *Let A and C be such matrices that AC exists, and let B and D be such matrices that BD exists. Then*

$$(A \otimes B)(C \otimes D) = AC \otimes BD. \tag{6.2}$$

Proof. If B is a $p \times q$ matrix and D is a $q \times r$ matrix, then the left-hand side of (6.2) is

$$\begin{bmatrix} Ab_{11} & Ab_{12} & \cdots & Ab_{1q} \\ Ab_{21} & Ab_{22} & \cdots & Ab_{2q} \\ \vdots & \vdots & \ddots & \vdots \\ Ab_{p1} & Ab_{p2} & \cdots & Ab_{pq} \end{bmatrix} \begin{bmatrix} Cd_{11} & Cd_{12} & \cdots & Cd_{1r} \\ Cd_{21} & Cd_{22} & \cdots & Cd_{2r} \\ \vdots & \vdots & \ddots & \vdots \\ Cd_{q1} & Cd_{q2} & \cdots & Cd_{qr} \end{bmatrix}.$$

Theorem 1.1.1 now yields

$$\begin{bmatrix} AC(b_{11}d_{11} + \cdots + b_{1q}d_{q1}) & \cdots & AC(b_{11}d_{1r} + \cdots + b_{1q}d_{qr}) \\ \vdots & \ddots & \vdots \\ AC(b_{p1}d_{11} + \cdots + b_{pq}d_{q1}) & \cdots & AC(b_{p1}d_{1r} + \cdots + b_{pq}d_{qr}) \end{bmatrix}$$

$$= AC \otimes BD.$$

$\square$

This theorem has a number of important corollaries.

Theorem 6.1.2 *If A and B are nonsingular matrices, then their tensor product is again a nonsingular matrix and*

$$(A \otimes B)^{-1} = A^{-1} \otimes B^{-1}. \tag{6.3}$$

Proof. Calculate the product of the matrices $A \otimes B$ and $A^{-1} \otimes B^{-1}$ (which, of course, does exist). By (6.2), we have

$$(A \otimes B)(A^{-1} \otimes B^{-1}) = (AA^{-1}) \otimes (BB^{-1}) = I_1 \otimes I_2 = I,$$

where I_1, I_2, and I are the identity matrices of the same orders as the matrices A, B and $A \otimes B$, respectively. Thus the matrices $A \otimes B$ and $A^{-1} \otimes B^{-1}$ are mutual inverses. $\square$

Theorem 6.1.3 *If the matrix A_1 is similar to the matrix A, $A_1 = PAP^{-1}$, and if B_1 is similar to B, $B_1 = QBQ^{-1}$, then the matrix $A_1 \otimes B_1$ is similar to the matrix $A \otimes B$:*

$$A_1 \otimes B_1 = (P \otimes Q)(A \otimes B)(P \otimes Q)^{-1}.$$

Proof. We employ (6.2) twice and use the formula (6.3) for $(P \otimes Q)^{-1}$. □

The next theorem is also of importance.

Theorem 6.1.4 *If A is an upper triangular matrix of order m and B an upper triangular matrix of order n, then the matrix $A \otimes B$ is upper triangular as well. If $a_1, \ldots, a_m$ and $b_1, \ldots, b_n$ are the diagonal entries of A and B, respectively, then $a_1b_1, a_2b_1, \ldots, a_mb_1, a_1b_2, a_2b_2, \ldots, a_mb_2, \ldots, a_1b_n, a_2b_n, \ldots, a_mb_n$ are (in this order) the diagonal entries of the matrix $A \otimes B$.*

We now prove the fundamental theorem on the eigenvalues of tensor product.

Theorem 6.1.5 *If A is a square matrix of order m with the eigenvalues $\alpha_1, \ldots, \alpha_m$ (some of them may be identical) and B is a square matrix of order n with the eigenvalues $\beta_1, \ldots, \beta_n$ (some of them may also be identical), then all the eigenvalues of the matrix $A \otimes B$ are $\alpha_i\beta_j$, $i = 1, \ldots, m$, $j = 1, \ldots, n$. If I_m and I_n are the identity matrices of orders m and n, and λ and μ are numbers, then the eigenvalues of the matrix*

$$\lambda(A \otimes I_n) + \mu(I_m \otimes B)$$

are the numbers $\lambda\alpha_i + \mu\beta_j$, $i = 1, \ldots, m$, $j = 1, \ldots, n$.

If we denote by e an eigenvector of A associated with the eigenvalue α_i and by f an eigenvector of B associated with β_j, then $e \otimes f$ is an eigenvector of the matrix $A \otimes B$ associated with the eigenvalue $\alpha_i\beta_j$ as well as an eigenvector of the matrix $\lambda(A \otimes I_n) + \mu(I_m \otimes B)$ associated with the eigenvalue $\lambda\alpha_1 + \mu\beta_j$.

Proof. By Theorem 2.4.3, there exist a nonsingular matrix P and an upper triangular matrix A_1 (possibly complex) such that $A = PA_1P^{-1}$. Similarly, there exist a nonsingular matrix Q and an upper triangular matrix B_1 such that $B = QB_1Q^{-1}$. Using the formula (6.2) twice and the formula (6.3), we have

$$A \otimes B = (P \otimes Q)(A_1 \otimes B_1)(P \otimes Q)^{-1}.$$

By Theorem 1.7.2, A_1 has the same eigenvalues (including their multiplicity) as A, B_1 has the same eigenvalues as B, and $A_1 \otimes B_1$ has the same eigenvalues as $A \otimes B$. But the eigenvalues of the matrices A_1, B_1, and $A_1 \otimes B_1$ are their diagonal entries. Since $\alpha_1, \ldots, \alpha_m$ are the diagonal entries of A_1 and $\beta_1, \ldots, \beta_n$ are the diagonal entries of B_1 the diagonal entries of $A_1 \otimes B_1$ are exactly the numbers $\alpha_i \beta_j$, $i = 1, \ldots, m$, $j = 1, \ldots, n$, by Theorem 6.1.4. Similarly,

$$\lambda(A \otimes I_n) + \mu(I_m \otimes B) =$$
$$(P \otimes Q)(\lambda(A_1 \otimes I_n) + \mu(I_m \otimes B_1))(P \otimes Q)^{-1}.$$

The matrix $\lambda(A_1 \otimes I_n) + \mu(I_m \otimes B_1)$ is upper triangular and its diagonal entries are

$$\lambda\alpha_1 + \mu\beta_1, \lambda\alpha_2 + \mu\beta_1, \ldots, \lambda\alpha_m + \mu\beta_1, \lambda\alpha_1 + \mu\beta_2, \ldots,$$
$$\ldots, \lambda\alpha_m + \mu\beta_2, \ldots, \lambda\alpha_1 + \mu\beta_n, \ldots, \lambda\alpha_m + \mu\beta_n.$$

These numbers are thus the eigenvalues not only of the matrix $\lambda(A_1 \otimes I_n) + \mu(I_m \otimes B_1)$ but also of the similar matrix $\lambda(A \otimes I_n) + \mu(I_m \otimes B)$.

If now $Ae = \alpha_i e$ and $Bf = \beta_j f$, then (6.2) implies that

$$(A \otimes B)(e \otimes f) = Ae \otimes Bf = \alpha_i \beta_j (e \otimes f)$$

(note that $e \otimes f \neq 0$ since $e \neq 0$ as well as $f \neq 0$). Analogously,

$$(\lambda(A \otimes I_n) + \mu(I_m \otimes B))(e \otimes f) = (\lambda\alpha_i + \mu\beta_j)(e \otimes f).$$

$\square$

The theorem can thus be employed for determining e.g. all the eigenvalues of the matrix (6.1) of Example 6.1 if the eigenvalues of the matrices Z_3 and Z_5 are known (see Exercise 6.3).

Theorem 6.1.6 *If A is a square matrix of order m and B a square matrix of order n, then*

$$\det(A \otimes B) = (\det A)^n (\det B)^m. \tag{6.4}$$

Proof. Follows from the preceding theorem and from the fact that the determinant of a matrix equals the product of all its eigenvalues by (1.39). Indeed, if $\alpha_1, \alpha_2, \ldots, \alpha_m$ are the eigenvalues of A and $\beta_1, \beta_2, \ldots, \beta_n$ are the eigenvalues of B, then

$$\det(A \otimes B) = \prod_{i=1}^{m}\prod_{j=1}^{n}(\alpha_i\beta_j) = \left(\prod_{i=1}^{m}\alpha_i\right)^n\left(\prod_{j=1}^{n}\beta_j\right)^m$$
$$= (\det A)^n(\det B)^m.$$

$\square$

Theorem 6.1.7 *A necessary and sufficient condition for the square matrices A and B of orders m and n, respectively, to have no common eigenvalue is*

$$\det(A \otimes I_n - I_m \otimes B) \neq 0. \tag{6.5}$$

Proof. Immediate since the left-hand side of (6.5) is the product of all the eigenvalues of the matrix $A \otimes I_n - I_m \otimes B$, i.e. the product of the numbers $\alpha_i - \beta_k$. $\square$

Remark. The fact that the left-hand side of (6.5) includes only the entries of A and B can be also used for the computation of the resultant of two polynomials (cf. Theorems 7.2.4 and 7.3.1, and Exercise 7.5).

The tensor product of matrices plays an important role in the investigation of solutions of linear matrix equations.

Theorem 6.1.8 *The system of linear equations in unknown entries x_{ik} of the matrix X*

$$A_1 X B_1 + A_2 X B_2 + \cdots + A_r X B_r = C,$$

where A_i, B_i, and C are matrices of proper dimensions, is identical with the system

$$(A_1 \otimes B_1^T + A_2 \otimes B_2^T + \cdots + A_r \otimes B_r^T)\text{vec}X = \text{vec}C,$$

where $\text{vec}X$ and $\text{vec}C$ are column vectors formed from all the columns of X and C, respectively, in such a way that these columns are written below each other.

Proof. Let $X = [X_1, X_2, \ldots, X_n]$ be the matrix X expressed in terms of its columns $X_1, X_2, \ldots, X_n$. Then $(\text{vec } X)^T = [X_1^T, X_2^T, \ldots, X_n^T]$. It is sufficient to prove the equivalence of both the systems for $r = 1$. Let

$$
B = \begin{bmatrix} b_{11} & \cdots & b_{1p} \\ \vdots & \ddots & \vdots \\ b_{n1} & \cdots & b_{np} \end{bmatrix}.
$$

Then

$$
\begin{aligned}
\text{vec } AXB &= \text{vec } [[AX_1, AX_2, \ldots, AX_n]B] \\
&= \begin{bmatrix} AX_1 b_{11} + AX_2 b_{21} + \cdots + AX_n b_{n1} \\ AX_1 b_{12} + AX_2 b_{22} + \cdots + AX_n b_{n2} \\ \vdots \\ AX_1 b_{1p} + AX_2 b_{2p} + \cdots + AX_n b_{np} \end{bmatrix} \\
&= \begin{bmatrix} Ab_{11} & Ab_{21} & \cdots & Ab_{n1} \\ Ab_{12} & Ab_{22} & \cdots & Ab_{n2} \\ \vdots & \vdots & \ddots & \vdots \\ Ab_{1p} & Ab_{2p} & \cdots & Ab_{np} \end{bmatrix} \begin{bmatrix} X_1 \\ X_2 \\ \vdots \\ X_n \end{bmatrix} \\
&= (A \otimes B^T) \text{vec } X.
\end{aligned}
$$

$\square$

Theorem 6.1.9 *Let A and B be square matrices of orders m and n, respectively, that have no eigenvalue in common. Then to any $m \times n$ matrix C there exists one and only one $m \times n$ matrix X for which*

$$
AX - XB = C.
$$

This matrix X can be determined from the relation

$$
(A \otimes I_n - I_m \otimes B) \text{vec} X = \text{vec} C
$$

as

$$
\text{vec} X = (A \otimes I_n - I_m \otimes B)^{-1} \text{vec} C.
$$

Proof. Follows readily from the preceding theorem and Theorem 6.1.7. $\square$

To complete our exposition we present a few more theorems on tensor products. Their proofs are left to the reader as an exercise (Exercises 6.1 and 6.2).

Theorem 6.1.10 *For any two matrices A and B we have*

$$(A \otimes B)^T = A^T \otimes B^T, \tag{6.6}$$

$$(A \otimes B)^* = A^* \otimes B^*. \tag{6.7}$$

Theorem 6.1.11 *If both A and B are unitary (orthogonal, Hermitian, symmetric) matrices, then their tensor product $A \otimes B$ is unitary (orthogonal, Hermitian, symmetric).*

6.2 Compound matrices

In the second part of this chapter we study the so called *compound matrices*. They are defined, roughly speaking, as matrices associated with a given matrix and an integer k in such a way that their entries are determinants of all the submatrices of order k formed from the matrix given.

In order to express the entries of compounds, we introduce the following notation: Let k and n be integers, $1 \leq k \leq n$, and $N = \{1, 2, \ldots, n\}$. Then $N^{(k)}$ denotes the set of all k-tuples of elements of N. Hence $N^{(k)}$ contains exactly $\binom{n}{k}$ elements. The elements of N are ordered in the natural order $1, 2, \ldots, n$ and, moreover, we shall assume that the elements of $N^{(k)}$ are ordered *lexicographically*. It means that the k-tuples containing 1 are taken first. When these k-tuples are exhausted, the k-tuples containing 2 (but no longer 1, of course) follow, etc. Among those containing 1, the k-tuples containing, in addition, 2 are in the beginning, etc. Moreover, the components of the k-tuples themselves are ordered in the natural ordering (i.e., in the same way as N). Hence, e.g. for $n = 4$, $k = 2$ we have

$$N^{(k)} = \{(1, 2), (1, 3), (1, 4), (2, 3), (2, 4), (3, 4)\}.$$

If $M = \{1, 2, \ldots, m\}$, then the lexicographically ordered set of k-tuples is denoted by $M^{(k)}$, etc.

Now we can turn to the compound matrices. Let $A = [a_{ik}]$ be an $m \times n$ matrix. Let k be a positive integer satisfying $k \leq \min(m, n)$. Define $M = \{1, 2, \ldots, m\}$ and $N = \{1, 2, \ldots, n\}$. Then the *kth compound* of A is the $\binom{m}{k} \times \binom{n}{k}$ matrix $A^{(k)}$ whose rows correspond to the elements of $M^{(k)}$ and columns to the elements of $N^{(k)}$ and

$$A^{(k)} = (\det A(P, Q)),$$

i.e., the entry in the "Pth" row and "Qth" column, where $P \in M^{(k)}$ and $Q \in N^{(k)}$, is just $\det A(P, Q)$.

Example 6.2. For $m = 4$, $n = 3$, $k = 2$, and

$$A = \begin{bmatrix} a_{11} & a_{12} & a_{13} \\ a_{21} & a_{22} & a_{23} \\ a_{31} & a_{32} & a_{33} \\ a_{41} & a_{42} & a_{43} \end{bmatrix},$$

we have

$$A^{(2)} = \begin{bmatrix} \begin{vmatrix} a_{11} & a_{12} \\ a_{21} & a_{22} \end{vmatrix} & \begin{vmatrix} a_{11} & a_{13} \\ a_{21} & a_{23} \end{vmatrix} & \begin{vmatrix} a_{12} & a_{13} \\ a_{22} & a_{23} \end{vmatrix} \\[2ex] \begin{vmatrix} a_{11} & a_{12} \\ a_{31} & a_{32} \end{vmatrix} & \begin{vmatrix} a_{11} & a_{13} \\ a_{31} & a_{33} \end{vmatrix} & \begin{vmatrix} a_{12} & a_{13} \\ a_{32} & a_{33} \end{vmatrix} \\[2ex] \begin{vmatrix} a_{11} & a_{12} \\ a_{41} & a_{42} \end{vmatrix} & \begin{vmatrix} a_{11} & a_{13} \\ a_{41} & a_{43} \end{vmatrix} & \begin{vmatrix} a_{12} & a_{13} \\ a_{42} & a_{43} \end{vmatrix} \\[2ex] \begin{vmatrix} a_{21} & a_{22} \\ a_{31} & a_{32} \end{vmatrix} & \begin{vmatrix} a_{21} & a_{23} \\ a_{31} & a_{33} \end{vmatrix} & \begin{vmatrix} a_{22} & a_{23} \\ a_{32} & a_{33} \end{vmatrix} \\[2ex] \begin{vmatrix} a_{21} & a_{22} \\ a_{41} & a_{42} \end{vmatrix} & \begin{vmatrix} a_{21} & a_{23} \\ a_{41} & a_{43} \end{vmatrix} & \begin{vmatrix} a_{22} & a_{23} \\ a_{42} & a_{43} \end{vmatrix} \\[2ex] \begin{vmatrix} a_{31} & a_{32} \\ a_{41} & a_{42} \end{vmatrix} & \begin{vmatrix} a_{31} & a_{33} \\ a_{41} & a_{43} \end{vmatrix} & \begin{vmatrix} a_{32} & a_{33} \\ a_{42} & a_{43} \end{vmatrix} \end{bmatrix}$$

where we write $|A(P, Q)|$ instead of $\det A(P, Q)$.

Our choice of the lexicographical ordering has the following important consequence.

Theorem 6.2.1 *If the matrix A is upper triangular (lower triangular, diagonal), then the matrix $A^{(k)}$ is upper triangular (lower triangular, diagonal) as well. If $a_{11}, \ldots, a_{nn}$ are the diagonal entries of the matrix A of order n, then the diagonal entries of $A^{(k)}$ are*

$$a_{11}a_{22} \cdots a_{kk}, \quad a_{11}a_{22} \cdots a_{k-1,k-1}a_{k+1,k+1}, \quad \ldots,$$

$$a_{n-k+1,n-k+1}a_{n-k+2,n-k+2} \cdots a_{nn}$$

(in this order), i.e. all possible products of k diagonal entries of A.

Proof. Let $P \in N^{(k)}$, $Q \in N^{(k)}$, $P = (p_1, \ldots, p_k)$, and $Q = (q_1, \ldots, q_k)$. P then precedes Q in the lexicographical ordering if the relations $p_1 = q_1$, $p_2 = q_2, \ldots, p_s = q_s$, and $p_{s+1} \neq q_{s+1}$ imply $p_{s+1} < q_{s+1}$. (The case of $s = 0$, i.e. $p_1 \neq q_1$, may also occur.)

Let thus the set P precede Q. We shall show that $\det A(Q, P) = 0$ if A is an upper triangular matrix, i.e., if $a_{ik} = 0$ for $i > k$. But then

$$A(P, Q) = \begin{bmatrix} A_{11} & A_{12} \\ A_{21} & A_{22} \end{bmatrix},$$

where A_{11} is an $s \times s$ matrix which corresponds to the indices $p_1, \ldots, p_s$ (if, however, $s = 0$, then $A(Q, P) = A_{22}$). Hence A_{11} is upper triangular. Further, A_{21} is a zero matrix since the indices $p_1, \ldots, p_s$ in the first block column and the indices $q_{s+1}, \ldots, q_k$ in the second block row fulfil the inequalities

$$p_1 < p_2 < \cdots < p_s < p_{s+1} < q_{s+1} < q_{s+2} < \cdots < q_k$$

and thus $a_{q_j p_i} = 0$ for $i \leq s$ and $j \geq s + 1$. Therefore $\det A(Q, P) = \det A_{11} \det A_{22}$. But the first column of A_{22} is zero since also $a_{q_j p_{s+1}} = 0$ for $j \geq s + 1$. We can prove in an analogous way that if A is a lower triangular matrix, then $A^{(k)}$ is lower triangular as well. If A is diagonal, it is both upper and lower triangular and so is $A^{(k)}$; $A^{(k)}$ is thus diagonal, too.

The second statement of the theorem follows readily from the first statement just proved and from the fact that $\det A(P, P)$ is the diagonal entry of the matrix $A^{(k)}$ for $P \in N^{(k)}$ and, moreover, that $A(P, P)$ are triangular matrices. Therefore $\det A(P, P) = a_{p_1 p_1} \cdots a_{p_k p_k}$ if $P = (p_1, \ldots, p_k)$. $\qquad\Box$

Now we are going to prove the fundamental theorem on the multiplication of compound matrices.

Theorem 6.2.2 *Let A be an $m \times n$ matrix and B and $n \times p$ matrix. For any positive integer k, $k \le \min(m, n, p)$, we have*

$$(AB)^{(k)} = A^{(k)} B^{(k)}. \tag{6.8}$$

Proof. Let $Q \in N^{(k)}$, where $N = \{1, 2, \ldots, n\}$, and let $S \in P^{(k)}$, where $P = \{1, 2, \ldots, p\}$. Put $\tilde{A} = A(Q, N)$ and $\tilde{B} = B(N, S)$. The value of $\det \tilde{A}\tilde{B}$ is then the entry of $(AB)^{(k)}$ in the position (Q, S). In the same position, the product $A^{(k)} B^{(k)}$ has the entry $\sum_{R \in N^{(k)}} \det A(Q, R) \det B(R, S)$. But the Binet-Cauchy formula (1.16) implies that both numbers are equal. $\qquad\Box$

Note that for $m = n = p = k$ we obtain the theorem on the determinant of the product of two square matrices: $\det(AB) = \det A \det B$.

The formula (6.8) allows us, as in the case of tensor product, to study spectral properties of compound matrices. However, we first present some simple corollaries.

Theorem 6.2.3 *If I_n is the identity matrix of order n and k is a positive integer, $k \le n$, then $I_n^{(k)}$ is the identity matrix of order $\binom{n}{k}$.*

Theorem 6.2.4 *If A^T is the transpose of an $m \times n$ matrix A and if k is a positive integer, $k \le \min(m, n)$, then*

$$(A^{(k)})^T = (A^T)^{(k)}.$$

Similarly,

$$(A^{(k)})^* = (A^*)^{(k)},$$

where $A^* = \bar{A}^T$.

Theorem 6.2.5 *If A is a unitary (orthogonal, Hermitian, real symmetric) matrix of order n and $1 \le k \le n$, then $A^{(k)}$ is unitary (orthogonal, Hermitian, real symmetric) as well.*

Theorem 6.2.6 *If A is a nonsingular matrix and if k is a positive integer not exceeding the order of A, then also $A^{(k)}$ is nonsingular and*

$$(A^{(k)})^{-1} = (A^{-1})^{(k)}. \tag{6.9}$$

Proof. By (6.8) and Theorem 6.2.3,

$$A^{(k)}(A^{-1})^{(k)} = (AA^{-1})^{(k)} = I_1^{(k)} = I_2,$$

where I_1 and I_2 are identity matrices of corresponding orders. Therefore, $A^{(k)}$ is nonsingular and $(A^{-1})^{(k)}$ is the inverse of $A^{(k)}$. $\square$

Theorem 6.2.7 *If $\alpha_1, \ldots, \alpha_n$ are the eigenvalues of a square matrix A of order n and if k is a positive integer, $k \leq n$, then the eigenvalues of $A^{(k)}$ are all the possible products $\alpha_{i_1}, \alpha_{i_2}, \ldots, \alpha_{i_k}$, where $1 \leq i_1 < i_2 < \cdots < i_k \leq n$. (Some of these eigenvalues, however, may be identical.)*

Proof. Given the matrix A, by Schur's Theorem 2.4.4 there exist a nonsingular matrix P and an upper triangular matrix A_1 (they both may be complex) such that

$$A = PA_1P^{-1}.$$

Using the formula (6.8) twice and the formula (6.9), we obtain

$$A^{(k)} = P^{(k)}A_1^{(k)}(P^{(k)})^{-1}.$$

Hence the eigenvalues of $A^{(k)}$ are the diagonal entries of $A_1^{(k)}$. Since the diagonal entries of A_1 are equal to the eigenvalues $\alpha_1, \ldots, \alpha_n$ of A, the second statement of Theorem 6.2.1 implies that the eigenvalues of $A^{(k)}$ are just all the numbers $\alpha_{i_1}, \alpha_{i_2}, \ldots, \alpha_{i_k}$, $1 \leq i_1 < i_2 < \cdots < i_k \leq n$. $\square$

Theorem 6.2.8 *If A is a square matrix of order n and if k is a positive integer, $k \leq n$, then*

$$\det A^{(k)} = (\det A)^{\binom{n-1}{k-1}}. \tag{6.10}$$

Proof. By (1.39), the determinant of a matrix equals the product of the eigenvalues of this matrix. Hence

$$\det A^{(k)} = \prod_{1 \leq i_1 < i_2 < \cdots < i_k \leq n} \alpha_{i_1}, \alpha_{i_2}, \ldots, \alpha_{i_k}.$$

Each of the numbers α_i (here we distinguish the identical eigenvalues!) does not occur in exactly $\binom{n-1}{k}$ terms on the right-hand side; it thus occurs just in

$$\binom{n}{k} - \binom{n-1}{k} = \binom{n-1}{k-1}$$

terms. Hence, we may easily rewrite the right-hand side as

$$\left(\prod_{i=1}^{n} \alpha_i\right)^{\binom{n-1}{k-1}} = (\det A)^{\binom{n-1}{k-1}}.$$

□

In order to find the eigenvectors of the matrix $A^{(k)}$ and to be able to handle such matrices more easily, we introduce – in our case only for matrices – a notation which is used in multilinear algebra.

Let $u_1, \ldots, u_k$ be n-component column vectors, $k \leq n$. The *exterior product* of the vectors $u_1, \ldots, u_k$, denoted by $u_1 \wedge u_2 \wedge \cdots \wedge u_k$, is introduced as the $\binom{n}{k}$-component column vector equal to the compound $U^{(k)}$ of the $n \times k$ matrix $U = [u_1, \ldots, u_k]$.

The following properties of the operation of exterior product follow readily from the definition:

EP1. If

$$P = \begin{pmatrix} 1 & 2 & \ldots & k \\ i_1 & i_2 & \ldots & i_k \end{pmatrix}$$

is a permutation, then we have for all the vectors $u_1, \ldots, u_k$ in X_n that

$$u_{i_1} \wedge u_{i_2} \wedge \cdots \wedge u_{i_k} = \sigma(P)(u_1 \wedge u_2 \wedge \cdots \wedge u_k),$$

where $\sigma(P)$ is the sign of P as in (1.5).

EP2. $(u_1 + v_1) \wedge u_2 \wedge \cdots \wedge u_k = u_1 \wedge u_2 \wedge \cdots \wedge u_k + v_1 \wedge u_2 \wedge \cdots \wedge u_k$
for all $u_1, \ldots u_k$ and v_1 in X_n.

EP3. $(\lambda u_1) \wedge u_2 \wedge \cdots \wedge u_k = \lambda(u_1 \wedge u_2 \wedge \cdots \wedge u_k)$ for all $u_1, \ldots, u_k$ in X_n and any real λ (if X_n is real) or any complex λ (if X_n is complex).

EP4. $u_1 \wedge u_2 \wedge \cdots \wedge u_k = 0$ if and only if $u_1, \ldots, u_k$ are linearly dependent.

EP5. Set $e_s = [0, \ldots, 0, 1, 0, \ldots, 0]^T$, where 1 is the sth component, $s = 1, \ldots, n$. Then the vectors

$$e_{i_1} \wedge e_{i_2} \wedge \cdots \wedge e_{i_k}, \qquad 1 \le i_1 < i_2 < \cdots < i_k \le n,$$

are linearly independent and form a basis of the space of $\binom{n}{k}$-component column vectors.

Let us show how an arbitrary exterior product $u_1 \wedge u_2 \wedge \cdots \wedge u_k$ can be expressed as a linear combination of vectors of the basis in EP5.

Theorem 6.2.9 *If $u_i = [u_{i_1}, \ldots, u_{i_n}]^T$, $i = 1, \ldots, k$, then*

$$u_1 \wedge u_2 \wedge \cdots \wedge u_k =$$
$$\sum_{I=(i_1,\ldots,i_k),1\le i_1<\cdots<i_k\le n} \det U(I,K)(e_{i_1} \wedge e_{i_2} \wedge \cdots \wedge e_{i_k}), \quad (6.11)$$

where $U = [u_{ij}]$ is the matrix $[u_1, \ldots u_k]$, $K = (1, 2, \ldots, k)$, and $e_1, \ldots e_n$ are the vectors of EP5.

Proof. It can certainly be performed directly with help of the definition of the exterior product. As an exercise we shall prove the theorem using the properties EP1 to EP5.

EP1 implies that analogies of EP2 and EP3 hold also for operands other than the first. Hence,

$$u_1 \wedge u_2 \wedge \cdots \wedge u_k =$$
$$= \left(\sum_{j_1} u_{1j_1} e_{j_1}\right) \wedge \left(\sum_{j_2} u_{2j_2} e_{j_2}\right) \wedge \cdots \wedge \left(\sum_{j_k} u_{kj_k} e_{j_k}\right)$$
$$= \sum_{j_1,\ldots,j_k=1}^{n} u_{1j_1} u_{2j_2} \ldots u_{kj_k}(e_{j_1} \wedge e_{j_2} \wedge \cdots \wedge e_{j_k})$$
$$= \sum_{I} \left(\sum_{P(I,J)} \sigma(P) u_{1j_1} u_{2j_1} \ldots u_{kj_k}\right)(e_{i_1} \wedge e_{i_2} \wedge \cdots \wedge e_{i_k})$$

where I is the set of all k-tuples $(i_1, \ldots, i_k)$ satisfying $1 \leq i_1 < \cdots < i_k \leq n$ and $P(I, J)$ is the permutation $\binom{i_1, \ldots, i_k}{j_1, \ldots, j_k}$ since it is sufficient to consider only k-tuples of distinct indices $j_1, \ldots, j_k$ in the sum involving $e_{j_1} \wedge e_{j_2} \wedge \cdots \wedge e_{j_k}$ by EP4 and, further, to group the terms in which the indices $j_1, \ldots, j_k$ differ only by their order with help of EP1. But since

$$\sum_{P=\binom{i_1, \ldots, i_k}{j_1, \ldots, j_k}} \sigma(P) u_{1j_1} u_{2j_1} \ldots u_{kj_k} = \det U(I, K)$$

for $I = (i_1, \ldots, i_k)$, $1 \leq i_1 < i_2 < \cdots < i_k \leq n$, the formula of Theorem 6.2.9 follows. □

Using exterior products, we can characterize the compound $A^{(k)}$.

Theorem 6.2.10 *Let A be an $m \times n$ matrix and let k be an integer, $1 \leq k \leq \min(m, n)$. Then the following properties of an $\binom{m}{k} \times \binom{n}{k}$ matrix C are equivalent:*

1. $C = A^{(k)}$, *where $A^{(k)}$ is the compound of A.*

2. $C(u_1 \wedge u_2 \wedge \cdots \wedge u_k) = Au_1 \wedge Au_2 \wedge \cdots \wedge Au_k$ *for all n-component column vectors $u_1, \ldots, u_k$.*

3. *The matrix C satisfies*

$$C(e_{i_1} \wedge e_{i_2} \wedge \cdots \wedge e_{i_k}) = Ae_{i_1} \wedge Ae_{i_2} \wedge \cdots \wedge Ae_{i_k})$$

for all indices $i_1, \ldots, i_k$, $1 \leq i_1 < i_2 < \cdots < i_k \leq n$.

Proof. We shall prove the implications 1. → 2., 2. → 3., and 3. → 1.

1. → 2. We are to show that

$$A^{(k)}(u_1 \wedge u_2 \wedge \cdots \wedge u_k) = Au_1 \wedge Au_2 \wedge \cdots \wedge Au_k.$$

Let $U = [u_1, u_2, \ldots u_k]$ be an $n \times k$ matrix having $u_1, u_2, \ldots u_k$ as its columns. By Theorem 1.1.1 $AU = [Au_1, Au_2, \ldots, Au_k]$. Hence $Au_1 \wedge Au_2 \wedge \cdots \wedge Au_k = (AU)^{(k)}$ by the definition of exterior product. But, on the other hand,

$$A^{(k)}(u_1 \wedge u_2 \wedge \cdots \wedge u_k) = A^{(k)} U^{(k)}$$

and the implication thus follows from (6.8).

2. → 3. It is sufficient to take $e_{i_1}, e_{i_2}, \ldots, e_{i_k}$, $1 \leq i_1 < i_2 < \cdots < i_k \leq n$, for $u_1, u_2, \ldots u_k$.

3. → 1. As the exterior product $e_{i_1} \wedge e_{i_2} \wedge \cdots \wedge e_{i_k}$, $1 \leq i_1 < i_2 < \cdots < i_k \leq n$, is an $\binom{n}{k}$-component column vector whose all components are zero except for the one corresponding to the "index" $I = (i_1, i_2, \ldots, i_k) \in N^{(k)}$, $N = \{1, 2, \ldots, n\}$, it follows from 3. that the column of C corresponding to the "index" I is $Ae_{i_1} \wedge Ae_{i_2} \wedge \cdots \wedge Ae_{i_k}$. By (6.11) this can be expressed as

$$\sum_{J=(j_1,\ldots,j_k),1 \leq j_1 < \cdots < j_k \leq m} \det A(J,I) e_{j_1} \wedge e_{j_2} \wedge \cdots \wedge e_{j_k}.$$

Hence the entry of C with the row index J and the column index I is just $\det A(J, I)$, i.e., $C = A^{(k)}$. □

Now we are in a position to study the eigenvectors of $A^{(k)}$.

Theorem 6.2.11 *Let A be a square matrix of order n and let $u_1, u_2, \ldots u_k$ be linearly independent eigenvectors of A associated with the eigenvalues $\alpha_1, \ldots, \alpha_k$. Then $u_1 \wedge u_2 \wedge \cdots \wedge u_k$ is an eigenvector of $A^{(k)}$ associated with the product $\alpha_1 \alpha_2 \ldots \alpha_k$.*

Proof. By 2. and 1. of Theorem 6.2.10,

$$\begin{aligned}
A^{(k)}(u_1 \wedge u_2 \wedge \cdots \wedge u_k) &= Au_1 \wedge Au_2 \wedge \cdots \wedge Au_k) \\
&= (\alpha_1 u_1) \wedge (\alpha_2 u_2) \wedge \cdots \wedge (\alpha_k u_k) \\
&= (\alpha_1 \alpha_2 \ldots \alpha_k)(u_1 \wedge u_2 \wedge \cdots \wedge u_k),
\end{aligned}$$

which proves the theorem since $(u_1 \wedge u_2 \wedge \cdots \wedge u_k) \neq 0$ by EP4. □

In the first part of this chapter we saw that with two square matrices it is possible to associate not only a matrix whose eigenvalues are products of two eigenvalues (i.e., we take one eigenvalue of each matrix), but also a matrix whose eigenvalues are sums of two eigenvalues (we again take one eigenvalue of each matrix). There is now a question whether for a given square matrix A it is possible to construct in an easy way (i.e., using only the entries of A) a matrix whose eigenvalues are sums of two or, in general, of k eigenvalues of A, that is,

$$\alpha_{i_1} + \alpha_{i_2} + \cdots + \alpha_{i_k}, \qquad 1 \leq i_1 < i_2 < \cdots < i_k \leq n.$$

In the conclusion of this chapter we shall give the affirmative answer to this question and, moreover, we shall present a method [18] for finding a matrix whose eigenvalues are arbitrary symmetric polynomial functions of k eigenvalues of A.

First we note that for a square matrix A of order n and the identity matrix I of the same order, the compound $(A + tI)^{(k)}$ (where k is a positive integer, $k \le n$, and t is a variable) is a matrix polynomial of degree k in t since each entry of this matrix is the determinant of order k of some submatrix of $A + tI$:

$$(A+tI)^{(k)} = A^{(k,k)} + A^{(k,k-1)}t + \cdots + A^{(k,1)}t^{k-1} + A^{(k,0)}t^k, \quad (6.12)$$

where $A^{(k,k)} = A^{(k)}, \ldots, A^{(k,1)}, A^{(k,0)}I^{(k)}$ are square matrices of order $\binom{n}{k}$. These matrices $A^{(k,s)}$ will be called the *generalized compounds* of the matrix A.

Theorem 6.2.12 *Let A be a square matrix of order n. All the generalized compounds $A^{(k,s)}$ of A with fixed k commute. Further, the matrix $A^{(k,s)}$ can be characterized as a matrix for which*

$$A^{(k,s)}(x_1 \wedge x_2 \wedge \cdots \wedge x_n) =$$
$$\sum (A^{\varepsilon_1}x_1) \wedge A^{\varepsilon_2}x_2) \wedge \cdots \wedge A^{\varepsilon_k}x_k), \quad (6.13)$$

where the sum is taken over all possible k-tuples $(\varepsilon_1, \ldots, \varepsilon_n)$ such that $\varepsilon_i = 0$ or $\varepsilon_i = 1$ and $\sum_{i=1}^{n} \varepsilon_i = s$.

Moreover, if $\alpha_1, \ldots, \alpha_n$ are the eigenvalues of A, then the eigenvalues of $A^{(k,s)}$ are just all the numbers $E_s(\alpha_{i_1}, \alpha_{i_2}, \ldots, \alpha_{i_k})$ for $1 \le i_1 < i_2 < \cdots < i_k \le n$, where $E_s(z_1, \ldots, z_k)$ denotes the sth elementary symmetric function in the variables $z_1, \ldots, z_k$.

Proof. Choose $k + 1$ arbitrary distinct real numbers $t_1, \ldots, t_{k+1}$. Let s be a fixed integer, $0 \le s \le k$. We shall show that there are real numbers $c_1, \ldots, c_{k+1}$ such that

$$A^{(k,s)} = c_1(A + t_1I)^{(k)} + c_2(A + t_2I)^{(k)} + \cdots + c_{k+1}(A + t_{k+1}I)^{(k)}.$$

The system of $k + 1$ linear equations in $k + 1$ unknowns $c_1, \ldots, c_{k+1}$,

$$c_1 t_1^s + c_2 t_2^s + \cdots + c_{k+1} t_{k+1}^s = 1,$$
$$c_1 t_1^j + c_2 t_2^j + \cdots + c_{k+1} t_{k+1}^j = 0, \quad (6.14)$$
$$j = 0, 1, \ldots, s-1, s+1, \ldots, k,$$

has one and only one solution since the matrix of this system (the so-called Vandermonde matrix) is nonsingular as will be shown in Theorem 7.1.3. Forming now the sum

$$c_1(A+t_1 I)^{(k)} + c_2(A+t_2 I)^{(k)} + \cdots + c_{k+1}(A+t_{k+1} I)^{(k)} \quad (6.15)$$

and expressing $(A+t_j I)^{(k)}$ as in (6.12), we obtain $A^{(k,s)}$ by (6.15). But all the matrices of the form (6.15) commute since

$$\begin{aligned}
(A + t_i I)^{(k)}(A + t_j I)^{(k)} &= ((A + t_i I)(A + t_j I))^{(k)} \\
&= ((A + t_j I)(A + t_i I))^{(k)} \\
&= ((A + t_j I)^{(k)}(A + t_i I))^{(k)}
\end{aligned}$$

holds for $i, j = 1, \ldots, k + 1$ by (6.8). Hence also all the matrices $A^{(k,s)}$ commute.

To prove the second statement, it is sufficient to observe that

$$\begin{aligned}
(A + tI)^{(k)}(x_1 \wedge x_2 \wedge \cdots \wedge x_k) &= \\
&= (A + tI)x_1 \wedge (A + tI)x_2 \wedge \cdots \wedge (A + tI)x_k \\
&= \sum_{r=0}^{k} t^r \sum (A^{\varepsilon_1} x_1) \wedge (A^{\varepsilon_2} x_2) \wedge \cdots \wedge (A^{\varepsilon_k} x_k),
\end{aligned}$$

where the sum on the right-hand side is taken over the same k-tuples $(\varepsilon_1, \ldots, \varepsilon_n)$ as on the right-hand side of (6.13). On the other hand,

$$(A + tI)^{(k)}(x_1 \wedge x_2 \wedge \cdots \wedge x_k) = \sum_{r=0}^{k} t^r A^{(k,r)}(x_1 \wedge x_2 \wedge \cdots \wedge x_k).$$

Now, let us prove the third statement. By Schur's Theorem 2.1.3, A is similar to an upper triangular matrix A_1, i.e., $A = TA_1 T^{-1}$ with a nonsingular (or even unitary or orthogonal) matrix T. Moreover, the diagonal entries of A are just the eigenvalues of A, i.e. the numbers $\alpha_1, \ldots, \alpha_n$. But

$$(A + tI)^{(k)} = (T(A_1 + tI)T^{-1})^{(k)} = T^{(k)}(A_1 + tI)^{(k)}(T^{(k)})^{-1}.$$

Hence $(A + tI)^{(k)}$ has the same eigenvalues as $(A_1 + tI)^{(k)}$ and thus any matrix of the form (6.15) has the same eigenvalues as the matrix

$$c_1(A_1 + t_1 I)^{(k)} + c_2(A_1 + t_2 I)^{(k)} + \cdots + c_{k+1}(A_1 + t_{k+1} I)^{(k)}.$$

Therefore, each matrix $A^{(k,s)}$ has the same eigenvalues as $A_1^{(k,s)}$. But by Theorem 6.2.1 the matrix

$$(A_1 + tI)^{(k)} = \sum_{r=0}^{k} t^r A_1^{(k,r)}$$

is again upper triangular. Hence also the matrices $A_1^{(k,s)}$ are upper triangular. The eigenvalues of $(A_1 + tI)^{(k)}$ are thus the diagonal entries of this matrix, i.e., the numbers

$$(\alpha_{i_1} + t)(\alpha_{i_2} + t)\ldots(\alpha_{i_k} + t) = \sum_{r=0}^{k} t^r E_r(\alpha_{i_1}, \alpha_{i_2}, \ldots, \alpha_{i_k}).$$

Since, as we have seen, $A^{(k,s)}$ may be expressed in the form (6.15), where the numbers c_i satisfy (6.15), the eigenvalues of this matrix are

$$\sum_{p=1}^{k+1} c_p t_p^r E_r(\alpha_{i_1}, \alpha_{i_2}, \ldots, \alpha_{i_k}) = E_s(\alpha_{i_1}, \alpha_{i_2}, \ldots, \alpha_{i_k})$$

by (6.15). $\qquad\square$

To study the matrix $A^{(k,1)}$ in more detail, we shall denote it by $A^{[k]}$ and call it the *additive compound* of A. Its properties are summarized in the following theorem [18].

Theorem 6.2.13 *Let A be a square matrix of order n and let k be a positive integer, $k \leq n$. Then there exists a square matrix $A^{[k]}$ of order $\binom{n}{k}$ such that*

$$A^{[k]}(x_1 \wedge x_2 \wedge \cdots \wedge x_k) = Ax_1 \wedge x_2 \wedge \cdots \wedge x_k \qquad (6.16)$$
$$+ x_1 \wedge Ax_2 \wedge \cdots \wedge x_k + \cdots + x_1 \wedge x_2 \wedge \cdots \wedge Ax_k.$$

The eigenvalues of this matrix are $\alpha_{i_1} + \alpha_{i_2} + \cdots + \alpha_{i_k}$, $1 \leq i_1 < i_2 < \cdots < i_k \leq n$, where $\alpha_1, \ldots, \alpha_n$ are all the eigenvalues of A. (Some of these eigenvalues may be identical.)

The entries of $A^{[k]}$ (for $I = (i_1, \ldots, i_k)$ and $J = (j_1, \ldots, j_k)$, $1 \leq i_1 < \cdots < i_k \leq n$, $1 \leq j_1, < \cdots < j_k \leq n$) are

$$A^{[k]}_{IJ} = \sum_{p \in I} a_{pp} \quad for \quad J = I \ (i.e.\ |I \cap J| = k);$$

$$A^{[k]}_{IJ} = (-1)^\sigma a_{pq} \quad if \quad |I \cap J| = k - 1, p = I \backslash (I \cap J),$$
$$q = J \backslash (I \cap J), and \ \sigma \ is \ the$$
$$number \ of \ elements \ of \ I \cap J$$
$$between \ p \ and \ q;$$

$$A^{[k]}_{IJ} = 0 \qquad if \quad |I \cap J| \leq k - 2.$$

Proof. The first and second statements follow from the general Theorem 6.2.12. We shall now prove the third statement. By (6.17), the number $A^{[k]}_{IJ}$ is the coefficient of $e_{i_1} \wedge e_{i_2} \wedge \cdots \wedge e_{i_k}$ in the expansion

$$Ae_{j_1} \wedge e_{j_2} \wedge \cdots \wedge e_{j_k} + e_{j_1} \wedge Ae_{j_2} \wedge \cdots \wedge e_{j_k} +$$
$$+ \cdots + e_{j_1} \wedge e_{j_2} \wedge \cdots \wedge Ae_{j_k}. \qquad (6.17)$$

Thus if $I = J$, then the first term contributes with $a_{i_1 i_1}$, the second with $a_{i_2 i_2}$ and, finally, the k-th with $a_{i_k i_k}$.

If I and J have less than $k - 1$ elements in common, then the coefficient $A^{[k]}_{IJ}$ is zero since at most one index of a factor may be different from $i_1, i_2, \ldots, i_k$ in each term.

The remaining case is that of I and J having $k - 1$ elements in common. Let p and q have the meaning as in the statement of the theorem. Let further

$$I = (s_1, \ldots, s_v, p, s_{v+1}, \ldots, s_{k-1}),$$
$$J = (s_1, \ldots, s_w, q, s_{w+1}, \ldots, s_{k-1}).$$

The coefficient $A^{[k]}_{IJ}$ is obtained from the only factor of

$$e_{s_1} \wedge \cdots \wedge e_{s_w} \wedge Ae_q \wedge e_{s_{w+1}} \wedge \cdots \wedge e_{s_{k-1}},$$

namely from the term $a_{pq} e_p$ in the expansion of Ae_q. To obtain the "normal form" $e_{s_1} \wedge \cdots \wedge e_{s_v} \wedge e_p \wedge e_{s_{v+1}} \wedge \cdots \wedge e_{s_{k-1}}$ from the term

$$a_{pq}(e_{s_1} \wedge \cdots \wedge e_{s_w} \wedge e_p \wedge e_{s_{w+1}} \wedge \cdots \wedge e_{s_{k-1}}),$$

it suffices to perform σ transpositions, where σ is the number of indices s_t between p and q. Hence,

$$A_{IJ}^{[k]} = (-1)^\sigma a_{pq},$$

which was to be proved. $\qquad\square$

Theorem 6.2.14 *The number of nonzero off-diagonal entries of $A^{[k]}$ equals $\binom{n-2}{k-1}$ times the number of nonzero off-diagonal entries of A.*

Proof. Let $A = [a_{ik}]$, $p \neq q$, and $a_{pq} \neq 0$. By the preceding theorem, the entry $\pm a_{pq}$ is then in such positions (I, J) which satisfy the second condition. This case occurs if and only if $I \cap J$ has $k - 1$ elements extracted from $n - 2$ elements (all n elements except p and q). It therefore occurs exactly $\binom{n-2}{k-1}$ times for every $a_{pq} \neq 0$, $p \neq q$. $\qquad\square$

The final two theorems of this chapter employ generalized compound matrices to construct matrices with prescribed properties.

Theorem 6.2.15 *Let A be a square matrix of order n having the eigenvalues $\alpha_1, \ldots, \alpha_n$. Let $\varphi(x_1, x_2, \ldots, x_k)$ be a symmetric polynomial in $k \leq n$ variables, i.e., a polynomial satisfying the equality $\varphi(x_{i_1}, x_{i_2}, \ldots, x_{i_k}) = \varphi(x_1, x_2, \ldots, x_k)$ for any permutation $(i_1, i_2, \ldots, i_k)$ of the indices $1, 2, \ldots, k$. Let $\Phi(z_1, z_2, \ldots, z_k)$ be a polynomial satisfying*

$$\varphi(x_1, x_2, \ldots, x_k) = $$
$$\Phi(E_1(x_1, \ldots, x_k), E_2(x_1, \ldots, x_k), \ldots, E_k(x_1, \ldots, x_k)),$$

where $E_j(x_1, \ldots, x_n)$ is the jth elementary symmetric function. Such a polynomial does always exist by virtue of the well-known theorem on symmetric polynomials[1]. Then the eigenvalues of the matrix polynomial

$$\Phi(A^{(k,1)}, A^{(k,2)}, \ldots, A^{(k,k)})$$

(which is a matrix of order $\binom{n}{k}$) are exactly all the numbers

$$\varphi(\alpha_{j_1}, \alpha_{j_2}, \ldots, \alpha_{j_k}), \qquad 1 \leq j_1 < j_2 < \cdots < j_k \leq n.$$

---

[1]See, for example, [60]

Proof. Schur's Theorem 2.1.3 implies that, as in the proof of Theorem 6.2.12, there is a nonsingular matrix T such that the matrix $A_1 = T^{-1}AT$ is upper triangular with diagonal entries $\alpha_1, \ldots, \alpha_n$ in some order. We may suppose that this order is exactly $\alpha_1, \ldots, \alpha_n$. All the matrices $A_1^{(k,s)} = (T^{(k)})^{-1}A^{(k,s)}T^{(k)}$ are thus also upper triangular and their diagonal entries are

$$E_s(\alpha_{i_1}, \alpha_{i_2}, \ldots, \alpha_{i_k}), \qquad 1 \leq i_1 < i_2 < \cdots < i_k \leq n,$$

in the lexicographical ordering. Now, the numbers

$$\Phi(E_1(\alpha_{i_1}, \ldots, \alpha_{i_k}), E_2(\alpha_{i_1}, \ldots, \alpha_{i_k}), \ldots, E_k(\alpha_{i_1}, \ldots, \alpha_{i_k}))$$

in the lexicographical ordering, i.e. the numbers

$$\varphi(\alpha_{i_1}, \alpha_{i_2}, \ldots, \alpha_{i_k}),$$

are the diagonal entries of the matrix

$$\Phi(A_1^{(k,1)}, A_1^{(k,2)}, \ldots, A_1^{(k,k)}) =$$
$$(T^{(k)})^{-1}\Phi(A^{(k,1)}, A^{(k,2)}, \ldots, A^{(k,k)})T^{(k)}.$$

These are all the eigenvalues of $\Phi(A_1^{(k,1)}, A_1^{(k,2)}, \ldots, A_1^{(k,k)})$ and thus those of $\Phi(A^{(k,1)}, A^{(k,2)}, \ldots, A^{(k,k)})$ as well, which was to be proved.   □

Theorem 6.2.16 *Let A be a square matrix of order n, $n \geq 2$, having the eigenvalues $\alpha_1, \ldots, \alpha_n$. Then the eigenvalues of the matrix*

$$(A^{[2]})^2 - 4A^{(2)},$$

which is of order $\binom{n}{2}$, are just all the numbers $(\alpha_i - \alpha_j)^2$, $1 \leq i < j \leq n$.

Proof. Follows readily from the preceding theorem for $\varphi(x_1, x_2) = (x_1 - x_2)^2$ since the polynomial $\Phi(z_1, z_2) = z_1^2 - 4z_2$ is such that

$$\Phi(E_1(x_1, x_2), E_2(x_1, x_2)) =$$
$$(x_1 + x_2)^2 - 4x_1 x_2 = (x_1 - x_2)^2 = \varphi(x_1, x_2).$$

□

The matrix $(A^{[2]})^2 - 4A^{(2)}$ can thus be considered as the *discriminant* of A. Indeed, it is singular if and only if an eigenvalue of A is multiple.

Exercises

6.1. Prove Theorem 6.1.10.

6.2. Prove Theorem 6.1.11.

6.3. Express all the eigenvalues of the matrix (6.1) in terms of the eigenvalues $\alpha_1, \alpha_2, \alpha_3$ of the matrix Z_3 and the eigenvalues $\beta_1, \beta_2, \beta_3, \beta_4, \beta_5$ of Z_5.

6.4 Prove Theorem 6.2.5.

6.5 Show that the exterior products of EP5 are exactly all the $\binom{n}{k}$-component vectors

$$[1, 0, \ldots, 0]^T, [0, 1, 0, \ldots, 0]^T, \ldots, [0, 0, \ldots, 1]^T.$$

Use this result to prove Theorem 6.2.9.

6.6 Write the additive compound $A^{[2]}$ of a square matrix $A = [a_{ik}]$ of order 4 explicitly.

6.7 Find the discriminant $(A^{[2]})^2 - 4A^{(2)}$ of the matrix

$$A = \begin{bmatrix} \alpha_1 & \beta_1 & 0 \\ \beta_1 & \alpha_2 & \beta_2 \\ 0 & \beta_2 & \alpha_3 \end{bmatrix}.$$

6.8.* If the rank of the matrix A is r, prove that the rank of its compound $A^{(r)}$ is 1. (Hint: Cf. Exercise 1.13.)

6.9.* If any r rows and any r columns of a matrix of rank r are linearly independent, prove that all its square submatrices of order r are nonsingular. (Hint: Cf. Exercises 6.8 and 1.14.)

6.10.* For a positive integer m and any complex 2×2 matrix

$$C = \begin{bmatrix} c_{11} & c_{12} \\ c_{21} & c_{22} \end{bmatrix},$$

let $C^{\langle m \rangle}$ be the $(m+1) \times (m+1)$ matrix defined by (x_1 and x_2 are variables)

$$C^{\langle m \rangle} X^{(m)} = Y^{(m)},$$

where

$$X^{(m)} = (x_1^m, x_1^{m-1} x_2, \ldots, x_2^m)^T$$

and

$$Y^{(m)} = ((c_{11} x_1 + c_{12} x_2)^m,$$
$$(c_{11} x_1 + c_{12} x_2)^{m-1} (c_{21} x_1 + c_{22} x_2), \ldots,$$
$$(c_{21} x_1 + c_{22} x_2)^m)^T.$$

Establish the main properties $((CD)^{\langle m \rangle} = C^{\langle m \rangle} D^{\langle m \rangle}$ for 2×2 matrices C, D etc.) in order to express the spectrum of $C^{\langle m \rangle}$ in terms of the spectrum of C.

6.11. Find the spectral properties of the additive version of $C^{\langle m \rangle}$.

6.12.* A real matrix is said to be *totally nonnegative (positive)* if all its minors are nonnegative (positive). Show that the product of totally nonnegative (positive) matrices is totally nonnegative (positive) as well.

6.13. The *Hadamard* (or *Schur*) *product* of matrices $A = [a_{ik}]$ and $B = [b_{ik}]$ of same dimensions is the matrix $A \circ B = [a_{ik} b_{ik}]$. Prove that for A and B square of the same order, $A \circ B$ is a principal submatrix of the tensor product $A \otimes B$.

6.14. Show that for A and B positive definite, $A \circ B$ is positive definite as well.

7

Matrices and polynomials. Stable Matrices

In this chapter, we investigate mutual relations between matrices and polynomials. To start with, we prove two important theorems on the characteristic polynomial.

7.1 Characteristic polynomial

Theorem 7.1.1 (Cayley-Hamilton) *Let A be a square matrix of order n and $f(x)$ its characteristic polynomial $\det(A - xI)$. Then it follows that $f(A) = 0$ (in the sense of equality for matrices). In other words, every square matrix A substituted into its own characteristic polynomial yields the zero matrix.*

Proof. For a while, we set $G = A - xI$, $G = [g_{ik}]$. Suppose that $f(x) = a_0 + a_1 x + \cdots + a_n x^n$. Let $H = [h_{ik}]$ be the matrix whose entry h_{ik} equals the cofactor of the entry g_{ki} in G (the so-called adjoint of G). The formulae (1.10) and (1.14) then imply that

$$HG = (\det G)I. \tag{7.1}$$

Since $G = A - xI$, each entry in H is a polynomial of degree at most $n - 1$ in x. Therefore,

$$H = H_0 + H_1 x + \cdots + H_{n-1} x^{n-1},$$

where H are matrices whose entries are numbers. The left-hand side of (7.1) may thus be written as

$$(H_0 + H_1 x + \cdots + H_{n-1} x^{n-1})(A - xI),$$

whereas the right-hand side is

$$f(x)I = a_0 I + a_1 xI + \cdots + a_n x^n I.$$

If we perform the multiplication on the left and compare the matrices at x^k on the two sides of (7.1) we obtain

$$
\begin{aligned}
H_0 A &&&= a_0 I, \\
H_1 A &&- H_0 &= a_1 I, \\
H_2 A &&- H_1 &= a_2 I, \\
&&\vdots& \\
H_{n-1} A &&- H_{n-2} &= a_{n-1} I, \\
&&- H_{n-1} &= a_n I.
\end{aligned}
$$

We now right-multiply the second equation by A, the third by A^2, and so on. Finally, the $(n+1)$th equation is right-multiplied by A^n. Summing up the equations we get the zero matrix on the left whereas on the right we obtain

$$a_0 I + a_1 A + a_2 A^2 + \cdots + a_{n-1} A^{n-1} + a_n A^n.$$

Hence $f(A) = 0$. $\qquad\qquad\qquad\qquad\qquad\qquad\qquad\square$

Theorem 7.1.2 *Every polynomial of degree n with complex co-efficients and the leading coefficient $(-1)^n$ is the characteristic polynomial of an $n \times n$ complex matrix. In particular, if*

$$\varphi(x) = x^n + a_{n-1} x^{n-1} + a_{n-2} x^{n-2} + \cdots + a_1 x + a_0,$$

then $(-1)^n \varphi(x)$ is the characteristic polynomial of the matrix

$$
A = \begin{bmatrix}
0 & 1 & 0 & \cdots & 0 & 0 \\
0 & 0 & 1 & \cdots & 0 & 0 \\
\vdots & \vdots & \vdots & \ddots & \vdots & \vdots \\
0 & 0 & 0 & \cdots & 0 & 1 \\
-a_0 & -a_1 & -a_2 & \cdots & -a_{n-2} & -a_{n-1}
\end{bmatrix}. \tag{7.2}
$$

Remark. The matrix A from (7.2) is called the *companion matrix* of the polynomial $\varphi(x)$.

Proof. Right-multiplying $A - xI$ by the matrix

$$\begin{bmatrix} 1 & 0 & 0 & \cdots & 0 \\ x & 1 & 0 & \cdots & 0 \\ x^2 & x & 1 & \cdots & 0 \\ \vdots & \vdots & \vdots & \ddots & \vdots \\ x^{n-1} & x^{n-2} & x^{n-3} & \cdots & 1 \end{bmatrix}$$

whose determinant equals 1, we obtain the matrix

$$C = \begin{bmatrix} 0 & 1 & 0 & \cdots & 0 \\ 0 & 0 & 1 & \cdots & 0 \\ \vdots & \vdots & \vdots & \ddots & \vdots \\ 0 & 0 & 0 & \cdots & 1 \\ -\varphi(x) & \times & \times & \cdots & \times \end{bmatrix},$$

where $\times$ indicates an unspecified entry. Therefore,

$$\det(A - xI) = \det C = (-1)^n \varphi(x).$$

$\square$

We now introduce two special matrices which will be often used later. Given an ordered r-tuple $(s_1, \ldots, s_r)$ of numbers, we define

$$V(s_1, \ldots, s_r) = \begin{bmatrix} 1 & s_1 & s_1^2 & \cdots & s_1^{r-1} \\ \vdots & \vdots & \vdots & \ddots & \vdots \\ 1 & s_r & s_r^2 & \cdots & s_r^{r-1} \end{bmatrix}, \quad (7.3)$$

$$W(s_1, \ldots, s_r) = \begin{bmatrix} f_1(0) & \frac{1}{1!}f_1'(0) & \cdots & \frac{1}{(r-1)!}f_1^{(r-1)}(0) \\ \vdots & \vdots & \ddots & \vdots \\ f_r(0) & \frac{1}{1!}f_r'(0) & \cdots & \frac{1}{(r-1)!}f_r^{(r-1)}(0) \end{bmatrix}, \quad (7.4)$$

where

$$f_i(x) = \prod_{j=1, j \neq i}^{r} (x - s_j), \quad i = 1, \ldots, r. \quad (7.5)$$

Remark. The matrix $V(s_1, \ldots, s_r)$ is called a *Vandermonde matrix*. $W(s_1, \ldots, s_r)$ is the matrix whose entries are the coefficients of the polynomial from (7.5).

Theorem 7.1.3 *The   matrix $V(s_1, \ldots, s_r)$ and   the   matrix $W(s_1, \ldots, s_r)$ satisfy*

$$W(s_1, \ldots, s_r)V^T(s_1, \ldots, s_r) = \operatorname{diag}(f_1(s_1), \ldots, f_r(s_r)), \quad (7.6)$$

$$\det V(s_1, \ldots, s_r) = \prod_{1 \le i < j \le r} (s_j - s_i). \quad (7.7)$$

Therefore, if the numbers $s_1, \ldots, s_r$ are distinct, then both $V(s_1, \ldots, s_r)$ and $W(s_1, \ldots, s_r)$ are nonsingular matrices.

Proof. To prove (7.6) it suffices to observe that the left-hand side equals the matrix $[f_i(s_j)]$. The formula (7.7) is well known [51]. The last statement is an easy consequence of (7.6).   □

7.2   Matrices associated with polynomials

We shall now consider a special type of matrices associated with polynomials and show that they can be applied in studying the relative primeness of polynomials.

Let $a(x)$ be a polynomial of degree n,

$$a(x) = a_0 + a_1 x + a_2 x^2 + \cdots + a_n x^n$$

($a_n \ne 0$). Given any integer $s \ge 1$, we associate to $a(x)$ the $s \times (n + s)$ matrix $[a]^{(s)}$ defined by

$$[a]^{(s)} = \begin{bmatrix} a_0 & a_1 & a_2 & \ldots & a_n & 0 & \ldots & 0 & 0 \\ 0 & a_0 & a_1 & \ldots & a_{n-1} & a_n & \ldots & 0 & 0 \\ \vdots & \ddots & \ddots & \ddots & \vdots & \ddots & \ddots & \vdots & \vdots \\ \vdots & \vdots & \ddots & \ddots & \ddots & \vdots & \ddots & \ddots & \vdots \\ 0 & 0 & 0 & \ldots & a_0 & a_1 & \ldots & a_{n-1} & a_n \end{bmatrix}. \quad (7.8)$$

Furthermore, we introduce the vectors $X^{(p)}$ defined for positive integers p by

$$X^{(p)} = [1, x, x^2, \ldots, x^{p-1}]^T. \quad (7.9)$$

One easily obtains the following result.

Theorem 7.2.1 *If* $[a]^{(s)}$ *is the matrix from* (7.8), *then*

$$[a]^{(s)} X^{(n+s)} = a(x) X^{(s)}, \qquad (7.10)$$

where the vectors $X^{(n+s)}$ *and* $X^{(s)}$ *are given by* (7.9). *Conversely, if a matrix* A *satisfies* $AX^{(n+s)} = a(x)X^{(s)}$, *then* $A = [a]^{(s)}$.

Next, for a polynomial $b(x) = b_0 + b_1 x + b_2 x + \cdots + b_m x^m$ we introduce the $s \times (m + s)$ matrix $[b]^{(s)}$ analogously to (7.8) and similarly we define the matrix $[a\,b]^{(s)}$ for the product $a(x)b(x)$.

Theorem 7.2.2 *Given a polynomial* $a(x)$ *of degree* n *and a polynomial* $b(x)$ *of degree* m, *then for every positive integer* s *we have*

$$[a\,b]^{(s)} = [a]^{(s)} [b]^{(n+s)} \qquad (7.11)$$

and also

$$[a\,b]^{(s)} = [b]^{(s)} [a]^{(m+s)} \qquad (7.12)$$

Proof. We have

$$[a]^{(s)} [b]^{(n+s)} X^{(m+n+s)} = [a]^{(s)} b(x) X^{(n+s)}$$
$$= b(x) [a]^{(s)} X^{(n+s)} = b(x) a(x) X^{(s)}$$

in view of (7.10) and using Theorem 7.2.1 we obtain (7.11). The proof of (7.12) is analogous. $\qquad \square$

The introduced notation finds its use in the investigation of certain properties of polynomials (see also Exercise 7.1). We first consider the effect of substituting $x = y + h$ into a polynomial. For a number h and a positive integer k we define the matrix $S^{(k)}(h)$ by

$$S^{(k)}(h) =$$

$$\begin{bmatrix}
1 & 0 & 0 & 0 & \cdots & 0 \\
h & 1 & 0 & 0 & \cdots & 0 \\
h^2 & 2h & 1 & 0 & \cdots & 0 \\
h^3 & 3h^2 & 3h & 1 & \cdots & 0 \\
\vdots & \vdots & \vdots & \vdots & \ddots & \vdots \\
h^{k-1} & \binom{k-1}{1} h^{k-2} & \binom{k-1}{2} h^{k-3} & \binom{k-1}{3} h^{k-4} & \cdots & 1
\end{bmatrix} . \qquad (7.13)$$

Then we may state the following result.

Theorem 7.2.3 *Let $a(x)$ be a polynomial of degree n. Let h be a number and s a positive integer. If $[c]^{(s)}$ is the matrix associated with the polynomial $c(y) = a(y + h)$, then*

$$[c]^{(s)} = (S^{(s)}(h))^{-1}[a]^{(s)}S^{(n+s)}(h). \qquad (7.14)$$

Proof. For each positive t, set $Y^{(t)} = [1, y, y^2, \ldots, y^{t-1}]^T$ and $Z^{(t)} = [1, z, z^2, \ldots, z^{t-1}]^T$, where $z = y + h$. Clearly

$$Z^{(t)} = S^{(t)}(h) Y^{(t)},$$

from which

$$[a]^{(s)} Z^{(n+s)} = [a]^{(s)} S^{(n+s)}(h) Y^{(n+s)}.$$

On the other hand,

$$[a]^{(s)} Z^{(n+s)} = a(z)Z^{(s)} = a(y + h)S^{(s)}(h) Y^{(s)}$$

by (7.10). Altogether we have

$$(S^{(s)}(h))^{-1}[a]^{(s)} S^{(n+s)}(h) Y^{(n+s)} = c(y) Y^{(s)}$$

and using Theorem 7.2.1 we conclude that (7.14) holds.   □

Now, we shall consider relatively prime polynomials.

Theorem 7.2.4 *If $a(x)$ is a polynomial of degree n and $b(x)$ a polynomial of degree m, $m \leq n$, then the following statements are equivalent:*

1. *The polynomials $a(x)$ and $b(x)$ are relatively prime.*

2. *For every integer $p \geq 0$, the rank of the $(m + n + 2p) \times (m + n + p)$ matrix*

$$\begin{bmatrix} [a]^{(m+p)} \\ [b]^{(n+p)} \end{bmatrix}$$

   is $m + n + p$.

3. *There exists an integer $p \geq 0$ such that the matrix*

$$\begin{bmatrix} [a]^{(m+p)} \\ [b]^{(n+p)} \end{bmatrix}$$

   is of rank $m + n + p$.

4.

$$\det \begin{bmatrix} [a]^{(m)} \\ [b]^{(n)} \end{bmatrix} \neq 0.$$

5.

$$\det \begin{bmatrix} [a]^{(n)} \\ [\tilde{b}]^{(n)} \end{bmatrix} \neq 0,$$

where $\tilde{b}(x)$ is the polynomial $b(x)$ viewed as a polynomial of degree at most n.[1]

Proof. We shall prove the implications 1. → 2., 2. → 3., 3. → 1., 2. → 4., 4. → 1., and the equivalence 4. ↔ 5.

1. → 2. Suppose that 1. holds and that for some $p \geq 0$ the rank of the matrix

$$C = \begin{bmatrix} [a]^{(m+p)} \\ [b]^{(n+p)} \end{bmatrix}$$

is less than $m+n+p$. Hence, by Theorem 1.5.1 the set of the first m and last $n + p$ rows of C is linearly dependent. Thus there exist numbers $c_0, c_1, \ldots, c_{m-1}, d_0, d_1, \ldots, d_{n+p-1}$, not all zero, such that

$$zC = 0,$$

where $z = [c_0, c_1, \ldots, c_{m-1}, 0, 0, \ldots, 0, d_0, d_1, \ldots, d_{n+p-1}]$. But this equation may be rewritten as

$$[c]^{(1)}[a]^{(m+p)} = -[d]^{(1)}[b]^{(n+p)},$$

where

$$c(x) = c_0 + c_1 x + \cdots + c_{m+p-1} x^{m+p-1},$$
$$d(x) = d_0 + d_1 x + \cdots + d_{n+p-1} x^{n+p-1}$$

(and $c_m = c_{m+1} = \cdots = c_{m+p-1} = 0$).

In view of (7.11) the above relation implies

$$[ca]^{(1)} = -[db]^{(1)},$$

---

[1] Hence, for instance, the matrix $[b]^{(1)}$ consists of one row and $n + 1$ columns, the last $n - m$ entries being zero.

or

$$a(x)c(x) = -b(x)d(x). \qquad (7.15)$$

If $d(x)$ were identically zero, then $c(x)$ would vanish identically as well, which contradicts $z \neq 0$. Therefore, the polynomial $b(x)$ of degree m is a divisor of the product $a(x)c(x)$. Since $b(x)$ and $a(x)$ are relatively prime, $b(x)$ divides the polynomial $c(x)$ of degree at most $m - 1$. But this is possible only if $c(x)$ is identically zero. Then, however, $d(x)$ must be identically zero and we again arrive at a contradiction. Hence the rank of C is $m+n+p$.

2. $\rightarrow$ 3. Obvious.

3. $\rightarrow$ 1. Suppose that there is some $p \geq 0$ such that the rank of

$$C = \begin{bmatrix} [a]^{(m+p)} \\ [b]^{(n+p)} \end{bmatrix}$$

is $m+n+p$; then the columns of C are linearly independent. Further suppose that $a(x)$ and $b(x)$ are not relatively prime. Then there exists (real or complex) number x_0 such that

$$a(x_0) = 0, \qquad b(x_0) = 0.$$

Set $X_0^{(t)} = [1, x_0, \ldots, x_0^{(t-1)}]^T$. Clearly $X_0^{(m+n+p)} \neq 0$ and we have

$$
\begin{aligned}
C X_0^{(m+n+p)} &= \begin{bmatrix} [a]^{(m+p)} X_0^{(m+n+p)} \\ [b]^{(n+p)} X_0^{(m+n+p)} \end{bmatrix} \\
&= \begin{bmatrix} a(x_0) X_0^{(m+p)} \\ b(x_0) X_0^{(n+p)} \end{bmatrix} \\
&= 0.
\end{aligned}
$$

This contradiction proves the desired implication.

Since 2. $\rightarrow$ 4. is obvious and 4. $\rightarrow$ 1. has been proved along with 3. $\rightarrow$ 1. it remains to prove the equivalence of 4. and 5. only.

The equivalence is clear for $m = n$. If $m < n$, then the matrix whose determinant is in 5. may be partitioned into the form

$$\begin{bmatrix} [a]^{(m)} & 0 \\ P & Q \\ [b]^{(n)} & 0 \end{bmatrix},$$

where Q is a square matrix of order $n - m$ which is lower triangular with nonzero diagonal entries equal to a_n, the leading coefficient of the polynomial $a(x)$. Hence Q is nonsingular and applying Laplace Theorem 1.2.1 to the last $n - m$ columns of Q we conclude that the determinant of 5. is equal to that of 4. multiplied by $\pm(a_n)^{n-m}$. The two determinants are thus either both zero or both nonzero. $\qquad \square$

Remark. The determinant of 4. is called the *resultant* of the polynomials $a(x)$ and $b(x)$. Hence, the resultant is nonzero if and only if $a(x)$ and $b(x)$ are relatively prime.

7.3 Bézout matrices

In what follows, we shall say that $a(x)$ and $b(x)$ are *polynomials of maximum degree n* if both the polynomials are of degree n or less and at least one of them is exactly of degree n.

Before defining the so-called Bézout form of the resultant, we denote, for polynomials $a(x)$ and $b(x)$ of maximum degree n, by $B(a, b)$ the $n \times m$ matrix $[\beta_{ik}]$, $i, k = 0, \ldots, n - 1$, satisfying

$$\frac{a(x)b(y) - a(y)b(x)}{x - y} = \sum_{i,k=0}^{n-1} \beta_{ik} x^i y^k. \qquad (7.16)$$

$B(a, b)$ is called the *Bézout matrix* for the pair of polynomials $a(x), b(x)$.

Theorem 7.3.1 *Let $a(x)$ and $b(x)$ be polynomials of maximum degree n. Consider the matrices $[a]^{(n)}$ and $[b]^{(n)}$ partitioned in the form*

$$[a]^{(n)} = [A_1, A_2], \qquad [b]^{(n)} = [B_1, B_2], \qquad (7.17)$$

where all the blocks are square matrices of order n.

If J denotes the $n \times n$ matrix

$$J = \begin{bmatrix} 0 & \cdots & 0 & 1 \\ 0 & \cdots & 1 & 0 \\ \vdots & \ddots & \vdots & \vdots \\ 1 & \cdots & 0 & 0 \end{bmatrix}, \qquad (7.18)$$

then

$$B(a, b) = J(A_2B_1 - B_2A_1). \tag{7.19}$$

Moreover, the polynomials $a(x)$ and $b(x)$ are relatively prime if and only if $B(a, b)$ is nonsingular or, equivalently, if

$$\det(A_2B_1 - B_2A_1) \neq 0. \tag{7.20}$$

Remark. The determinant $\det B(a, b)$ is called the *Bézout form* of the resultant of the polynomials $a(x), b(x)$. If (i, j) means the determinant

$$\begin{vmatrix} b_i & b_j \\ a_i & a_j \end{vmatrix}$$

for

$$a(x) = \sum_{i=0}^{n} a_i x^i, \qquad b(x) = \sum_{i=0}^{n} b_i x^i,$$

then

$$\det B(a, b) =$$

$$\det \begin{bmatrix} (0,1) & (0,2) & (0,3) & \cdots \\ (0,2) & (1,2)+(0,3) & (1,3)+(0,4) & \cdots \\ (0,3) & (1,3)+(0,4) & (2,3)+(1,4)+(0,5) & \cdots \\ \vdots & \vdots & \vdots & \cdots \end{bmatrix}.$$

Proof. Let

$$a(x) = \sum_{i=0}^{n} a_i x^i, \qquad b(x) = \sum_{i=0}^{n} b_i x^i.$$

Multiplying both the sides of (7.16) by $x - y$ we get

$$\sum_{i,k=0}^{n-1} \beta_{ik}(x^{i+1}y^k - x^i y^{k+1}) = a(x)b(y) - a(y)b(x)$$

and equating the coefficients of like powers, we arrive at

$$\beta_{i-1,k} - \beta_{i,k-1} = a_i b_k - a_k b_i, \quad i,k = -1,0,\ldots,n,$$

where we have put

$$\beta_{-1,k} = \beta_{i,-1} = \beta_{nk} = \beta_{in} = 0.$$

For $i, k = 0, \ldots, n - 1$ we thus have

$$\beta_{i+p,k-p} - \beta_{i+p+1,k-p-1} = a_{i+p+1}b_{k-p} - a_{k-p}b_{i+p+1},$$

where $p = 0, 1, \ldots, m$ and $m = \min(k, n - 1 - i)$. Summing we find

$$\beta_{ik} = \sum_{p=0}^{m} (a_{i+p+1}b_{k-p} - a_{k-p}b_{i+p+1}),$$

or

$$
B = \begin{bmatrix} a_1 & a_2 & \cdots & a_n \\ a_2 & a_3 & \cdots & 0 \\ \vdots & \vdots & \ddots & \vdots \\ a_n & 0 & \cdots & 0 \end{bmatrix} \begin{bmatrix} b_0 & b_1 & \cdots & b_{n-1} \\ 0 & b_0 & \cdots & b_{n-2} \\ \vdots & \vdots & \ddots & \vdots \\ 0 & 0 & \cdots & b_0 \end{bmatrix} -
$$

$$
- \begin{bmatrix} b_1 & b_2 & \cdots & b_n \\ b_2 & b_3 & \cdots & 0 \\ \vdots & \vdots & \ddots & \vdots \\ b_n & 0 & \cdots & 0 \end{bmatrix} \begin{bmatrix} a_0 & a_1 & \cdots & a_{n-1} \\ 0 & a_0 & \cdots & a_{n-2} \\ \vdots & \vdots & \ddots & \vdots \\ 0 & 0 & \cdots & a_0 \end{bmatrix},
$$

which is $B = JA_2B_1 - JB_2A_1$, i.e. (7.19), in the notation of the theorem.

This also shows the form in the remark. To prove the last assertion, we may assume that the degree of $a(x)$ is exactly n so that A_2 is a nonsingular lower triangular matrix. In view of the equivalence of 5. and 1. of Theorem 7.2.4, it is sufficient to show that the rank of the matrix

$$\begin{bmatrix} A_1 & A_2 \\ B_1 & B_2 \end{bmatrix}$$

is by n greater than the rank of $A_2B_1 - B_2A_1$.

But the matrices A_1 and B_1 commute, i.e. we have $A_1B_1 = B_1A_1$, since both are matrix polynomials in $J_n(0)$, the Jordan block defined by (1.43) (cf. Exercise 1.5). Similarly, A_2 and B_2

commute and A_2 is nonsingular, which implies $B_2 A_2^{-1} = A_2^{-1} B_2$. We thus have

$$\operatorname{rank} \begin{bmatrix} A_1 & A_2 \\ B_1 & B_2 \end{bmatrix} = \operatorname{rank} \begin{bmatrix} A_2 & A_1 \\ B_2 & B_1 \end{bmatrix}$$

$$= \operatorname{rank} \left(\begin{bmatrix} A_2 & 0 \\ B_2 & I \end{bmatrix} \begin{bmatrix} I & A_2^{-1} \\ 0 & B_1 - B_2 A_2^{-1} A_1 \end{bmatrix} \right)$$

$$= n + \operatorname{rank}(B_1 - B_2 A_2^{-1} A_1)$$

$$= n + \operatorname{rank}(B_1 - A_2^{-1} B_2 A_1)$$

$$= n + \operatorname{rank}(A_2 B_1 - B_2 A_1),$$

as was to be proved. $\square$

7.4   Hankel matrices

The Bézout matrices are closely related to the so-called Hankel matrices. The *Hankel matrix* of order n is an $n \times n$ complex matrix of the form $A = [a_{ik}]$, $i, k = 0, 1, \ldots, n - 1$, where $a_{ik} = \alpha_{i+k}$ for some given numbers $\alpha_0, \ldots, \alpha_{2n-2}$. Hence,

$$A = \begin{bmatrix}
\alpha_0 & \alpha_1 & \alpha_2 & \cdots & \cdots & \alpha_{n-1} \\
\alpha_1 & \alpha_2 & \cdots & \cdots & \alpha_{n-1} & \alpha_n \\
\alpha_2 & \cdots & \cdots & \cdots & \alpha_n & \alpha_{n+1} \\
\vdots & \vdots & \vdots & \vdots & \vdots & \vdots \\
\vdots & \alpha_{n-1} & \alpha_n & \cdots & \cdots & \cdots \\
\alpha_{n-1} & \alpha_n & \alpha_{n+1} & \cdots & \cdots & \alpha_{2n-2}
\end{bmatrix}. \qquad (7.21)$$

We now present a fundamental theorem on Hankel matrices and their relation to Bézout matrices.

Theorem 7.4.1 *Let V_t be the Vandermonde matrix of order n for the distinct complex numbers $t_1, \ldots, t_n$. Let C be a nonsingular diagonal matrix, $C = \operatorname{diag}(c_1, \ldots, c_n)$. Then the matrix*

$$A = V_t^T C V_t \qquad (7.22)$$

is a nonsingular Hankel matrix of the form (7.21) *with*

$$\alpha_r = \sum_{i=1}^{n} c_i t_i^r, \qquad r = 0, \ldots, 2n - 2. \qquad (7.23)$$

*Conversely, if A is a nonsingular matrix of the form (7.21),
then there exist numbers $t_1, \ldots, t_n$, all distinct, and numbers
$c_1, \ldots, c_n$ such that (7.23) and (7.22) hold. In fact, there are
infinitely many such n-tuples of numbers $t_1, \ldots, t_n$ (which, in
their turn, determine the numbers c_i). All these n-tuples can be
obtained as the n-tuples of roots of those polynomials*

$$
\det \begin{bmatrix}
\alpha_0 & \cdots & \alpha_{n-2} & \alpha_{n-1} & 1 \\
\alpha_1 & \cdots & \alpha_{n-1} & \alpha_n & x \\
\vdots & \ddots & \vdots & \vdots & \vdots \\
\alpha_{n-1} & \cdots & \alpha_{2n-3} & \alpha_{2n-2} & x^{n-1} \\
\alpha_n & \cdots & \alpha_{2n-2} & \xi & x^n
\end{bmatrix} \tag{7.24}
$$

*(where ξ is a parameter) which possess only simple roots. If $f(x)$
and $g(x)$ are polynomials of the form (7.24) corresponding to two
different values of ξ, $\xi = \xi_1$ and $\xi = \xi_2$ (we allow also $\xi = \infty$,
which corresponds to the polynomial equal to the coefficient of ξ
in (7.24)), then there is a nonzero number ρ such that*

$$
B(f, \rho g) = A^{-1}. \tag{7.25}
$$

Proof. The first statement as well as the nonsingularity of the
matrix (7.22) are evident.

We now assume that the numbers $\alpha_0, \ldots, \alpha_{2n-2}$ satisfy (7.23)
and show that then $t_1, \ldots, t_n$ are roots of (7.24) for some ξ.

Indeed, if we put

$$
\xi = \sum_{i=1}^{n} c_i t_i^{2n-1},
$$

then

$$
\det \begin{bmatrix}
\alpha_0 & \cdots & \alpha_{n-2} & \alpha_{n-1} & 1 \\
\alpha_1 & \cdots & \alpha_{n-1} & \alpha_n & x \\
\vdots & \ddots & \vdots & \vdots & \vdots \\
\alpha_n & \cdots & \alpha_{2n-2} & \xi & x^n
\end{bmatrix}
$$

may be written as the determinant of the matrix

$$
[\hat{V}_t^T C V_t, X], \tag{7.26}
$$

where $X = [1, x, \ldots, x^n]^T$, V_t is the nth order Vandermonde matrix for $t_1, \ldots, t_n$, and $\hat{V}^t$ is an analogous $n \times (n + 1)$ matrix with one additional column. The matrix (7.26), however, can be expressed as the product

$$[\hat{V}_t^T, X] \begin{bmatrix} C & 0 \\ 0 & 1 \end{bmatrix} \begin{bmatrix} V_t & 0 \\ 0 & 1 \end{bmatrix},$$

where $C = \mathrm{diag}(c_1, \ldots, c_n)$. The first factor here is the transpose of the Vandermonde matrix of order $n + 1$ for the numbers $t_1, \ldots, t_n, x$. Hence, by (7.7), the determinant in question equals

$$\rho \sum_{i=1}^{n} (t_i - x),$$

where ρ is a nonzero number. The polynomial (7.24) with the above given ξ thus really has just the roots $t_1, \ldots, t_n$. On the other hand, if for some ξ the distinct numbers $t_1, \ldots, t_n$ are roots of the polynomial (7.24), which is not identically zero, then there exist numbers $u_0, \ldots, u_n$ satisfying

$$\sum_{i=0}^{n} \alpha_{i+s} u_i = 0, \qquad s = 0, \ldots, n - 2, \qquad (7.27)$$

$$u_n \neq 0, \qquad (7.28)$$

and, moreover,

$$\sum_{i=0}^{n} u_i t_k^i = 0 \quad \text{for} \quad k = 1, \ldots, n. \qquad (7.29)$$

Let $c_1, \ldots, c_n$ be the solution of the linear system

$$\sum_{k=1}^{n} c_k t_k^r = \alpha_r, \qquad r = 0, \ldots, n - 1. \qquad (7.30)$$

Referring to (7.27) and (7.28), we have

$$\sum_{k=1}^{n} c_k t_k^r = u_n^{-1} c_k \left(-\sum_{i=0}^{n-1} u_i t_k^i \right)$$

$$= -u_n^{-1} \sum_{i=0}^{n-1} u_i \sum_{k=1}^{n} c_k t_k^i$$

$$= -u_n^{-1} \sum_{i=0}^{n-1} \alpha_i u_i$$

$$= \alpha_n.$$

It is easy to show by induction that (7.30) holds also for $r = n+1, \ldots, 2n-2$.

To prove the existence of the distinct numbers $t_1, \ldots, t_n$ we need three lemmas. For brevity, we shall use the term *leading submatrix* for any upper-left-corner submatrix.

Lemma 7.4.2 [27] *If $C = [\gamma_{i+k}]$ is a singular Hankel matrix of order n and r is a nonnegative integer, then the following two statements are equivalent:*

1. *The first r columns of C are linearly independent, whereas the first $r+1$ columns of C are already linearly dependent.*

2. *The order of the maximal nonsingular leading principal submatrix of C is r.*

Proof. Denote the number r of 2. by r'. Clearly, $r' \leq r$. To prove that also $r \leq r'$ we denote the columns of C by $C_0, \ldots, C_{n-1}$. Further, let $\tilde{C}$ be the submatrix of C formed by its first r rows and denote by $\tilde{C}_0, \ldots, \tilde{C}_{n-1}$ the columns of $\tilde{C}$. By 1., there exist numbers $\lambda_0, \ldots, \lambda_{r-1}$, not all zero, such that

$$C_r = \lambda_0 C_0 + \lambda_1 C_1 + \cdots + \lambda_{r-1} C_{r-1}. \qquad (7.31)$$

Hence. we have

$$\gamma_{s+r} = \lambda_0 \gamma_s + \lambda_1 \gamma_{s+1} + \cdots + \lambda_{r-1} \gamma_{s+r-1}$$

for $s = 0, 1, \ldots, n-1$ so that

$$\tilde{C}_{r+k} = \lambda_0 \tilde{C}_k + \lambda_1 \tilde{C}_{k+1} + \cdots + \lambda_{r-1} \tilde{C}_{r+k-1}$$

holds for $k = 0, 1, \ldots, n - r - 1$. Since the rank of $\tilde{C}$ is r, this is possible only if the columns $\tilde{C}_0, \ldots, \tilde{C}_{r-1}$ are linearly independent. Therefore, the leading principal submatrix of order r in $\tilde{C}$ is nonsingular, i.e., $r \leq r'$.   □

Lemma 7.4.3 *If* $A = [\alpha_{i+k}]$ *is a nonsingular Hankel matrix, then any two polynomials of the form* (7.24) *corresponding to different values of* ξ *(including* $\xi = \infty$*) are relatively prime.*

Proof. Suppose that two such polynomials are not relatively prime. Then they have a common root, say t, and it follows that

$$
\det \begin{bmatrix}
\alpha_0 & \cdots & \alpha_{n-2} & \alpha_{n-1} & 1 \\
\alpha_1 & \cdots & \alpha_{n-1} & \alpha_n & t \\
\vdots & \ddots & \vdots & \vdots & \vdots \\
\alpha_n & \cdots & \alpha_{2n-2} & \xi & t^n
\end{bmatrix} = 0
$$

for all ξ. After obvious elementary row operations we find that also $\det M = 0$ for any ξ, where

$$
M = \begin{bmatrix}
\alpha_1 - \alpha_0 t & \cdots & \alpha_n - \alpha_{n-1} t \\
\alpha_2 - \alpha_1 t & \cdots & \alpha_{n+1} - \alpha_n t \\
\vdots & \ddots & \vdots \\
\alpha_n - \alpha_{n-1} t & \cdots & \xi - \alpha_{2n-2} t
\end{bmatrix}.
$$

Thus M is a singular Hankel matrix, for which the number r from 2. of Lemma 7.4.2 satisfies

$$r \leq n - 2.$$

By Lemma 7.4.2, it means that the $(r + 1)$th column of M is a linear combination of the first r columns. Adding the column $[\alpha_0, \alpha_1, \ldots, \alpha_{n-1}]^T$ to the $r + 1$ columns of M, we obtain $r + 2$ linearly dependent vectors. Using elementary column operations we now find that also the first $r + 2$ ($\leq n$) columns of the matrix A are linearly dependent, contrary to the nonsingularity of A.   □

Lemma 7.4.4 *Let* $n \geq 1$. *Let* $f_1(x)$ *and* $f_2(x)$ *be polynomials of maximum degree* n *which are relatively prime. Then there exists a number* λ *such that* $f_1(x) + \lambda f_2(x)$ *has* n *simple roots. In fact, there are even infinitely many such numbers* λ.

Proof. A simple consideration shows that it is sufficient to give the proof for $f_1(x)$ of degree n and $f_2(x)$ of degree less than n. To prove the first statement, suppose that for every λ the polynomial $f_1(x) + \lambda f_2(x)$ has a multiple root. Then for every λ the polynomial $f_1(x) + \lambda f_2(x)$ and its derivative $f_1'(x) + \lambda f_2'(x)$ have a common non-constant factor. However, since $f_1(x)$ and $f_2(x)$ are relatively prime and $f_1(x) + \lambda f_2(x)$ is linear in λ, the polynomial $f_1(x) + \lambda f_2(x)$ considered as a polynomial in two variables, x and λ, is irreducible. Thus, $f_1'(x) + \lambda f_2'(x)$ is divisible by $f_1(x) + \lambda f_2(x)$. But this is impossible as $f_1'(x) + \lambda f_2'(x)$ is an $(n-1)$th degree polynomial which is not identically zero.

The second statement is a consequence of the first since there is a neighborhood of the number $\tilde{\lambda}$ whose existence we have proved above such that for every $\tilde{\lambda}$ from this neighborhood all the roots of $f_1(x) + \tilde{\lambda} f_2(x)$ are simple. □

We now turn back to the proof of Theorem 7.4.1. By Lemma 7.4.3, any two polynomials (7.24) with distinct values of ξ are relatively prime. Therefore, by Lemma 7.4.4 there are infinitely many ξ for which (7.24) has n distinct roots.

It remains to prove the last statement. Since $B(f + \lambda g, g) = B(f, g)$, it is sufficient to give the proof for the case when $f(x)$ has n distinct roots; we denote them by $t_1, \ldots, t_n$. As also $B(f, g + \mu f) = B(f, g)$, we may assume the degree of $g(x)$ to be less than n. By Lemma 7.4.3, $f(x)$ and $g(x)$ are relatively prime so that $g(t_i) \neq 0$, $i = 1, \ldots, n$.

Consider the Bézout matrix $B(f, g) = [\beta_{ij}]$. For $j = 1, \ldots, n$ define the polynomials $f_j(x)$ by

$$f_j(x) = \prod_{k=1, k \neq j}^{n} (x - t_k) \qquad (7.32)$$

and set

$$d_j = \prod_{k=1, k \neq j}^{n} (t_j - t_k), \, j = 1, \ldots, n \qquad (7.33)$$

$(d_j = f_j(t_j))$. Then

$$g(x) = \sum_{k=1}^{n} d_k^{-1} g(t_k) f_k(x)$$

by Theorem 7.2.1. Also,

$$\sum_{i,j=0}^{n-1} \beta_{ij} x^i y^j = \frac{f(x)g(y) - f(y)g(x)}{x - y}$$

$$= \frac{1}{x - y} \sum_{k=1}^{n} d_k^{-1} g(t_k)(f(x)f_k(y) - f(y)f_k(x)).$$

Since $f(x) = (x - t_k)f_k(x)$ we have

$$\sum_{i,j=0}^{n-1} \beta_{ij} x^i y^j = \sum_{k=1}^{n} d_k^{-1} g(t_k) f_k(x) f_k(y). \qquad (7.34)$$

Now, assume that $c_1, \ldots, c_n$ are numbers satisfying (7.23). Since in view of (7.24) we have

$$g(x) = \mu \det \begin{bmatrix} \alpha_0 & \cdots & \alpha_{n-2} & 1 \\ \alpha_1 & \cdots & \alpha_{n-1} & x \\ \vdots & \ddots & \vdots & \vdots \\ \alpha_{n-1} & \cdots & \alpha_{2n-3} & x^{n-1} \end{bmatrix}$$

for some $\mu \neq 0$, the coefficients $v_0, \ldots, v_{n-1}$ of

$$g(x) = \sum_{i=0}^{n-1} v_i x^i$$

satisfy

$$\sum_{i=0}^{n-1} \alpha_{i+s} v_i = 0, \qquad s = 0, \ldots, n - 2.$$

Using (7.23) we may write this as

$$\sum_{i=0}^{n-1} \sum_{k=1}^{n} c_k t_k^{i+s} v_i = 0,$$

or

$$\sum_{k=1}^{n} c_k t_k^s g(t_k) = 0, \qquad s = 0, \ldots, n - 2. \qquad (7.35)$$

Furthermore, we define

$$\omega = \sum_{k=1}^{n} c_k t_k^{n-1} g(t_k). \tag{7.36}$$

Equations (7.35) and (7.36) imply that

$$\omega = \sum_{k=1}^{n} c_k h(t_k) g(t_k)$$

for every polynomial $h(x)$ of degree $n-1$ whose leading coefficient is 1. Hence

$$\omega = \sum_{k=1}^{n} c_k f_i(t_k) g(t_k)$$

for $i = 1, \ldots, n$ so that

$$\omega = c_i d_i g(t_i). \tag{7.37}$$

From (7.34)

$$\sum_{i,j=o}^{n-1} \beta_{ij} t_r^i t_s^j = \sum_{k=1}^{n} d_k^{-1} g(t_k) f_k(t_r) f_k(t_s)$$

$$= \begin{cases} 0 & \text{for } r \neq s, \\ d_r g(t_r) & \text{for } r = s. \end{cases} \tag{7.38}$$

Therefore, if V_t is defined as in (7.22) and $C = \operatorname{diag}(c_1, \ldots, c_n)$ we have

$$V_t B(f, g) V_t^T = L,$$

where $L = \operatorname{diag}(d_1 g(t_1), \ldots, d_n g(t_n))$. Applying (7.22), we get

$$V_t B(f, g) A = V_t B(f, g) V_t^T C V_t = L C V_t.$$

But

$$LC = \omega I$$

in view of (7.37) so that

$$V_t B(f, g) A = \omega V_t.$$

Since V_t is nonsingular we have

$$B(f,g)A = \omega I.$$

By Theorem 7.3.1, $B(f,g)$ is nonsingular so that $\omega \neq 0$. Thus, putting $\rho = 1/\omega$, we have

$$B(f,\rho g) = A^{-1}$$

as was to be proved.   □

Theorem 7.4.1 shows that the inverse of a nonsingular Hankel matrix is a Bézout matrix. We shall prove that also the converse is true [45].

Theorem 7.4.5 *If B is a nonsingular matrix, then the following two statements are equivalent:*
  1. B is a Bézout matrix.
  2. B^{-1} is a Hankel matrix.

Proof. Since 2. → 1. holds by Theorem 7.4.1, it suffices to prove 1. → 2.

Hence, suppose that $B = B(f,g)$ where $f(x)$ is a polynomial of degree n and $g(x)$ is a polynomial of degree n or less. Since B is nonsingular $f(x)$ and $g(x)$ are relatively prime. Clearly, $B(f,g) = B(f + \lambda g, g)$ for any number λ. In view of Lemma 7.4.4 we may, therefore, assume that $f(x)$ has n distinct roots. We denote then by $t_1, \ldots, t_n$. Since $B(f,g) = B(f, g + \mu f)$ for any number μ as well, we may also assume that the degree of $g(x)$ is less than n. We now introduce the polynomials $f_i(x)$ and numbers d_i by (7.32) and (7.33), respectively.

If we put

$$c_i = \frac{1}{d_i g_i(t)},$$

then the Hankel matrix $A = [\alpha_{i+k}]$ with entries given by (7.23) has the property that the numbers c_i, t_i satisfy (7.38) and (7.37) with $\omega = 1$. Thus, A satisfies (7.25) with $\rho = 1$.   □

Theorems 7.4.5 and 7.4.1, and the expression of the Bézout matrix in the form (7.20) can be used to invert a nonsingular Hankel matrix A by solving only two systems of linear equations (cf. [35] for Toeplitz matrices).

Theorem 7.4.6 *Let $A = [\alpha_{i+k}]$, $i, k = 0, \ldots, n-1$, be a nonsingular Hankel matrix. Let $u = [u_0, \ldots, u_{n-1}]^T$ and $v = [v_0, \ldots, v_{n-1}]^T$ be solutions of the systems*

$$Au = p, \quad Av = e_n,$$

*where $p = [\alpha_n, \alpha_{n+1}, \ldots, \alpha_{2n-2}, 0]^T$ and $e_n = [0, 0, \ldots, 0, 1]^T$.
Then*

$$A^{-1} = B(f, g),$$

where

$$f(x) = x^n - u_{n-1}x^{n-1} - \ldots - u_0,$$
$$g(x) = \qquad\quad v_{n-1}x^{n-1} + \ldots + v_0.$$

Thus, A^{-1} can be expressed as in (7.19).

Another theorem [42] characterizes definite Bézout matrices.

Theorem 7.4.7 *The Bézout matrix $B(f, g)$ for two real polynomials, $f(x)$ and $g(x)$, is positive or negative definite if all the roots of $f(x)$ and $g(x)$ are real and simple, every two of them are distinct, and they interlace. By this last condition we mean – for a polynomial $f(x)$ whose order is exactly n and a polynomial $g(x)$ of order at most n – that each open interval whose endpoints are two neighbouring roots of $f(x)$ contains one and only one root of $g(x)$.*

Remark. By Theorem 7.4.1, an analogous statement holds for the Hankel matrix and arbitrary two different polynomials (7.24).

Lemma 7.4.2 implies that with each singular Hankel matrix A whose first column is not the zero vector we can associate the polynomial

$$\varphi_A(x) \equiv x^r - \lambda_0 x^{r-1} - \cdots - \lambda_{r-1}, \tag{7.39}$$

whose (positive) degree r is equal to r of Lemma 7.4.2 and whose coefficients $\lambda_0, \ldots, \lambda_{r-1}$ are the numbers uniquely determined by (7.31). The polynomial (7.39) will be called the H-polynomial of the (singular) Hankel matrix A. If the first column of A is zero we say that the Hankel matrix A is *degenerate*, putting $\varphi_A(x) \equiv 1$

in such a case. A Hankel matrix whose rank equals the number r of Lemma 7.4.2 is said to be *proper*.

For the class H of complex Hankel matrices, the problem of quasidirect decomposition (Exercise 2.17) can be solved completely. This problem shows to be of fundamental importance for investigating the properties of Hankel matrices. In order to be able to present (without proof, see [23]) a more lucid formulation of the result we define, for a complex number t and integers m, n such that $1 \leq m \leq n$, $m \times n$ matrices $P_{mn}(t)$, which are generalizations of the transpose of the matrix (7.13). We put

$$P_{mn}(t) = \begin{bmatrix} 1 & t & t^2 & \cdots & t^{n-1} \\ 0 & 1 & 2t & \cdots & (n-1)t^{n-2} \\ \vdots & \vdots & \vdots & \ddots & \vdots \\ 0 & 0 & 0 & \cdots & \binom{n-1}{m-1}t^{n-m} \end{bmatrix} = [p_{ij}] \qquad (7.40)$$

so that

$$p_{ij} = \binom{j}{i} t^{j-i}, \qquad i = 0, \ldots, m-1, \ j = 0, \ldots, n-1.$$

Theorem 7.4.8 *Let A be a singular Hankel matrix. Suppose its H-polynomial is factored as follows ($t_i \neq t_j$ whenever $i \neq j$):*

$$\varphi_A(x) = \prod_{i=1}^{s}(x - t_i)^{n_i}, \quad n_i \geq 1, \quad \sum_{i=1}^{s} n_i = r. \qquad (7.41)$$

Further, put

$$n_0 = h - r,$$

where h is the rank of A.

Then the following holds:

a) *The matrix A is (quasidirectly) H-indecomposable if and only if either $s = 0$ and A is degenerate, or $s = 1$, $n_0 = 0$ and A is of the form*

$$A = P_{n_1 n}(t_1) C_1 P_{n_1 n}^T(t_1), \qquad (7.42)$$

where

$$C_1 = \begin{bmatrix} \gamma_0 & \gamma_1 & \cdots & \gamma_{h-1} \\ \gamma_1 & \gamma_2 & \cdots & 0 \\ \vdots & \vdots & \ddots & \vdots \\ \gamma_{h-1} & 0 & \cdots & 0 \end{bmatrix}, \gamma_{h-1} \neq 0. \qquad (7.43)$$

b) *If A is not H-indecomposable, then it may be written as*

$$A = \sum_{i=1}^{s} P_{n_1 n}(t_i) C_i P_{n_1 n}^T(t_i) + C_0, \qquad (7.44)$$

where C_i, $i = 1, \ldots, s$, are nonsingular $n_i \times n_i$ matrices of the form similar to that of C_1 in (7.43) and for $n_0 > 0$ the matrix C_0 is a degenerate Hankel matrix of order n and rank n_0. Moreover, (7.44) is a quasidirect sum, each summand is an H-indecomposable Hankel matrix, and (7.44) is a unique (except for the order of terms) expression of A as a quasidirect sum of H-indecomposable matrices.

Remark. Every nonsingular Hankel matrix of order greater than 1 is always H-decomposable (exercise 7.8). All the decompositions of such a Hankel matrix A into a quasidirect sum of H-indecomposable matrices are of the form

$$A = \sum_{i=1}^{t} A_i$$

where $t > 2$. The product of the H polynomials of the summands A_i (these are always singular Hankel matrices) is then a polynomial of degree n or less and every polynomial resulting in this way belongs (after dividing by its leading coefficient) to the family of polynomials (7.24) of Theorem 7.4.1.

We now turn to other related classes of matrices.

7.5 Toeplitz and Löwner matrices

The *Toeplitz matrices* of order n are the matrices having the form

$$C = [\gamma_{i-k}], \qquad i, k = 0, \ldots, n-1,$$

where $\gamma_{-(n-1)}, \ldots, \gamma_{-1}, \gamma_0, \gamma_1, \ldots, \gamma_{n-1}$ are, in general, complex numbers.

Denoting by J_n the matrix introduced in (7.18), we clearly have the following theorem.

Theorem 7.5.1 *If A is a Hankel matrix of order n, then AJ_n and J_nA are Toeplitz matrices. If C is a Toeplitz matrix of order n, then CJ_n and J_nC are Hankel matrices.*

Employing Theorem 7.5.1 and multiplying by J_n, we can reformulate Theorems 7.4.1 and 7.4.8 to get results for Toeplitz matrices.

The *Löwner matrices* of order n ([46, 12]) have the form

$$\left[\frac{c_i - d_j}{y_i - z_j}\right], \qquad i, j = 1, \ldots, n,$$

where $y_1, \ldots, y_n, z_1, \ldots, z_n$ are distinct (real or complex) numbers. Considering these numbers fixed, we then speak of the class $\mathcal{L}(y, z)$ of Löwner matrices.

It is clear that the Hankel matrices as well as the Toeplitz and Löwner matrices (in $\mathcal{L}(y, z)$) of order n form a linear subspace of the linear space of $n \times n$ matrices. Its dimension is $2n - 1$ in all the three cases.

Theorem 7.5.2 *Let $A = [\alpha_{i+k}]$, $i, k = 0, \ldots, n - 1$, be a Hankel matrix. Then*

$$L = W(y_1, \ldots, y_n)AW^T(z_1, \ldots, z_n),$$

where W are the matrices defined by (7.4), is a Löwner matrix in $\mathcal{L}(y, z)$,

$$L = \left[\frac{c_i - d_j}{y_i - z_j}\right].$$

The numbers c_i, d_j (determined uniquely except for an additive constant) satisfy

$$\begin{bmatrix} c \\ d \end{bmatrix} = W(y_1, \ldots, y_n, z_1, \ldots, z_n)[\alpha],$$

where we have put $c = [c_1, \ldots, c_n]^T, d = [d_1, \ldots, d_n]^T$, and

$$[\alpha] = [\alpha_0, \ldots, \alpha_{2n-1}, \xi]^T$$

for a certain ξ.

Conversely, if L is a Löwner matrix, then

$$A = (W(y_1, \ldots, y_n))^{-1} L (W^T(z_1, \ldots, z_n))^{-1}$$

is a Hankel matrix.

The proof is left to the reader (Exercise 7.9).
Löwner matrices are closely related to the interpolation by rational functions. The relationship is described in the following theorem, which we present without proof [24].

Theorem 7.5.3 *Let $x_1, \ldots, x_{2n-1}, y_1, \ldots, y_{2n-1}$ be (real or complex) numbers such that $x_i \neq x_j$ whenever $i \neq j$. A necessary and sufficient condition for the existence of polynomials $p(x)$ and $q(x)$ of maximum degree $n-1$ which are relatively prime and satisfy*

$$\frac{p(x_i)}{q(x_i)} = y_i, \qquad q(x_i) \neq 0, \quad i = 1, \ldots, 2n-1,$$

is that all the square submatrices of order $n-1$ in the $n \times (n-1)$ matrix

$$\hat{L} = \left[\frac{y_i - y_{n+j}}{x_i - x_{n+j}} \right], \quad i = 1, \ldots, n, \; j = 1, \ldots, n-1,$$

be nonsingular.
If this is the case we may take

$$p(x) = \det(\hat{L}, YF(x)),$$
$$q(x) = \det(\hat{L}, F(x)),$$

where

$$F(x) = [f_1(x), \ldots, f_n(x)]^T, \quad f_i(x) = \prod_{j=1, j \neq i}^{n} (x - x_j),$$
$$Y = \mathrm{diag}(y_1, \ldots, y_n).$$

7.6   Stable matrices

In the conclusion of this chapter we shall be concerned with the so-called stable matrices, which find their important applications in solving systems of differential equations $\dot{x} = Ax$, in certain economic problems, etc.

A square matrix A is said to be *stable* if all the eigenvalues of A have negative real parts, and it is said to be *positive stable* if all the eigenvalues of A have positive real parts. Obviously, A is stable if and only if $-A$ is positive stable.

The following theorem is a simple consequence of the definition.

Theorem 7.6.1 *If A is a stable (or positive stable) matrix, then A is nonsingular and A^{-1} is again a stable (or positive stable) matrix. If, in addition, T is a nonsingular matrix, then TAT^{-1} is a stable (or positive stable) matrix as well.*

The next theorem characterizes the stable matrices.

Theorem 7.6.2 (Fundamental Theorem on Stable Matrices) *Let A be a square (real or complex) matrix and set $B = (A + I)^{-1}(A - I)$. Then the following properties of A are equivalent:*

1. *A is positive stable.*

2. *The matrix B exists and $\rho(B) < 1$.*

3. *The matrix B exists and for every positive definite (in general Hermitian) matrix C there exists a (Hermitian) positive definite matrix P such that $P - BPB^* = C$.*

4. *The matrix B exists and the series*

   $$C + BCB^* + B^2C(B^*)^2 + B^3C(B^*)^3 + \dots$$

   converges for any positive definite matrix C.

5. *The matrix B exists and the series*

   $$I + BB^* + B^2(B^*)^2 + B^3(B^*)^3 + \dots$$

   converges.

6. *The matrix* B *exists and there is a positive definite matrix* Q *such that*
$$Q - BQB^* = I.$$

7. *The matrix* B *exists and there is a positive definite matrix* R *such that* $R - BRB^*$ *is positive definite.*

8. *There exists a positive definite matrix* S *such that* $AS + SA^*$ *is positive definite.*

9. *For every positive definite matrix* G *there exists a positive definite matrix* V *such that*
$$AV + VA^* = G.$$

10. *There exists a positive definite matrix* V *such that*
$$AV + VA^* = I.$$

11. *There exists a positive definite matrix* W *such that the matrix*
$$WAW^{-1} + W^{-1}A^*W$$
is positive definite.

12. *There exists a nonsingular matrix* T *such that the symmetric part of* TAT^{-1}, *i.e.,* $\mathrm{Re}(TAT^{-1}) = \frac{1}{2}(TAT^{-1} + (T^*)^{-1}A^*T^*)$, *is positive definite.*

Proof. We shall prove the implications 1. → 2., 2. → 3., 3. → 4., 4. → 5., 5. → 6., 6. → 7., 7. → 8., 8. → 1., 3. → 9., 9. → 10., 10. → 11., 11. → 12., and 12. → 1.

1. → 2. If 1. holds, then $A + I$ is nonsingular so that B does exist. By Theorem 1.7.12, the eigenvalues of B are $\beta_i = (\alpha_i - 1)/(\alpha_i + 1)$, where α_i, $i = 1, \ldots, n$, are all the eigenvalues of A. But since $\mathrm{Re}\,\alpha_i > 0$ we have

$$|\beta_i|^2 = \left| \frac{\alpha_i - 1}{\alpha_i + 1} \right|^2 = \frac{|\alpha_i|^2 + 1 - 2\,\mathrm{Re}\,\alpha_i}{|\alpha_i|^2 + 1 + 2\,\mathrm{Re}\,\alpha_i} < 1,$$

and thus $\rho(B) < 1$.

2. → 3. Suppose that the matrix B exists and $\rho(B) < 1$. By Theorem 6.1.8, the system of linear equations

$$P - BPB^* = C \tag{7.45}$$

in the unknown entries of the matrix P is equivalent to the system

$$(I - B \otimes \bar{B})\mathrm{vec}P = \mathrm{vec}C, \tag{7.46}$$

where $\mathrm{vec}P$ and $\mathrm{vec}C$ are column vectors formed – as in Chapter 6 – by writing all the columns of P and C, respectively, below each other as a single vector.

The eigenvalues of the matrix $B \otimes \bar{B}$ occuring in (7.46) are $\beta_i \bar{\beta}_j$, where β_i are the eigenvalues of B. Hence, $\rho(B \otimes \bar{B}) < 1$ and $I - B \otimes \bar{B}$ is a nonsingular matrix. For a given matrix C there exists, therefore, a unique matrix P satisfying (7.45). If C is Hermitian, P is Hermitian as well since both P and P^* satisfy (7.45). We shall prove that for a positive definite matrix C the matrix P is positive definite. By Theorem 1.7.14, the solution $\mathrm{vec}P$ of (7.46) may be written as

$$\mathrm{vec}P = (I + B \otimes \bar{B} + (B \otimes \bar{B})^2 + \dots)\mathrm{vec}C.$$

Since $(B \otimes \bar{B})^k = B^k \otimes (\bar{B})^k$ and the series in brackets converges, we have

$$P = C + BCB^* + B^2 C(B^*)^2 + \dots$$

by Theorem 6.1.8. However, $B^k C(B^*)^k$, $k = 1, 2, \dots$, are positive semidefinite matrices since for every x we have

$$\langle (B^k C(B^*)^k x, x \rangle = \langle C(B^*)^k x, (B^*)^k x \rangle = \langle Cy, y \rangle \geq 0,$$

where $y = (B^*)^k x$. But $\langle Cx, x \rangle > 0$ for any $x \neq 0$ so that $\langle Px, x \rangle > 0$ for any $x \neq 0$ and the desired implication has been proved.

3. → 4. Let C be a positive definite matrix. By 3., there exists a positive definite matrix P satisfying $P - BPB^* = C$. We now show that $\rho(B) < 1$. Indeed, if β_i is an eigenvalue of B, then $\bar{\beta}_i$ is an eigenvalue of B^* and there is a vector $y \neq 0$ such that

$B^*y = \bar{\beta}_i y$. For this y, we find that

$$
\begin{aligned}
\langle Cy, y \rangle &= \langle Py, y \rangle - \langle BPB^*y, y \rangle \\
&= \langle Py, y \rangle - \langle PB^*y, B^*y \rangle \\
&= \langle Py, y \rangle - |\beta_i|^2 \langle Py, y \rangle \\
&= (1 - |\beta_i|^2) \langle Py, y \rangle.
\end{aligned}
$$

Since $y \neq 0$, we have $\langle Py, y \rangle > 0$ and also $\langle Cy, y \rangle > 0$ so that $|\beta_i|^2 < 1$. This demonstrates that $\rho(B) < 1$. Further, the convergence of the series $C + BCB^* + B^2C(B^*)^2 + \dots$ follows similarly as in the proof of the implication $2. \to 3$.

$4. \to 5.$ It is sufficient to take $C = I$.

$5. \to 6.$ Denote by Q_k, $k = 1, 2, \dots$, the kth partial sum

$$
Q_k = I + BB^* + B^2(B^*)^2 + \dots + B^{k-1}(B^*)^{k-1}
$$

and put $Q = \lim_{k \to \infty} Q_k$. As

$$
Q_k - BQ_kB^* = I - B^k(B^*)^k
$$

and $\lim_{k \to \infty} B^k(B^*)^k = 0$, we have

$$
Q - BQB^* = I.
$$

Given a vector $x \neq 0$, we get

$$
\begin{aligned}
\langle Qx, x \rangle &= \langle x, x \rangle + \sum_{k=1}^{\infty} \langle B^k(B^*)^k x, x \rangle \\
&= \langle x, x \rangle + \sum_{k=1}^{\infty} \langle (B^*)^k x, (B^*)^k x \rangle \\
&\geq \langle x, x \rangle > 0.
\end{aligned}
$$

Hence, Q really is a positive definite matrix.

$6. \to 7.$ It is sufficient to take $R = Q$.

$7. \to 8.$ Since

$$
\begin{aligned}
R - BRB^* &= R - (A+I)^{-1}(A-I)R(A^*-I)(A^*+I)^{-1} \\
&= (A+I)^{-1}((A+I)R(A^*+I) \\
&\quad -(A-I)R(A^*-I))(A^*+I)^{-1} \\
&= 2(A+I)^{-1}(AR + RA^*)(A^*+I)^{-1},
\end{aligned}
$$

the positive definiteness of $C = R - BRB^*$ implies that the matrix $H = AR + RA^*$ is positive definite. Indeed, for $x \neq 0$ we get $\langle Hx, x \rangle = \frac{1}{2}\langle Cy, y \rangle$, where $y = (A^* + I)x \neq 0$, and thus $\langle Hx, x \rangle > 0$.

8. → 1. Assume that 8. holds. If λ is an eigenvalue of the matrix A, $Az = \lambda z$, $z \neq 0$, then for $y = S^{-1}z$ we have

$$\langle (AS + SA^*)y, y \rangle > 0.$$

On the other hand,

$$\begin{aligned} \langle (AS + SA^*)y, y \rangle &= \langle ASy, y \rangle + \langle y, ASy \rangle \\ &= \langle Ay, S^{-1}z \rangle + \langle S^{-1}z, Az \rangle \\ &= (\lambda + \bar{\lambda})\langle S^{-1}z, z \rangle. \end{aligned}$$

Since $\langle S^{-1}z, z \rangle > 0$, we have $(\lambda + \bar{\lambda}) = 2\,\mathrm{Re}\,\lambda > 0$. Therefore, A is positive stable.

3. → 9. Suppose 3. holds and let G be a positive definite matrix. Choose

$$C = 2(A + I)^{-1}G(A^* + I)^{-1};$$

the matrix C exists as B exists. Clearly, C is a positive definite matrix again. Hence, by 3. there exists a positive definite matrix V such that $V - BVB^* = C$. Proceeding similarly as in the proof of 7. → 8. we obtain

$$C = V - BVB^* = 2(A + I)^{-1}(AV + VA^*)(A^* + I)^{-1}$$

so that $AV + VA^* = G$.

9. → 10. It is sufficient to take $G = I$.

10. → 11. By 10., there exists a positive definite matrix V such that $AV + VA^* = I$. By Theorem 2.5.4 the matrix V^{-1} is positive definite as well. Using the result of Exercise 2.11 we can see that there is a positive definite matrix W such that $W^2 = V^{-1}$. But then

$$\begin{aligned} WAW^{-1} + W^{-1}AW &= W(A(W^2)^{-1} + (W^2)^{-1}A^*)W \\ &= W(AV + VA^*)W \\ &= W^2 = V^{-1} \end{aligned}$$

and 11. holds.

11. → 12. It is sufficient to put $T = W$.

12. → 1. Suppose that 12. is true and let λ be an eigenvalue of A, i.e., $Ax = \lambda x$ where $x \neq 0$. Choosing $y = Tx$ we then obtain

$$
\begin{aligned}
0 < \langle (\mathrm{Re}(TAT^{-1}))y, y \rangle &= \tfrac{1}{2}\langle TAT^{-1}y, y \rangle + \tfrac{1}{2}\langle y, TAT^{-1}y \rangle \\
&= \tfrac{1}{2}\langle TAx, y \rangle + \tfrac{1}{2}\langle y, TAx \rangle \\
&= \tfrac{1}{2}(\lambda + \bar{\lambda})\langle y, y \rangle \\
&= (\mathrm{Re}\,\lambda)\langle y, y \rangle.
\end{aligned}
$$

Hence, $\mathrm{Re}\,\lambda > 0$ and A is positive stable. $\qquad\square$

Remark. The equivalence 1. → 8. is known as Ljapunov's Theorem; the equivalence of this theorem and 10. was proved by O. Taussky-Todd in [56]. Stein's Theorem states the equivalence of 2. and 3.

It is worth noting that the conditions 2. to 7. themselves are equivalent for any matrix B. Indeed, in the formulation of the theorem A is primary, B derived. However, it can be easily verified that the roles of A and B can be interchanged so that B will be primary and A derived, $A = (I + B)(I - B)^{-1}$. (All the statements on A will then start with "A exists and ...".)

A reasonable question to ask at this point is how to find whether A is stable (or positive stable) or not without computing its eigenvalues. One possibility is to construct the matrix V of 10. employing Theorem 6.1.8 and to investigate the positive definiteness of V. However, if A is of order n the linear system for the entries of V is of order n^2 (if A is real we may use symmetry to reduce the order roughly to $\frac{1}{2}n^2$, which may be a rather large number. A simpler direct method was described e.g. by Carlson and Datta [6]. Another possibility is to find the characteristic polynomial (for example, by the Krylov method – see Chapter 13) and use the Routh-Hurwitz theorem, which we present here without proof[2].

Theorem 7.6.3 (Routh-Hurwitz) *Let* $f(x) = a_0 + a_1x + \cdots + a_nx^n$ *($a_n \neq 0$) be a real polynomial. Suppose that $a_0 > 0$ and define $a_j = 0$ for $j > n$. Then, all the roots of $f(x)$ have negative*

---

[2]See, for example, [42].

real parts if and only if all the determinants

$$D_1 = a_1,$$

$$D_2 = \det \begin{bmatrix} a_1 & a_0 \\ a_3 & a_2 \end{bmatrix},$$

$$D_3 = \det \begin{bmatrix} a_1 & a_0 & 0 \\ a_3 & a_2 & a_1 \\ a_5 & a_4 & a_3 \end{bmatrix},$$

$$D_4 = \det \begin{bmatrix} a_1 & a_0 & 0 & 0 \\ a_3 & a_2 & a_1 & a_0 \\ a_5 & a_4 & a_3 & a_2 \\ a_7 & a_6 & a_5 & a_4 \end{bmatrix},$$

$$\vdots$$

$$D_n = \det \begin{bmatrix} a_1 & a_0 & 0 & \ldots & 0 \\ a_3 & a_2 & a_1 & \ldots & 0 \\ \vdots & \vdots & \vdots & \ddots & \vdots \\ a_{2n-1} & a_{2n-2} & a_{2n-3} & \ldots & a_n \end{bmatrix}$$

are positive.

Let us present without proof another instance of the application of matrices to the investigation of roots of real polynomials.

Consider a polynomial

$$f(x) = a_n x^n + a_{n-1} x^{n-1} + \cdots + a_1 x + a_0$$

with the roots $z_1, \ldots, z_n$ (some of which may coincide). We first note that the sums of the kth powers of the roots

$$s_k = \sum_{i=1}^{n} z_i^k, \qquad k = 1, 2, \ldots, \quad s_0 = n,$$

can be obtained from the recurrence relations[3]

$$a_n s_k + a_{n-1} s_{k-1} + \cdots + a_{n-k+1} s_1 + k a_{n-k} = 0,$$

for $k = 1, \ldots, n$; for $i = 1, 2, \ldots$

$$a_n s_{n+1} + a_{n-1} s_{n+i-1} + \cdots + a_0 s_i = 0. \qquad (7.47)$$

We now have the following theorem.

---

[3]See, for example, [60].

Theorem 7.6.4 *Let* $f(x) = a_n x^n + a_{n-1} x^{n-1} + \cdots + a_0$ *(*$a_n \neq 0$*)* *be a polynomial with real coefficients. If* $s_1, s_2, \ldots, s_{2n-2}$ *satisfy the system* (7.47), *then the number of distinct roots of* $f(x)$ *is equal to the rank of the Hankel matrix*

$$
S = \begin{bmatrix}
s_0 & s_1 & \cdots & s_{n-1} \\
s_1 & s_2 & \cdots & s_n \\
s_2 & s_3 & \cdots & s_{n+1} \\
\vdots & \vdots & \ddots & \vdots \\
s_{n-1} & s_{n-2} & \cdots & s_{2n-2}
\end{bmatrix}.
$$

Further, the number of distinct real roots of $f(x)$ *equals the number of positive eigenvalues diminished by the number of negative eigenvalues of* S *(this is called the signature of* S*).*

Remark. There are other theorems in this field, which enable us to determine, for example, also the number of distinct real roots of $f(x)$ lying in a given interval[4].

We conclude the chapter by presenting (without proof, cf. [44]) the main theorem on matrices whose entries are polynomials in one variable. Such matrices are usually called *lambda-matrices* or *λ-matrices* as the variable has been denoted by λ traditionally. Let us remark that an analogous theorem holds also for integral matrices (whose entries are integers) and, more generally, for matrices with entries in a Euclidean ring [60].

We need the notion of the *Smith form*, which is a λ-matrix of the form

$$
\begin{bmatrix}
D & 0 \\
0 & 0
\end{bmatrix},
$$

where D (possibly void) is a diagonal matrix $D = \mathrm{diag}(d_1(\lambda), \ldots, d_r(\lambda))$ and $d_i(\lambda)$ are monic polynomials (i.e., polynomials with the leading coefficient 1) such that $d_{i+1}(\lambda)$ is divisible by $d_i(\lambda)$, $i = 1, \ldots, r - 1$.

Theorem 7.6.5 *Let* $A(\lambda)$ *be a* λ-*matrix. Then there exist square* λ-*matrices* $P(\lambda)$, $Q(\lambda)$ *whose determinants are non-zero numbers*

---

[4]See [51].

(units in the more general case) and a Smith form $S_A(\lambda)$ such that
$$A(\lambda) = P(\lambda)S_A(\lambda)Q(\lambda).$$

The Smith form
$$S_A(\lambda) = \begin{bmatrix} D & 0 \\ 0 & 0 \end{bmatrix},$$

$D = \mathrm{diag}(d_1(\lambda), \ldots, d_r(\lambda))$ *is uniquely determined by $A(\lambda)$ and can be obtained e.g. using the fact that $d_k(\lambda)/d_{k-1}(\lambda)$, $k = 1, \ldots,$ r, $(d_0(\lambda) = 1$ by definition) is the greatest common divisor of all the minors of order k of the matrix $A(\lambda)$.*

Exercises

7.1. Denote (for $s \geq 1$) by $D^{(s)}$ the $s \times s$ matrix
$$D^{(s)} = \begin{bmatrix} 0 & 0 & 0 & \cdots & 0 & 0 \\ 1 & 0 & 0 & \cdots & 0 & \\ 0 & 2 & 0 & \cdots & 0 & 0 \\ \vdots & \vdots & \vdots & \ddots & \vdots & \vdots \\ 0 & 0 & 0 & \cdots & s-1 & 0 \end{bmatrix}.$$

If $a(x)$ is a polynomial of degree n and $a'(x)$ is its derivative (viewed as a polynomial of degree n or less with the coefficient of x^n equal to zero), prove that $[a']^{(s)} = [a]^{(s)}D^{(n+s)} - D^{(s)}[a]^{(s)}$ for $s \geq 1$.

7.2.* Given n distinct numbers $t_1, \ldots, t_n$, prove that there exists a unique polynomial $\varphi(x)$ of degree $n - 1$ or less which assumes prescribed values k_i, $i = 1, \ldots, n$, at the points t_i: $\varphi(t_i) = k_i$, $i = 1, \ldots, n$. If $f_j(x)$ are defined as in (7.32), show that $\varphi(x)$ may be written as
$$\varphi(x) = \sum_{i=1}^{n} \frac{f_i(x)}{f_i(t_i)} k_i.$$

7.3.* The *modified Chebyshev polynomials* $S_k(x)$, $k = 0, 1, \ldots,$ *of the second kind* can be defined, for example, by the recurrence relation

$$S_{i+1}(x) = xS_i(x) - S_{i-1}(x), \quad i = 1, 2, \ldots,$$

$$S_0(x) = 1, \qquad S_1(x) = x.$$

Show that they possess the following properties:

1. If we put $x = 2\cos t$, then $S_k(x) = (\sin t)^{-1}\sin(k+1)t$, $k = 0, 1, \ldots$.

2. $S_k(x) = \prod_{i=1}^{k}\left(x - 2\cos\frac{i\pi}{k+1}\right)$.

3. $S_i(x)S_{j-k}(x) = S_j(x)S_{i-k}(x) - S_{k-1}(x)S_{i-j-1}(x)$ for $i > j \geq k \geq 1$.

7.4. Suppose that $\lambda_1, \lambda_2, \ldots, \lambda_n$ are all the roots of the polynomial $f(x) = a_0 + a_1x + \cdots + a_nx^n$ ($a_n \neq 0$). Express the coefficients of the polynomial whose all roots are $\lambda_1^2, \lambda_2^2, \ldots, \lambda_n^2$ in terms of $a_0, a_1, \ldots, a_n$.

7.5. Let $f(x) = (-1)^n(x^n + a_{n-1}x^{n-1} + \cdots + a_1x + a_0)$ and let $g(x)$ be another polynomial which is not identically zero. Consider the matrix A given by formula (7.2). Prove that $\det(g(A)) = 0$ if and only if $f(x)$ and $g(x)$ have a common root. (Hence, $\det(g(A))$ is also a resultant of $f(x)$ and $g(x)$, that is, a polynomial in the coefficients of both f and g which vanishes if and only if $f(x)$ and $g(x)$ have a common root.)

7.6. Write explicitly the Bézout form of the resultant for polynomials $a(x)$ and $b(x)$ of degree 3.

7.7. Let A be a Hankel matrix of order $n + 1$ and let C be a nonsingular square matrix of order 2. Define the matrix $C^{\langle n \rangle}$ as in Exercise 6.10. Show that $C^{\langle n \rangle}A(C^{\langle n \rangle})^T$ is again a Hankel matrix.

7.8. Prove that every nonsingular Hankel matrix of order at least two is always H-decomposable.

7.9. Prove Theorem 7.5.2.

7.10. Let A be a singular Hankel matrix of order n and rank r. Prove that the Smith form $S_A(x)$ of the matrix $[A, X]$, $X = [1, x, \ldots, x^{n-1}]^T$ is

$$S_A(x) = \begin{bmatrix} I & 0 & 0 \\ 0 & \varphi_A(x) & 0 \\ 0 & 0 & 0 \end{bmatrix},$$

where I_r is the identity matrix of order r and $\varphi_A(x)$ is the H-polynomial of A.

7.11. Prove that the quasidirect sum of two proper Hankel matrices is again a proper Hankel matrix. If, in addition, the sum is singular, prove that the H-polynomial of the sum is the product of the H-polynomials of the summands and that these polynomials are relatively prime.

7.12. Let n and r be two integers, $n > r \geq 1$. If A is a proper Hankel matrix of order $r + 1$ and rank r, then there is a unique Hankel matrix of order n and rank r such that A is its leading principal submatrix.

7.13. Every Hankel matrix may be expressed as a quasidirect sum of a proper and a degenerate Hankel matrices. If A is singular, then the decomposition is unique. Therefore, the sum of the ranks of the "proper" and "degenerate" parts equals the rank of the matrix.

7.14. Prove that a degenerate Hankel matrix is H-indecomposable.

7.15. Let $A = [\alpha_{i+k}]$, $i = 0, \ldots, m_1 - 1$, $k = 0, \ldots, m_2 - 1$, be an $m_1 \times m_2$ rectangular Hankel matrix. Let $f_i(x)$, $i = 1, 2$, be polynomials of degrees n_i, $n_i < m_i$. Show that then

$$B = [f_1]^{(m_1 - n_1)} A([f_2]^{(m_2 - n_2)})^T$$

is an $(m_1 - n_1) \times (m_2 - n_2)$ Hankel matrix,

$$B = [\beta_{i+k}]. \qquad i = 0, \ldots, m_1 - n_1 - 1,$$
$$k = 0. \ldots, m_2 - n_2 - 1,$$

where for $f(x) = f_1(x)f_2(x)$ and $\hat{\alpha} = [\alpha_0, \ldots, \alpha_{m_1+m_2-2}]^T$, $\hat{\beta} = [\beta_0, \ldots, \beta_{m_1+m_2-n_1-n_2-2}]^T$ we have

$$\hat{\beta} = [f]^{(m_1+m_2-n_1-n_2-1)}\hat{\alpha}.$$

7.16. Let

$$L = \left[\frac{c_i - d_j}{y_i - z_j}\right]$$

be an $n \times n$ Löwner matrix in $\mathcal{L}(y, z)$. Put

$$f_i(x) = \prod_{j=1, j \neq i} (x - y_i), \quad i = 1, \ldots, n,$$

$F(x) = [f_1(x), \ldots, f_n(x)]^T$. If L is singular of rank r, then the *L-polynomial* of L is defined as the (monic) polynomial $\psi_L(x)$ for which the Smith form of $(L, F(x))$ is

$$\begin{bmatrix} I & 0 & 0 \\ 0 & \psi_L(x) & 0 \\ 0 & 0 & 0 \end{bmatrix}.$$

Show that if the matrix A in Theorem 7.5.2 is singular, then the H-polynomial of A coincides with the L-polynomial of the corresponding matrix L.

7.17. Let $t_1, \ldots, t_n$ be real numbers, $0 < t_1 < t_2 < \cdots < t_n$. Show that the Vandermonde matrix $V(t_1, \ldots, t_n)$ is totally positive (cf. Exercise 6.12).

7.18. Let $a_1 < a_2 < \cdots < a_n$ and $b_1 < b_2 < \cdots < b_n$. Show that the matrix $[e^{a_i b_j}]$ is totally positive.

7.19. Show that a nonnegative tridiagonal matrix in P_0 (cf. Chapter 5) is totally nonnegative.

7.20. A square $n \times n$ matrix $C = [c_{ik}]$ is said to be *circulant* if $c_{ik} = \gamma_j$ for $j = 0, 1, \ldots, n-1$, $j \equiv (i - k) \mod n$. We write $C = C(\gamma_0, \gamma_1, \ldots, \gamma_{n-1})$. Show that such C is a polynomial in $\hat{C} = C(0, 1, 0, \ldots, 0)$ and find the eigenvalues of the general circulant matrix $C(\gamma_0, \gamma_1, \ldots, \gamma_{n-1})$.

7.21. Find a necessary and sufficient condition for an $n \times n$ circulant matrix to belong to (a) class K, (b) class K_0.

8

Band Matrices

In this chapter, we consider matrices of a certain type which occur very frequently today, especially in numerical analysis when solving numerically ordinary differential or partial differential equations by discretization techniques, in the algorithms for the computation of matrix eigenvalues, etc.

8.1 Band matrices and graphs

If $A = [a_{ik}]$ is a square matrix of order n, we put

$$p = \max(p_0, 0), \quad p_0 = \max\{d; d = k - 1, \text{ where } a_{ik} \neq 0\}, \quad (8.1)$$
$$q = \max(q_0, 0), \quad q_0 = -\min\{d; d = k - 1, \text{ where } a_{ik} \neq 0\}, \quad (8.2)$$

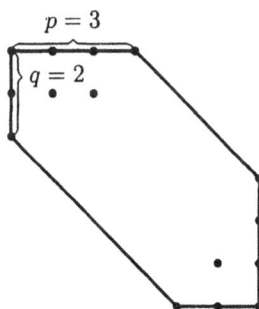

Fig. 12

The numbers p and q have clearly the following meaning (Fig. 12): They determine the dimensions of the minimal band along the main diagonal containing all nonzero off-diagonal entries of the matrix.

We therefore say that A is a (p, q)-*band matrix* (A is (p, q)-*banded*) if the numbers p and q are defined by (8.1) and (8.2). We are, of course, interested only in the case when at least one of the numbers p and q is much less than the order n of the matrix.

Important particular cases are: the lower triangular matrix, for which $p = 0$ obviously, the upper triangular matrix with $q = 0$, and the diagonal matrix, which is (0, 0)-banded. A (1, 1)-band matrix is also called *tridiagonal*; a matrix with $q = 1$ (sometimes $q \leq 1$) is called a *Hessenberg matrix*.

Theorem 8.1.1 *Let A_1 and A_2 be square matrices of order n. If A_1 is (p_1, q_1)-banded and A_2 is (p_2, q_2)-banded, then A_1^T is (q_1, p_1)-banded, $A_1 + A_1$ is (p_3, q_3)-banded, where $p_3 \leq \max(p_1, p_2)$, $q_3 \leq \max(q_1, q_2)$, and $A_1 A_2$ is (p_4, q_4)-banded, where*

$$p_4 \leq \min(n - 1, p_1 + p_2), \tag{8.3}$$
$$q_4 \leq \min(n - 1, q_1 + q_2). \tag{8.4}$$

Proof. The first two statements are obvious. The third statement follows from the fact that the numbers p_{40}, p_{10}, and p_{20} defined by (8.1) satisfy

$$
\begin{aligned}
p_{40} &= \max\{k - i; \sum_{j=1}^{n} a_{ij}^{(1)} a_{jk}^{(2)} \neq 0\} \\
&\leq \max\{k - j; a_{jk}^{(2)} \neq 0\} + \max\{j - i; a_{ij}^{(1)} \neq 0\} \\
&= p_{10} + p_{20},
\end{aligned}
$$

and, of course, also $p_{40} \leq n - 1$, and from the analogous inequalities for q_4.   □

The importance of band matrices is implied mainly by the next theorem.

Theorem 8.1.2 *Let A be a square matrix which is (p, q)-banded and strongly nonsingular. Then there exist a $(0, q)$-band matrix B and a $(p, 0)$-band matrix C such that $A = BC$.*

Proof. By Theorem 1.4.3, any strongly nonsingular matrix A can be factored as a product BC of a lower triangular matrix

B and an upper triangular matrix C. Suppose that B is $(0, q_1)$-banded and C is $(p_1, 0)$-banded. By Theorem 8.1.1, $p \leq p_1$ and $q \leq q_1$. Therefore, it is sufficient to prove $p \geq p_1$ and $q \geq q_1$. Put $A = [a_{ik}]$, $B = [b_{ik}]$ and $C = [c_{ik}]$. Then clearly $b_{ii} \neq 0$, $c_{ii} \neq 0$, $i = 1, \ldots, n$, as both B and C are nonsingular. Hence

$$p_1 = \max\{d; d = k - i, \text{ where } c_{ik} \neq 0\}$$

and thus there is an entry $c_{ik} \neq 0$, $i \leq k$, such that $p_1 = k - i$ and, in addition, $c_{jk} = 0$ for all j, $1 \leq j < i$. For this pair i, k of indices we then have

$$a_{ik} = \sum_{j=1}^{i} b_{ij} c_{jk} = b_{ii} c_{ik} \neq 0.$$

From (8.1). it follows that $p \geq k - i = p_1$, so that $p = p_1$ as required. $\square$

Theorem 8.1.2 implies – as well we shall see in Chapter 14 – that to perform Gaussian elimination on A (without row or column interchanges, of course) it is sufficient to store in the memory of the computer only the entries within the band. This reduction in storage is very important, especially when we solve large linear systems with relatively small dimensions of the band.

It is clear that the dimensions of the band may vary if we permute rows or columns, or even if we perform a simultaneous permutation of the rows and columns of A. This gives rise to the very important problem of finding an appropriate permutation transforming A to a (p, q)-band matrix with, for example, minimal $p + q$. To show the difficulty of this problem we shall consider the case of the simultaneous permutations of rows and columns of a symmetric matrix A. Since only the zero-nonzero structure of A, and not the values of the entries, is relevant for our considerations we represent A by its (undirected, in this case) graph $G(A)$ here. Its vertices are, for $n \times n$ matrix $A = [a_{ik}]$, the numbers $1, \ldots, n$, its edges those unordered pairs (i, k), $i \neq k$, for which $a_{ik} \neq 0$.

When A is a symmetric (p, q)-band matrix, we refer to p as to the *bandwidth* of A. The following theorem relates the bandwidth of A to the structure of $G(A)$.

Theorem 8.1.3 *Let A be a symmetric square matrix of order n which is (p,p)-banded. Let the vertices of its graph $G(A)$ be labelled by the indices $1, 2, \ldots, n$ of the corresponding rows (and columns) of A. If each edge (i, k) of $G(A)$ is assigned the number $|i - k|$, then p equals the maximum over all edges in $G(A)$ of the numbers assigned in this way.*

Proof. Follows immediately from the symmetry of A, the definition of $G(A)$, and from (8.1). □

Theorem 8.1.3 tells us that finding the optimal band form of a symmetric matrix is equivalent to the problem of labelling the points of a given graph with n vertices by distinct numbers $1, 2, \ldots, n$ in such a way that the maximal value of the differences $|i - k|$ over all the edges be minimum. We have thus arrived at a problem of the theory of undirected finite graphs.

For some graphs the solution is almost obvious. So, the solution for the rectangular lattice $M_{p,q}$ (cf. Example 3.10) is (for $p = 5$ and $q = 7$) in Fig. 13 (the resulting bandwidth is $\min(p, q)$). In applications, it is favourable to recognize the graph structure of A (say, by analyzing the situation from which A arose).

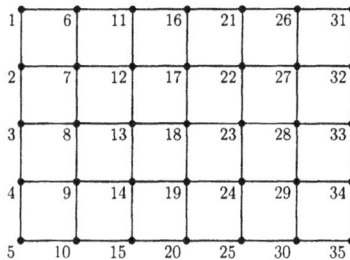

Fig. 13

A practicable solution of the above problem (not, for example, examining all the $n!$ possible labellings) is not known. It has been even proved that the problem of deciding whether there exists a labelling of a given graph that leads to a bandwidth p or less belongs to the class of the so-called NP-complete problems, which are mutually equivalent as regards their algorithmic complexity and comprise also the well known travelling salesman problem, etc. Several rather efficient algorithms leading to a reasonably good (suboptimal) solution have been developed. Here, we describe the so-called Reverse Cuthill-McKee algorithm (RCM) [9] and the basic ideas of the GPS algorithm due to Gibbs, Poole, and Stockmeyer [30].

The underlying idea of both the algorithms is finding a *level structure* of the graph $G = (V, H)$, i.e., a partition $R = [V_1, \ldots, V_r]$ of the set V of vertices such that the condensation of G with respect to R is a path with vertices $V_1, \ldots, V_r$ and edges (V_1, V_2), $(V_2, V_3), \ldots, (V_{r-1}, V_r)$ of G is connected. If G is disconnected the algorithm is applied to each component separately. The purpose is to find such a level structure R whose *width* $w(R)$ defined as

$$w(R) = \max_{i=1,\ldots,r} |V_i|$$

is minimum. For the corresponding matrix this means, roughly speaking, to find a simultaneous permutation giving the matrix a block tridiagonal form with the smallest possible order of the largest diagonal block. This interpretation shows that the bandwidth corresponding to a level structure is not greater than $2w(R) - 1$ provided the numbering of the vertices of G is done level by level, first numbering the vertices in V_1, then those in V_2, and so forth up to V_r.

We shall restrict ourselves to connected graphs since the generalization to the disconnected case is trivial. First, we note that for any connected graph having at least two vertices there exists a level structure $R = [V_1, \ldots, V_r]$ such that the number r, which is called the *depth* of R, is not less than 2. Indeed, it is sufficient to choose any vertex u in G, put $V_1 = \{u\}$, and define the set V_k, $k > 1$, as the set of all the vertices whose distance from u is just $k - 1$. (The *distance* between two vertices in a connected graph is the number of edges in the shortest path connecting these points.) Such a level structure will be said to be *rooted* at the point u. Moreover the depth r always satisfies the inequality $r \leq d+1$, where d is the *diameter* of G, i.e., the maximal distance between two vertices in G.

We now describe the *RCM algorithm*. Suppose that G has N vertices whose degrees are $s_1, s_2, \ldots, s_N$. Put

$$S = \max_{i=1,\ldots,N} s_i, \qquad s = \min_{i=1,\ldots,N} s_i, \qquad \sigma = \frac{1}{N} \sum_{i=1}^{N} s_i.$$

We shall say that a vertex in G is of "low degree" if the degree of the vertex is less than or equal to $\max\{\min[(S+s)/2,\ \sigma-1], s\}$. In the first step of the algorithm we generate the level structures R_u rooted at each vertex u of "low degree" and compute their widths $w(R_u)$. Further we consider only the "minimal" level structures, i.e. those whose width $w(R_u)$ is minimum. For each "minimal" level structure R_u we then number the vertices of G in the following way:

The root vertex u of R_u is assigned the number 1 so that V_1 has now been numbered. If the vertices in V_k are numbered, then for the successive level, V_{k+1}, we first number the vertices adjacent to the lowest numbered vertex in V_k, in the order of increasing degree. If we cannot choose the next point uniquely we decide arbitrarily. Then the remaining vertices adjacent to the next lowest numbered vertex in V_k are numbered, again in order of increasing degree, and so forth. We continue the process until all vertices in V_{k+1} are numbered; then, we begin again on the next level V_{k+2} and continue until all vertices of G are numbered.

For each numbering produced in the above way we compute the corresponding bandwidth and select the numbering which gives the smallest bandwidth; if there are several such numberings we decide arbitrarily. In the last step the vertices are numbered in reverse order (whence the attribute "Reverse" CM algorithm).

The *GPS algorithm* is much more complex but it seems to give somewhat better results. We describe concisely its basic idea only. Given a connected graph G, we first find a pair of vertices that are at nearly maximal distance apart. This is accomplished in such a way that we pick an arbitrary vertex of minimal degree, find the vertex which is farthest away from it (or we pick one of such vertices), then we again pick a vertex which is farthest away from the preceding one, and we continue the procedure until the distance "gets stabilized". The algorithm terminates with vertices u and v of a "pseudo-diameter" of G. For this "pseudo-diameter", a level structure $[V_1, \ldots, V_k]$ where $u \in V_1$ and $v \in V_k$ is then constructed in a rather complicated manner[1].

---

[1]Here, we refer to the description of the algorithm in [30].

Finally, the vertices in V_j, $j = 1, \ldots, k$, are numbered as in the preceding algorithm and the numbering is then reversed.

Now, we consider several computational problems involving band matrices.

Theorem 8.1.4 *Let A be a nonsingular (p, q)-band matrix of order n whose entries $a_{k,k+p}$, $k = 1, \ldots, n-p$, are nonzero. Then A may be partitioned into the form*

$$A = \begin{bmatrix} A_{11} & A_{12} \\ A_{21} & A_{22} \end{bmatrix},$$

where A_{12} is a nonsingular matrix of order $n - p$ which is lower triangular. The system $Ax = b$, which in the partitioned form may be written as

$$\begin{bmatrix} A_{11} & A_{12} \\ A_{21} & A_{22} \end{bmatrix} \begin{bmatrix} x_1 \\ x_2 \end{bmatrix} = \begin{bmatrix} b_1 \\ b_2 \end{bmatrix} \tag{8.5}$$

(x_1 is a p-component column vector, x_2 and $(n - p)$-component column vector, b_1 is an $(n - p)$-component column vector, and b_2 a p-component column vector), has then the solution

$$x_1 = (A_{21} - A_{22}A_{12}^{-1}A_{11})^{-1}(b_2 - A_{22}A_{12}^{-1}b_1),$$
$$x_2 = A_{12}^{-1}b_1 - A_{12}^{-1}A_{11}x_1.$$

Proof. The first statement is obvious. To prove the second statement, we write the system (8.5) as

$$A_{12}x_2 + A_{11}x_1 = b_1,$$
$$A_{22}x_2 + A_{21}x_1 = b_2$$

and substitute $x_2 = A_{12}^{-1}b_1 - A_{12}^{-1}A_{11}x_1$ into the second block equation. The matrix $A_{21} - A_{22}A_{12}^{-1}A_{11}$ is nonsingular by Theorem 1.4.2 and the required formula for x_1 follows. □

Remark. The method of Theorem 8.1.4, however, has a certain drawback sometimes. If the off-diagonal entries are small compared with the diagonal ones, it often leads to operations involving large numbers and the rounding errors may then seriously affect the accuracy of the calculation.

8.2 Eigenvalues and eigenvectors of tridiagonal matrices

The inversion of tridiagonal matrices will be treated in more detail in Chapter 12. In the following portion of the present chapter we study the eigenvalue problem for tridiagonal matrices. As we shall see in Chapter 13, the computation of the eigenvalues of a symmetric (or Hermitian) matrix often employs the reduction of the matrix to tridiagonal form using orthogonal (or unitary) similarity transformations. Therefore, we study methods for finding the eigenvalues of tridiagonal matrices here. One of the methods is based on the following two theorems.

Theorem 8.2.1 *Given a real tridiagonal matrix*

$$
A = \begin{bmatrix}
\alpha_1 & \gamma_1 & & & & & \\
\gamma_1 & \alpha_2 & \gamma_2 & & & & \\
& \gamma_2 & \alpha_3 & \gamma_3 & & & \\
& & \ddots & \ddots & \ddots & & \\
& & & \gamma_{n-2} & \alpha_{n-1} & \gamma_{n-1} \\
& & & & \gamma_{n-1} & \alpha_n
\end{bmatrix}, \qquad (8.6)
$$

then its principal minors $\det A(N_k)$, $N_k = \{1, 2, \ldots, k\}$, $k = 1, \ldots, n$, *can be computed recursively from the relations*

$$
\begin{aligned}
\det A(N_1) &= \alpha_1, \\
\det A(N_2) &= \alpha_2 \det A(N_1) - \gamma_1^2, \\
\det A(N_k) &= \alpha_k \det A(N_{k-1}) - \gamma_{k-1}^2 \det A(N_{k-2}), \\
&\qquad k = 3, \ldots, n.
\end{aligned}
$$

Proof. The first and the second relation may be verified directly. Expanding $\det A(N_k)$ by the last row of $A(N_k)$, one obtains the third relation. $\qquad \square$

Theorem 8.2.2 *Let ξ be a real number. Compute the numbers* $\delta_0, \delta_1, \ldots, \delta_n$ *recursively from*

$$
\begin{aligned}
\delta_0 &= 1, \\
\delta_1 &= \alpha_1 - \xi, \\
\delta_k &= (\alpha_k - \xi)\delta_{k-1} - \gamma_{k-1}^2 \delta_{k-2}, \ k = 2, \ldots, n.
\end{aligned}
$$

If $\delta_k \neq 0$, $k = 1, \ldots, n$, then the number of the eigenvalues of A from (8.6) which are greater than ξ equals the number of agreements in sign in the sequence $\delta_0, \delta_1, \ldots, \delta_n$. (An agreement in sign occurs if sgn δ_k = sgn δ_{k+1}.)

Proof. Clearly, the number of the eigenvalues of A which are greater than ξ equals the number of the positive eigenvalues of $A - \xi I$. But for a symmetric matrix P this number equals the number of coincidences in sign in the sequence $1, \det P(N_1)$, $\det P(N_2), \ldots, \det P(N_n)$ according to Theorem 2.6.2. $\square$

Theorem 8.2.2 can be used to compute the eigenvalues of a symmetric tridiagonal matrix by the *bisection method*. First, we find (for example, using the norms; cf. Chapter 9) an interval $[\alpha, \beta]$ containing all the eigenvalues of A. Then we put $\xi = \frac{1}{2}(\alpha + \beta)$ and determine how many eigenvalues of A are contained in the interval $(\xi, \beta]$ (and thus also the number of the eigenvalues in $[\alpha, \xi]$). If some of these intervals contains at least one eigenvalue, it is then again bisected, and so forth. The procedure is continued until approximations of the required eigenvalues are computed with prescribed accuracy.

Theorem 8.2.3 *Every irreducible tridiagonal matrix of order n can be transformed into a matrix of the form*

$$\begin{bmatrix} \alpha_1 & -1 \\ -\beta_1 & \alpha_2 & -1 \\ & -\beta_2 & \alpha_3 & -1 \\ & & \ddots & \ddots & \ddots \\ & & & -\beta_{n-2} & \alpha_{n-1} & -1 \\ & & & & -\beta_{n-1} & \alpha_n \end{bmatrix} \quad (8.7)$$

by a similarity transformation with a diagonal matrix. Every real symmetric irreducible tridiagonal matrix of order n can be transformed into a matrix of the form (8.6) with $\gamma_1 < 0, \gamma_2 < 0, \ldots, \gamma_{n-1} < 0$ by a similarity transformation, the transformation matrix being a diagonal matrix whose diagonal entries are ± 1.

Proof is left to the reader (Exercise 8.4).

Suppose now that we know an eigenvalue λ and its associated eigenvector of a given tridiagonal matrix of order n. We shall present without proof (see [22]) a *deflation formula*, i.e., a formula enabling us to construct a tridiagonal matrix of order $n - 1$ whose eigenvalues are exactly the eigenvalues of the original matrix except λ. Theorem 8.2.3 implies that Theorems 8.2.4 and 8.2.5 may be used also in more general cases.

Theorem 8.2.4 *Let A be a tridiagonal matrix of order n having the form (8.7) with $\beta_i \neq 0$, $i = 1, \ldots, n - 1$. Suppose that an eigenvector $y = [y_1, \ldots, y_n]^T$ of A satisfies*

$$\sum_{k=1}^{n} \beta_k \beta_{k+1} \cdots \beta_{n-1} y_k^2 \neq 0$$

(the coefficient of y_n^2 being 1), $y_j \neq 0, j = 1, \ldots, n$. Then y is associated with a simple eigenvalue λ of A and all the other eigenvalues of A are (multiplicities included) eigenvalues of the $(n - 1) \times (n - 1)$ tridiagonal matrix

$$\hat{A} = \begin{bmatrix} \hat{\alpha}_1 & -1 & & & & \\ -\hat{\beta}_1 & \hat{\alpha}_2 & -1 & & & \\ & \ddots & \ddots & \ddots & & \\ & & -\hat{\beta}_{n-3} & \hat{\alpha}_{n-2} & -1 & \\ & & & -\hat{\beta}_{n-2} & \hat{\alpha}_{n-1} \end{bmatrix},$$

where

$$\hat{\alpha}_i = \alpha_{i+1} + \frac{y_{i+1}}{y_i} - \frac{y_{i+2}}{y_{i+1}}, \quad i = 1, \ldots, n - 2,$$

$$\hat{\alpha}_{n-1} = \alpha_n + \frac{y_n}{y_{n-1}},$$

$$\hat{\beta}_i = \frac{\beta_i y_i y_{i+2}}{y_{i+1}^2}. \quad i = 1, \ldots, n - 2.$$

If $z = [z_1, \ldots, z_n]^T$ is an eigenvector of A associated with an eigenvalue different from λ, then $p = [p_1, \ldots, p_{n-1}]^T$,

$$p_i = y_{i+1} \left(\frac{z_i}{y_i} - \frac{z_{i+1}}{y_{i+1}} \right), \quad i = 1, \ldots, n - 1.$$

is an eigenvector of $\hat{A}$ associated with the same eigenvalue as z. Conversely, if $p = [p_1, \ldots, p_{n-1}]^T$ is an eigenvector of the matrix $\hat{A}$, then $z = [z_1, \ldots, z_n]^T$, where

$$z_j = y_j \sum_{k=1}^{j-1} \rho_k p_k + y_i \sum_{k=j}^{n-1} \sigma_k p_k,$$

$$\rho_k = -y_{k+1}^{-1} \sum_{j=1}^{k} \beta_j \beta_{j+1} \ldots \beta_{n-1} y_j^2,$$

$$\sigma_k = y_{k+1}^{-1} \sum_{j=k+1}^{n} \beta_j \beta_{j+1} \ldots \beta_{n-1} y_j^2,$$

is an eigenvector of A associated with the same eigenvalue as p.

Theorem 8.2.5 *Let A be a real symmetric tridiagonal matrix of order n having the form (8.6) with $\gamma_i < 0$, $i = 1, \ldots, n-1$. Let $y = [y_1, \ldots, y_n]^T$ be a positive eigenvector associated with the smallest eigenvalue λ (cf. Exercise 5.6). Then all the remaining eigenvalues of A coincide with the eigenvalues of the $(n-1) \times (n-1)$ real symmetric tridiagonal matrix*

$$\hat{A} = \begin{bmatrix} \hat{\alpha}_1 & \hat{\gamma}_1 & & & \\ \hat{\gamma}_1 & \hat{\alpha}_2 & \hat{\gamma}_2 & & \\ & \ddots & \ddots & \ddots & \\ & & \hat{\gamma}_{n-3} & \hat{\alpha}_{n-2} & \hat{\gamma}_{n-2} \\ & & & \hat{\gamma}_{n-2} & \hat{\alpha}_{n-1} \end{bmatrix},$$

where

$$\hat{\alpha}_i = \alpha_{i+1} - \frac{\gamma_i y_{i+1}}{y_i} + \frac{\gamma_{i+1} y_{i+2}}{y_{i+1}}, \quad i = 1, \ldots, n-2,$$

$$\hat{\alpha}_{n-1} = \alpha_n - \frac{\gamma_{n-1} y_n}{y_{n-1}},$$

$$\hat{\gamma}_i = -\sqrt{\frac{\gamma_i \gamma_{i+1} y_i y_{i+2}}{y_{i+1}^2}}, \quad i = 1, \ldots, n-2.$$

If $z = [z_1, \ldots, z_n]^T$ is an eigenvector of A associated with an eigenvalue different from λ, then $p = [p_1, \ldots, p_{n-1}]^T$,

$$p_i = \left(\frac{z_i}{y_i} - \frac{z_{i+1}}{y_{i+1}} \right) \sqrt{(-\gamma_i y_i y_{i+1})}, \quad i = 1, \ldots, n-1,$$

*is an eigenvector of $\hat{A}$ associated with the same eigenvalue as z.
Conversely, if $p = [p_1, \ldots, p_{n-1}]^T$ is an eigenvector of $\hat{A}$, then
$z = [z_1, \ldots, z_n]^T$, where*

$$z_j = -y_j \sum_{k=1}^{j-1} \frac{p_k}{\sqrt{(-\gamma_k y_k y_{k+1})}} \sum_{t=1}^{k} y_t^2$$

$$+ y_j \sum_{k=j}^{n-1} \frac{p_k}{\sqrt{(-\gamma_k y_k y_{k+1})}} \sum_{t=k+1}^{n} y_t^2,$$

is an eigenvector of A associated with the same eigenvalue as p.

Theorem 8.2.6 *If $A = [a_{ik}]$ is a real tridiagonal matrix of order
n satisfying $a_{k,k+1}a_{k+1,k} > 0$ for $k = 1, \ldots, n - 1$, then A has n
real simple eigenvalues.*

Proof. We use induction on n. The theorem is true for $n = 1$.
Let $n > 1$ and suppose the theorem holds for matrices or order
$n - 1$. By Theorem 8.2.3, the matrix A can be transformed into
a matrix $\tilde{A}$ of the form (8.7) by a similarity transformation, the
transformation matrix being diagonal. Moreover, it is easy to
show that $\beta_i = a_{i,i+1}a_{i+1,i}$, and thus $\beta_i > 0$, $i = 1, \ldots, n - 1$.
Since $\tilde{A} \in Z$ there exists a number k such that $kI + \tilde{A} \in K$.
Using the result of Exercise 5.6 we find that there exists a real
simple eigenvalue of the matrix $kI + \tilde{A}$ and a positive associated
eigenvector. This means that A has a real simple eigenvalue
ω associated with a positive eigenvector y. Using now Theorem
8.2.4 we find that the matrix $\tilde{A}$, which has no eigenvalue equal to
ω, satisfies again the hypothesis of the present theorem and is of
order $n-1$. By the induction assumption, $\tilde{A}$ has $n-1$ real simple
eigenvalues. However, by Theorem 8.2.4 these are eigenvalues of
A as well and taking also ω into account we conclude that A has
n real eigenvalues altogether (which are simple, of course). $\square$

Eigenvectors of real tridiagonal matrices possess some interesting sign properties described in the next theorem.

Theorem 8.2.7 *Let $A = [a_{ik}]$ be a real tridiagonal matrix of
order $n \geq 2$. Let $a_{i,i+1}a_{i+1,i} > 0$, $i = 1, \ldots, n - 1$. If $\lambda_1 >$*

$\lambda_2 > \cdots > \lambda_n$ are the real eigenvalues of A (which exist by Theorem 8.2.6), then every (real) eigenvector $z = [z_1, \ldots, z_n]^T$ of the matrix A has the following properties:

1. $z_1 \neq 0$, $z_n \neq 0$.
2. If $z_k = 0$, then $a_{k-1,1} a_{k,k+1} z_{k-1} z_{k+1} < 0$.
3. If we delete zeros from the sequence

$$z_1, \quad a_{12} z_2, \quad a_{12} a_{23} z_3, \quad \ldots, \quad a_{12} a_{23} \cdots a_{n-1,n} z_n$$

and if the vector z is associated with λ_r, then there are exactly $r - 1$ changes of sign in the sequence.

Proof. Let $Az = \lambda_r z$, $z \neq 0$. Supposing $z_i = 0$, then clearly $z_k = 0$ for all $k = 1, \ldots, n$, which is a contradiction. The proof of $z_n \neq 0$ is analogous. Hence 1. holds. If $z_k = 0$, then $1 \neq k \neq n$ and the kth equation of the system $Az = \lambda_r z$ then yields

$$a_{k,k-1} z_{k-1} + a_{k,k+1} z_{k+1} = 0,$$

which implies 2. since $z_{k-1} = 0$ or $z_{k+1} = 0$ would lead to $z = 0$.

To prove 3. suppose first that all the components of z are nonzero. It may be easily shown that there exists a real diagonal matrix D with positive diagonal entries such that $D(A - \lambda_r I) D^{-1} = B$ is symmetric and tridiagonal. Put $y = Dz$. Then y_i and z_i, $i = 1, \ldots, n$, have equal signs and the signs of $b_{k,k+1}$ and $a_{k,k+1}$, $k = 1, \ldots, n-1$, are equal as well. Since $By = 0$ we may easily verify that $B = YLY^T$, where

$$Y = \begin{bmatrix} y_1^{-1} & 0 & \cdots & 0 & 0 \\ -y_2^{-1} & y_2^{-1} & \cdots & 0 & 0 \\ 0 & -y_3^{-1} & \cdots & 0 & 0 \\ \vdots & \vdots & \ddots & \vdots & \vdots \\ 0 & 0 & \cdots & y_{n-1}^{-1} & 0 \\ 0 & 0 & \cdots & -y_n^{-1} & y_n^{-1} \end{bmatrix}$$

and $L = \mathrm{diag}(-b_{12} y_1 y_2, -b_{23} y_2 y_3, \ldots, -b_{n-1,n} y_{n-1} y_n, 0)$. Therefore, by Theorem 2.6.1 we have $\mathrm{In}\, B = \mathrm{In}\, L$, and hence $\mathrm{In}(A - \lambda_r I) = \mathrm{In}\, L$ as well. We thus conclude that the number of negative values among $b_{k,k+1} y_k y_{k+1}$ equals the number of the eigenvalues of A which are greater that λ_r, i.e., equals $r - 1$. But this is also the number of negative values among $a_{k,k+1}z_k z_{k+1}$, i.e., the number of changes of sign in the sequence given in 3.

By an easy continuity argument 3. holds even if one (or more than one) component of z is zero.   □

Tridiagonal symmetric matrices are a special case of the so-called *acyclic* symmetric matrices. These are symmetric matrices $A = [a_{ik}]$ satisfying

$$a_{k_1 k_2} \, a_{k_2 k_3} \, \cdots a_{k_s k_1} = 0$$

whenever $k_1, k_2, \ldots, k_s$ ($s \geq 3$) are distinct indices. It means that the undirected graph $G(A)$ of A contains no polygon or, in other words, that every component of $G(A)$ is a tree (representing a path if A is an irreducible tridiagonal matrix). An analogue of Theorem 8.2.7 may be shown to hold for acyclic matrices (see [20]).

Exercises

8.1. Use the RCM algorithm to label the rectangular lattice $M_{3,5}$ of Example 3.9 (Fig. 7). What is the bandwidth of the corresponding band matrix?

8.2. If A is a tridiagonal nonnegative matrix of order n prove that each its additive compound $\bar{A}^{[k]}$, $k = 1, \ldots, n$, introduced in Theorem 6.2.13 is nonnegative.

8.3. Show that the matrix $A^{[k]}$ of the preceding exercise is irreducible for every $k = 1, \ldots, n$ if and only if A is irreducible. What can you state about the eigenvectors of A using the Perron-Frobenius theorem?

8.4. Prove Theorem 8.2.3.

8.5. Let A be the matrix from (8.7) and denote its characteristic polynomial by $f(\lambda)$ so that $f(\lambda) = \det(A - \lambda I)$. Prove that the following is true:

1. If $Au = \hat{\lambda}u$, $u = [u_1, \ldots, u_n]^T \neq 0$, then $v = [v_1, \ldots, v_n]^T$, where $v_k = \beta_k \beta_{k+1} \ldots \beta_{n-1} u_k$ for $k = 1, \ldots, n-1$ and $v_n =$

u_n, is an eigenvector of A^T associated with the eigenvalue $\hat{\lambda}$.

2. The value of the derivative $f'(\lambda)$ at $\hat{\lambda}$ is

$$f'(\hat{\lambda}) = -\frac{\sum_{k=1}^n \beta_k \, \beta_{k+1} \cdots \beta_{n-1} u_k^2}{u_1 u_n}.$$

(Hint: Use the results of Exercises 1.16 and 1.17.)

9

Norms and Their Use for Estimation of Eigenvalues

In this chapter, we deal with the problem of "measuring the distance" both in the linear space of column vectors and in the linear space of matrices. Such a measurement is very important for error estimates, for planning the computation, for localization of eigenvalues of matrices, etc. The results on norms will be further used for the localization of matrix eigenvalues and, in Chapter 11, for the study of the convergence of iterative methods.

9.1 Norms

The concept of norm is well-known. We only recall that the *norm* in the linear space X_n of n-component column vectors whose components are, in general, complex is a function g that assigns a real number $g(x)$ to every vector x and satisfies the following conditions:

N1. $g(x) \geq 0$ for all $x \in X_n$.

N2. $g(\lambda x) = |\lambda| g(x)$ for all $x \in X_n$ and all complex λ.

N3. $g(x + y) \leq g(x) + g(y)$ for all $x \in X_n$ and $y \in X_n$.

N4. $g(x) = 0$ if and only if $x = 0$.

The norm in the linear space of real column vectors is defined in the same way; in such a case N2 is, of course, required to hold only for all real λ. The property N2 implies that the norm of the zero vector is zero and the properties N4 and N1 show that the norm of a vector different from the zero vector is positive.

Example 9.1 The function g_1 which assigns to any (complex)

235

vector $x = [x_1, \ldots, x_n]^T$ the number

$$g_1(x) = \sum_{i=1}^{n} |x_i|$$

is a norm called the *sum norm*.

Similarly, the function

$$g_2(x) = \left(\sum_{i=1}^{n} |x_i|^2 \right)^{\frac{1}{2}}$$

is a norm; it is the well-known *Euclidean norm*.

Also the function

$$g_3(x) = \max_i |x_i|$$

is a norm, it is called the *uniform* or *maximum norm*.

The norms g_1 and g_2 are particular cases of the so-called l_p*norms* defined for $p \geq 1$ by the formula

$$g(x) = \left(\sum_{i=1}^{n} |x_i|^p \right)^{\frac{1}{p}}. \tag{9.1}$$

Clearly, $g_1(x)$ is obtained for $p = 1$ and $g_2(x)$ for $p = 2$. The norm g_3 is a limit case for $p \to \infty$.

The norm of a vector can be viewed as its length in some "metric". To explain the situation in more detail we shall describe the case of $n = 2$ and a real vector space X_2. Then x_1 and x_2 can be considered as the rectangular Cartesian coordinates of the vector $\overrightarrow{OX}$ (see Fig. 14).

If a norm g is introduced in X_2 then the end points X of *unit vectors*, i.e. the vectors x for which $g(x) = 1$, form a plane curve C that is symmetric with respect to the origin by the property N2 and represents a boundary of a convex domain containing the origin as its interior point by N3. The curve C corresponding to the norm g_1 of Example 9.1 is the boundary of a square whose diagonals coincide with the axes x_1 and x_2. The curve C corresponding to g_2 is a circle with center at the origin and the curve corresponding to g_3 is the boundary of a square with center at the origin and the sides parallel to the axes x_1 and x_2.

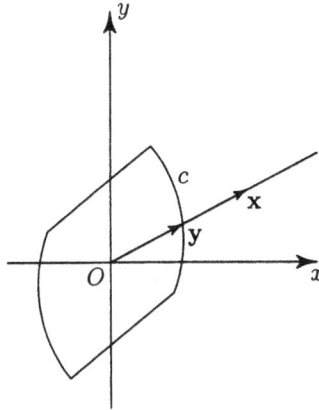

Fig. 14

Conversely, if a curve C which is symmetric about the origin and bounds a convex domain is given in the plane, then it defines a norm. Really, let $x \neq 0$ be the vector $\overrightarrow{OX}$. Denote by Y the intersection of the half-line OX with C. The norm of x is then the (positive) number by which we have to multiply the vector $\overrightarrow{OY}$ to obtain $\overrightarrow{OX}$:

$$\overrightarrow{OX} = g(x)\overrightarrow{OY}.$$

A similar situation occurs also in the three-dimensional case and, in general, in the n-dimensional case as well. For $n = 3$ we obtain, of course, a surface C instead of a curve. In the case of the norm g_1, C is the boundary of an octahedron, in the case of g_2, C is the spherical surface, and in the case of g_3 it is the boundary of a cube.

Analogously as we studied norms in the linear space of n-component column vectors we can introduce a norm also in the space of $m \times n$ matrices with certain fixed m and n.

It is, however, reasonable to define the norm in such a way that it might be related to the norms in the spaces of both n-component and m-component vectors since the result of the multiplication of A by and n-component vector is an m-component vector.

We shall therefore assume that a norm g_X is given in the linear space X_n of n-component vectors and a norm g_Y is given in the linear space Y_m of m-component vectors. We then introduce a function g_{YX} on the space Z of $m \times n$ matrices. (They may be complex or real; the spaces X_n and Y_m are, however, always considered either both complex or both real. The elements of Z can be viewed as mn-component column vectors formed from the columns of the corresponding matrix.) For $A \in Z$ we put

$$g_{YX}(A) = \sup\{g_Y(Ax); x \in X_n, \ g_X(x) = 1\},$$

or, which makes no difference in our case,

$$g_{YX}(A) = \max_{x \neq 0, x \in X_n} \frac{g_Y(Ax)}{g_X(x)}.$$

The function g_{YX} is a norm in the space Z, i.e., it fulfils N1 to N4 (Exercise 9.3). We shall call g_{YX} the *norm induced by the norms* g_X *and* g_Y. Clearly, it can also be characterized as the number $g_{YX}(A)$ satisfying the two conditions:

1. For any vector $x \in X_n$,

$$g_Y(Ax) \leq g_{YX}(A)g_X(x). \tag{9.2}$$

2. There exists a vector $x_0 \in X_n$, $x_0 \neq 0$, such that

$$g_Y(Ax_0) = g_{YX}(A)g_X(x_0). \tag{9.3}$$

The norms of matrices cannot usually be calculated in a closed form (using an explicit formula). An exception is the case when the same norm g_1 (or g_2, or g_3) is prescribed in both X and Y.

Theorem 9.1.1 *Let* $A = [a_{ik}]$ *be an* $m \times n$ *(in general complex) matrix. For the norm of* A *we then have*

$$g_1(A) = \max_k \sum_i |a_{ik}|$$

if g_1 *is defined in both* X *and* Y,

$$g_3(A) = \max_i \sum_k |a_{ik}|$$

if g_3 *is defined both* X *and* Y, *and*

$$g_2(A) = (\rho(A^*A))^{\frac{1}{2}}$$

if g_2 *is defined in both* X *and* Y.

Remark. We can, of course, write $\lambda_{\max}(A^*A)$ (i.e., the largest eigenvalue of A^*A) instead of $\rho(A^*A)$. Since A^*A and AA^* have the same nonzero eigenvalues (Exercise 2.16), it is also possible to write $\rho(AA^*)$ or $\lambda_{\max}(AA^*)$ instead of $\rho(A^*A)$.

Proof. We shall prove that the number

$$g_1(A) = \max_k \sum_i |a_{ik}|$$

possesses the properties 1. and 2. (i.e., that it fulfils the relations (9.2) and (9.3)). Let $x \in X_n$, $x = [x_1, \ldots, x_n]^T$. Then

$$
\begin{aligned}
g_1(Ax) &= \sum_{i=1}^{m} \left| \sum_{k=1}^{n} a_{ik} x_k \right| \\
&\leq \sum_{i=1}^{m} \sum_{k=1}^{n} |a_{ik}| |x_k| \\
&= \sum_{k=1}^{n} \left(\sum_{i=1}^{m} |a_{ik}| \right) |x_k| \\
&\leq \left(\max_{k=1,\ldots,n} \sum_{i=1}^{m} |a_{ik}| \right) \sum_{k=1}^{n} |x_k| \\
&= \left(\max_{k=1,\ldots,n} \sum_{i=1}^{m} |a_{ik}| \right) g_1(x).
\end{aligned}
$$

If further

$$\max_k \sum_i |a_{ik}| = \sum_i |a_{ip}|$$

we put $x_0 = [0, \ldots, 0, 1, 0, \ldots, 0]^T$ where 1 is the pth component. Clearly $x_0 \neq 0$ and, moreover,

$$g_1(Ax_0) = \sum_i |a_{ip}| = \left(\max_k \sum_i |a_{ik}| \right) g_1(x_0),$$

which proves the first formula.

We shall prove the second formula in a similar way. For any

$x \in X_n$ we have

$$g_3(Ax) = \max_i \left| \sum_k a_{ik} x_k \right| \le \max_i \sum_k |a_{ik}||x_k|$$

$$\le \left(\max_i \sum_k |a_{ik}| \right) \max_j |x_j| = \left(\max_i \sum_k |a_{ik}| \right) g_3(x).$$

Let

$$\max_i \sum_k |a_{ik}| = \sum_k |a_{pk}|.$$

Put $x_0 = [x_{01}, \dots, x_{0n}]^T$, where $x_{0k} = |a_{pk}|/a_{pk}$ if $a_{pk} \ne 0$ and $x_{0k} = 1$ if $a_{pk} = 0$. Clearly $x_0 \ne 0$ and, moreover,

$$g_3(Ax_0) = \max_i \left| \sum_k |a_{ik} x_{0k}| \right|$$

$$\ge \left| \sum_k a_{pk} x_{0k} \right|$$

$$= \sum_k |a_{pk}|$$

$$= \left(\max_i \sum_k |a_{ik}| \right) g_3(x_0)$$

since $|x_{0k}| = 1$ for all k. As by the above established formula

$$g_3(Ax_0) \le \left(\max_i \sum_k |a_{ik}| \right) g_3(x_0),$$

equality takes place here and the second formula has been proved.

   In order to prove the third formula we choose an arbitrary vector $x \in X_n$. By Theorems 2.1.3 and 2.7.3 then

$$g_2(Ax) = \langle Ax, Ax \rangle^{\frac{1}{2}}$$

$$= \langle A^*Ax, x \rangle^{\frac{1}{2}}$$

$$\le (\lambda(A^*A))^{\frac{1}{2}} \langle x, x \rangle^{\frac{1}{2}}$$

$$= (\lambda(A^*A))^{\frac{1}{2}} g_2(x),$$

where $\lambda(A^*A)$ is the maximal eigenvalue of the symmetric (or Hermitian) matrix (A^*A).

Finally, if x_0 is an eigenvector of (A^*A) associated with the eigenvalue $\lambda(A^*A)$, then

$$
\begin{aligned}
g_2(Ax_0) &= \langle Ax_0, Ax_0 \rangle^{\frac{1}{2}} \\
&= \langle A^*Ax_0 . x_0 \rangle^{\frac{1}{2}} \\
&= \langle \lambda(A^*A)x_0, x_0 \rangle^{\frac{1}{2}} \\
&= (\lambda(A^*A))^{\frac{1}{2}} g_2(x_0),
\end{aligned}
$$

which completes the proof of the theorem.   □

Any induced norm of a square matrix can be compared with the spectral radius. This is the subject of the next theorem.

Theorem 9.1.2 *If A is a square matrix and g a norm, then*

$$\rho(A) \leq g(A).$$

If A is symmetric, then $\rho(A) = g_2(A)$, where g_2 is the norm induced by the Euclidean norm.

Proof. Let λ be an eigenvalue of A such that $|\lambda| = \rho(A)$ and let $Au = \lambda u$, $u \neq 0$. Then

$$g(Au) \leq g(A)g(u)$$

and, at the same time,

$$g(Au) = g(\lambda u) = |\lambda| g(u) = \rho(A)g(u).$$

Comparing these two relations, we obtain the inequality of the theorem. If A is symmetric, then $g_2(A) = (\rho(A^*A))^{1/2} = (\rho(A^2))^{1/2} = \rho(A)$ by Theorem 1.7.12.   □

Remark. The norm $g_2(A)$ is often called the *spectral norm* of the matrix A.

Besides the norms $g_1(A)$, $g_2(A)$, and $g_3(A)$, the so-called *Schur* (or *Frobenius*) *norm* $N(A)$ is frequently used. It is defined as

$$N(A) = \left(\sum_{i=1}^{m} \sum_{k=1}^{n} |a_{ik}|^2 \right)^{\frac{1}{2}} \tag{9.4}$$

or

$$N(A) = (\operatorname{tr} AA^*)^{\frac{1}{2}} \tag{9.5}$$

(for an $m \times n$ matrix A). The Schur norm is not (for $n > 1$ or $m > 1$) a matrix norm in the sense of our definition, i.e., it is not a norm induced by norms in X and Y. For the Euclidean norms of vectors in X and Y and the Schur norm, however, we have the inequality

$$g_2(Ax) \le N(A)g_2(x). \tag{9.6}$$

Indeed, by Cauchy-Schwarz inequality,

$$
\begin{aligned}
(g_2(Ax))^2 &= \sum_i \left| \sum_k a_{ik} x_k \right|^2 \\
&\le \sum_i \left(\left(\sum_k |a_{ik}|^2 \right) \sum_j |x_j|^2 \right) \\
&= N^2(A)(g_2(x))^2.
\end{aligned}
$$

From this inequality the next theorem readily follows in virtue of (9.3).

Theorem 9.1.3 *If A is a square matrix, then*

$$g_2(A) \le N(A). \tag{9.7}$$

If now X, Y, and Z are spaces of column vectors (either all complex or all real) of dimensions m, n, and p, respectively, and if A is an $m \times n$ matrix and B is an $n \times p$ matrix, then we can multiply A by B to obtain the $m \times p$ matrix $C = AB$. If g_X, g_Y, and g_Z are norms in X, Y, and Z, respectively, we may ask how the norms $g_{XY}(A)$, $g_{YZ}(B)$, and $g_{XZ}(C)$ are related to each other. For the sake of simplicity we denote by $L(Y, X)$ the linear space of matrices that (by multiplication) transform column vectors in Y into column vectors in X, etc. Thus $A \in L(Y, X)$, $B \in L(Z, Y)$, and also $AB \in L(Z, X)$.

Theorem 9.1.4 *Let X, Y, and Z be spaces of column vectors with norms g_X, g_Y, and g_Z, respectively. If $A \in L(Y, X)$ and $B \in L(Z, Y)$, then*

$$g_{XZ}(AB) \le g_{XY}(A)g_{YZ}(B).$$

Proof. By (9.2), we have for $g_{XY}(A)$ and $g_{YZ}(B)$ that

$$g_X(Ay) \leq g_{XY}(A)g_Y(y) \quad \text{for all } y \in Y,$$
$$g_Y(Bz) \leq g_{YZ}(B)g_Z(z) \quad \text{for all } z \in Z.$$

It means that

$$g_X(ABz) \leq g_{XY}(A)g_Y(Bx) \leq g_{XY}(A)g_{YZ}(B)g_Z(z)$$

for all $z \in Z$. By (9.3), however, there exists a vector z_0, $z_0 \neq 0$, such that

$$g_X(Cz_0) = g_{XZ}(C)g_Z(z_0).$$

Therefore

$$g_X(ABz_0) = g_{XZ}(AB)g_Z(z_0)$$
$$\leq g_{XY}(A)g_{YZ}(B)g_Z(z_0),$$

from which the inequality $g_{XZ}(AB) \leq g_{XY}(A)g_{YZ}(B)$ follows.

$\square$

The Schur norm $N(A)$ possesses an analogous property. However, even a stronger theorem holds.

Theorem 9.1.5 *If A and B are matrices such that the product AB exists, then*

$$N(AB) \leq N(A)g_2(B) \tag{9.8}$$

and

$$N(AB) \leq g_2(A)N(B). \tag{9.9}$$

Proof. We shall first prove the following auxiliary statement.

Theorem 9.1.6 *If P and Q are positive semidefinite matrices of the same order, then*

$$\operatorname{tr} PQ \leq \operatorname{tr} P\rho(Q) \tag{9.10}$$

and

$$\operatorname{tr} PQ \leq \rho(P)\operatorname{tr}(Q). \tag{9.11}$$

Proof. The inequalities (9.10) and (9.11) can be transformed into analogous inequalities for matrices P_1 and Q_1 if we put $P = UP_1U^*$ and $Q = UQ_1U^*$, where U is an arbitrary unitary matrix. (Indeed, $\operatorname{tr} PQ = \operatorname{tr} UP_1U^*UQ_1U^*) = \operatorname{tr}(UP_1Q_1U^*) = \operatorname{tr} P_1Q_1$ by Theorem 1.7.2; similarly $\operatorname{tr} P = \operatorname{tr} P_1$, $\rho(Q) = \rho(Q_1)$.) We can thus suppose that the matrix P (of order n) has the diagonal form $P = \operatorname{diag}(\lambda_1, \ldots, \lambda_n)$, $\lambda_i \geq 0$, $i = 1, \ldots, n$, by Theorem 2.5.2. The left-hand side of (9.10) is then $\sum_{i=1}^{n} \lambda_i q_{ii}$. The diagonal entries of Q are

$$q_{ii} = \sum_{j=1}^{n} |v_{ij}|^2 \mu_j$$

if $Q = VMV^*$, $V = [v_{ij}]$ unitary, and $M = \operatorname{diag}(\mu_1, \ldots, \mu_n)$. The matrix $[|v_{ij}|^2]$ is doubly stochastic by Exercise 4.5. Therefore

$$
\begin{aligned}
\operatorname{tr} PQ &= \sum_{i=1}^{n} \lambda_i q_{ii} = \sum_{i,j=1}^{n} |v_{ij}|^2 \lambda_i \mu_j \\
&\leq (\max_k \mu_k) \sum_{i=1}^{n} \left(\sum_{j=1}^{n} |v_{ij}|^2 \right) \lambda_i = \max_k \mu_k \sum_{i=1}^{n} \lambda_i \\
&= \operatorname{tr} P\rho(Q)
\end{aligned}
$$

as asserted. The inequality (9.11) follows if we interchange P and Q since $\operatorname{tr} PQ = \operatorname{tr} QP \leq \operatorname{tr} Q\rho(P)$.   □

Let us return to the proof of Theorem 9.1.5. The square of the left-hand side of the inequality (9.8) is

$$\operatorname{tr}(ABB^*A^*) = \operatorname{tr}((A^*A)(BB^*))$$

by the result of Exercise 1.2. By Theorem 9.1.6, if we substitute the positive definite matrices A^*A and BB^* for P and Q, then

$$(N(AB))^2 \leq \operatorname{tr}(A^*A)\rho(BB^*) = (N(A))^2(g_2(B))^2,$$

i.e., (9.8) really holds.
Similarly

$$(N(AB))^2 \leq \rho(A^*A)\operatorname{tr}(BB^*) = (g_2(A))^2(N(B))^2,$$

i.e., (9.9) also holds.   □

By (9.7), we have the following statement.

Corollary 6 *If the product of the matrices A and B exists, then*

$$N(AB) \leq N(A)N(B), \tag{9.12}$$

We now turn to the special case when both the spaces are identical and so are both the norms g_X and g_Y. In this case we do not use subscripts and write simply g. By the preceding theorems, the induced norm (denoted again by g) is a norm in the linear space of square matrices M_n fulfilling, in addition to the assertions

N'1. $g(A) \geq 0$ for all matrices $A \in M_n$;

N'2. $g(\lambda A) = |\lambda| g(A)$ for all matrices $A \in M_n$ and all numbers λ (complex or real);

N'3. $g(A + B) \leq g(A) + g(B)$ for all matrices $A \in M_n$ and $B \in M_n$;

N'4. $g(A) = 0$ if and only if $A = 0$;

also the assertions

N'5. $g(AB) \leq g(A)g(B)$ for all matrices $A \in M_n$ and $B \in M_n$

and

N'6. $g(I) = 1$ (I is the identity matrix).

The property N'6, however, follows readily from the definition of norm.

The fact that for square matrices of order $n > 1$ the Schur norm $N(A)$ is not induced by any norm follows for example from the equality

$$N(I) = \sqrt{n}, \tag{9.13}$$

where I is the identity matrix.

9.2 Measure of nonsingularity. Dual norms

We now introduce the concept of the measure of nonsingularity of a square matrix with respect to a given norm. If A is a square matrix of order n and g is a norm in the linear space of n-component column vectors, then the *measure of nonsingularity* of A is is

$$\hat{g}(A) = \inf\{g(Ax); x \in X_n. g(x) \geq 1\}.$$

This can also be written as

$$\hat{g}(A) = \min_{x \neq 0, x \in X_n} \frac{g(Ax)}{g(x)}$$

or in this way: $\hat{g}(A)$ is the number fulfilling both the following conditions:

1. For every $x \in X_n$,

$$g(Ax) \geq \hat{g}(A)g(x). \tag{9.14}$$

2. There exists a vector $x_o \in X_n$, $x_0 \neq 0$, such that

$$g(Ax_0) = \hat{g}(A)g(x_0). \tag{9.15}$$

The relation

$$\hat{g}(A) \geq 0 \tag{9.16}$$

follows from the definition.

The next important theorem characterizes the measure of nonsingularity.

Theorem 9.2.1 *Let A be a square matrix of order n and let g be a norm in the linear space of n-component column vectors. Then*

$$\hat{g}(A) = 0 \qquad \text{if } A \text{ is singular,}$$
$$\hat{g}(A) = (g(A^{-1}))^{-1} > 0 \quad \text{if } A \text{ nonsingular.}$$

Proof. Let A be singular. Then there exists (Theorem 1.5.3) a vector $y \neq 0$ such that $Ay = 0$. By (9.14) thus

$$0 = g(Ay) \geq \hat{g}(A)g(y),$$

from which

$$\hat{g}(A) = 0$$

follows by (9.16) and the inequality $g(y) > 0$. Let now A be non-singular. We shall show that $(g(A^{-1}))^{-1}$ fulfils both the properties 1. and 2.

If $x \in X_n$, then for $x = 0$ the inequality

$$g(Ax) \geq (g(A^{-1}))^{-1}g(x) \tag{9.17}$$

holds. If $x \neq 0$ and $y = Ax$, then

$$g(A^{-1}y) \leq g(A^{-1})g(y)$$

by (9.2), from which

$$g(y) \geq (g(A^{-1}))^{-1}g(A^{-1}y),$$

and thus (9.17) again holds. Further, by (9.3) there exists a vector $y_0 \neq 0$ such that $g(A^{-1}y_0) = g(A^{-1})g(y_0)$. Putting further $A^{-1}y_0 = x_0$, we then have

$$g(Ax_0) = (g(A^{-1}))^{-1}g(x_0)$$

and (9.15) thus holds. Therefore

$$\hat{g}(A) = (g(A^{-1}))^{-1}.$$

$\square$

Remark. A matrix A is thus nonsingular if and only if its measure of nonsingularity with respect to some norm is positive. The greater the measure of nonsingularity, the "more nonsingular" is the matrix. Nevertheless, from the numerical point of view the measure of nonsingularity is not very advantageous since to calculate it, for example in the norms g_1 and g_3, it is necessary to find the inverse of A and its norm. In some cases, however, the measure of nonsingularity can be easily computed.

Theorem 9.2.2 *If D is a diagonal matrix, $D = \mathrm{diag}(d_1, \ldots, d_n)$, then $\hat{g}_1(D) = \hat{g}_2(D) = \hat{g}_3(D) = \min_i |d_i|$. Moreover,*

$$\hat{g}(cI) = |c| \tag{9.18}$$

holds for the matrix cI.

Theorem 9.2.3 *If A is a symmetric matrix, then $\hat{g}_2(A)$ equals the minimum of the moduli of the eigenvalues of A:*

$$\hat{g}_2(A) = \min\{|\lambda|; \lambda \in S(A)\}.$$

The following is an analogue of N′3.

Theorem 9.2.4 *If P and Q are square matrices of the same order and g is a norm, then*

$$\hat{g}(P - Q) \geq \hat{g}(P) - g(Q).$$

Proof. By (9.15), there exists a vector $x_0 \neq 0$ such that

$$g((P - Q)x_0) = \hat{g}(P - Q)g(x_0).$$

Then

$$
\begin{aligned}
\hat{g}(P)g(x_0) &\leq g(Px_0) \\
&= g((P - Q)x_0 + Qx_0) \\
&\leq g((P - Q)x_0) + g(Qx_0) \\
&= \hat{g}(P - Q)g(x_0) + g(Qx_0) \\
&\leq (\hat{g}(P - Q) + g(Q))g(x_0).
\end{aligned}
$$

$\square$

The next theorem is a consequence of Theorem 9.2.4.

Theorem 9.2.5 *Let A be a square matrix and suppose that there exist square matrices P and Q such that $A = P - Q$ and*

$$\hat{g}(P) > g(Q)$$

for a norm g. Then A is nonsingular.

Proof. By the preceding theorem,

$$\hat{g}(A) = \hat{g}(P - Q) \geq \hat{g}(P) - g(Q) > 0.$$

Thus A is nonsingular by Theorem 9.2.1. $\square$

The connection between the measure of nonsingularity and the eigenvalues of a matrix is shown in the next theorem.

Theorem 9.2.6 *If λ is any eigenvalue of a square matrix A and g is a norm, then*

$$|\lambda| \geq \hat{g}(A).$$

Proof. Let $u \neq 0$ be an eigenvector associated with the eigenvalue λ, i.e. $Au = \lambda u$. Then

$$g(Au) \geq \hat{g}(A)g(u);$$

but

$$g(Au) = g(\lambda u) = |\lambda|g(u).$$

Comparing these two relations we obtain the statement of the theorem. □

The concept of duality is very important in linear algebra (and, more generally, in functional analysis). The duality is, in general, realized by a bilinear form, for example $\langle x, x' \rangle$,[1] where x is an element of the linear space X and x' is an element of the (dual) space X'. The form $\langle x, x' \rangle$ is said to be *bilinear* if the numbers $\langle x, x' \rangle$ etc. fulfil the following relations (for all $x_1, x_2, x \in X$, $x_1', x_2', x' \in X'$, and all complex or real numbers λ and μ; these numbers are chosen either complex in both the spaces or real in both of them)

$$\langle x_1 + x_2, x' \rangle = \langle x_1, x' \rangle + \langle x_2, x' \rangle,$$
$$\langle \lambda x_1, x' \rangle = \lambda \langle x_1, x' \rangle,$$
$$\langle x, x_1' + x_2' \rangle = \langle x, x_1' \rangle + \langle x, x_2' \rangle,$$
$$\langle x, \mu x' \rangle = \mu \langle x, x' \rangle.$$

The spaces X and X' are said to be *dual* with respect to the bilinear form $\langle x, x' \rangle$ if both the following statements hold:
1. For any vector $x \in X$, $x \neq 0$, there exists a vector $x' \in X'$ such that $\langle x, x' \rangle \neq 0$.
2. For any vector $x' \in X'$, $x' \neq 0$, there exists a vector $x \in X$ such that $\langle x, x' \rangle \neq 0$.

It may be shown that the dual spaces have the same dimension and that there exist the so-called *dual bases*. In the case of finite-dimensional (say, n-dimensional) spaces, these are pairs of bases $e_1, \ldots, e_n$ in X and $e_1', \ldots, e_n'$ in X' such that $\langle e_i, e_j' \rangle = \delta_{ij}$ (where $\delta_{ij} = 1$ for $i = j$, $\delta_{ij} = 0$ for $i \neq j$).

In the case of the linear space X of (real or complex) column vectors, a dual space is e.g. the space of row vectors of the same

---

[1] We use the same symbol as for the inner product. However, it will not lead to misunderstanding.

dimension if we consider the bilinear form $\langle x, x' \rangle = x'x$; on the
right-hand side, there is the usual (matrix) product of the row x'
and the column x. To keep our study within the scope of column
vectors, we shall assume in what follows that the dual space X' is
again the space of column vectors and the corresponding bilinear
form is then $\langle x, x' \rangle = x'^T x$ for $x \in X$ and $x' \in X'$.

If now a norm g in the space X and a norm g' in the dual
space X' are given, then these norms are said to be *dual* if the
following statements hold:

D1. For all vectors $x \in X$ and $x' \in X'$,

$$|\langle x, x' \rangle| \le g(x)g'(x').$$

D2. For any vector $x \in X$, there exists a vector $x' \in X'$, $x' \ne 0$,
such that
$$|\langle x, x' \rangle| = g(x)g'(x').$$

D3. For any vector $x' \in X'$, there exists a vector $x \in X$, $x \ne 0$,
such that
$$|\langle x, x' \rangle| = g(x)g'(x').$$

Example 9.2 The norms g_1 in the space X_n of n-component
column vectors and g'_3 in the space X'_n of n-component column
vectors are dual with respect to the bilinear form $\langle x, x' \rangle = x'^T x$,
$x \in X_n$, $x' \in X'_n$. Here g_1 denotes the norm from Example 9.1
and g'_3 denotes the norm g_3 in X'_n from the same example.

Indeed,

$$\left| \sum_i x_i x'_i \right| \le \sum_i |x_i||x'_i| \le \left(\sum_i |x_i| \right) \max_i |x'_i| = g_1(x)g'_3(x')$$

for $x = [x_1, \ldots, x_n]^T$ and $x' = [x'_1, \ldots, x'_n]^T$. Let $x \in X$. If
$x = 0$, then it is certainly possible to fulfil D2. If $x \ne 0$, $x =
[x_1, \ldots, x_n]^T$, we put $x' = [x'_1, \ldots, x'_n]^T$, where $x'_i = \bar{x}_i/|x_i|$ ($\bar{x}_i$ is
the complex conjugate of x_i) for $x_i \ne 0$ and $x'_i = 0$ for $x_i = 0$.

Then $x'^T x = \sum_{i=1}^n |x_i|$, $g'_3(x') = 1$, $g_1(x) = \sum_{i=1}^n |x_i|$, and
D2 really holds.

If $x' \in X'$, then D3 can be fulfilled for $x' = 0$. Let $x' \ne 0$
and let k be the subscript (or one of the subscripts) such that
$|x'_k| = \max_i |x'_i|$. We then put $x = e_k$, where e_k is the vector

having 1 as its kth component, all other components being 0. Hence $|x'^T x| = |x'_k|$, $g_1(x) = 1$, $g'_3(x') = |x'_k|$, and D3 is thus also fulfilled.

In an analogous way we can show that the norms g_3 (in X_n) and g'_1 (in X'_n) are dual and that the norms g_2 (in X_n) and g'_2 (in X'_n) are dual as well (cf. Exercise 9.2).

For a given norm g in X_n, there exists always a *dual norm* g' in X'_n (if the duality is given by a fixed bilinear form $\langle x, x' \rangle$). Indeed, for a fixed x' it is sufficient to choose

$$g'(x') = \sup_{x \neq 0} \frac{|\langle x, x' \rangle|}{g(x)}$$

or, which is the same, the number $g'(x')$ for which the following two statements hold:

D'1. $|\langle x, x' \rangle| \leq g(x) g'(x')$ for all $x \in X$.

D'3. There exists a vector $x_0 \in X$, $x_0 \neq 0$, such that $|\langle x_0, x' \rangle| = g(x_0) g'(x')$.

It can be shown that the function $g'(x')$ defined in this way is a norm which together with $g(x)$ satisfies the properties D1 to D3.

Theorem 9.1.1 implies that $g_3(A)$ equals $g_1(A^T)$ and also $g_1(A)$ equals $g_3(A^T)$. This is not by chance. The general statement follows.

Theorem 9.2.7 *If g is a norm in the space of n-component column vectors and g' is the dual norm in the space of n-component column vectors with respect to the bilinear from $\langle x, x' \rangle = x'^T x$, then*

$$g(A) = g'(A^T).$$

Proof. If $A = 0$, then $A^T = 0$ as well and the theorem holds. Let thus $A \neq 0$. We know by (9.2) that for all vectors $x \in X_n$, the number $g(A)$ satisfies the inequality

$$g(Ax) \leq g(A) g(x)$$

and, by (9.3), there exists $x_0 \neq 0$ for which the equality takes place:

$$g(Ax_0) = g(A) g(x_0). \qquad (9.19)$$

If x' is an arbitrary vector, then

$$
\begin{aligned}
g'(A^T x') &= \max_{x \neq 0} \frac{|x'^T A x|}{g(x)} \\
&\leq \max_{x \neq 0} \frac{g'(x')g(Ax)}{g(x)} \\
&= g(A)g'(x').
\end{aligned}
$$

Since $g(A) > 0$, (9.19) implies $g(Ax_0) > 0$ and $Ax_0 \neq 0$. By D2, there exists a vector $x'_0 \neq 0$ such that

$$
|x_0'^T Ax_0| = g'(x'_0)g(Ax_0).
$$

Altogether,

$$
\begin{aligned}
g'(A^T x'_0)g(x_0) &\geq |x_0'^T Ax_0| \\
&= g'(x'_0)g(Ax_0) \\
&= g'(x'_0)g(A)g(x_0),
\end{aligned}
$$

whence

$$
g'(A^T x'_0) \geq g(A)g'(x'_0).
$$

Thus

$$
g'(A^T x'_0) = g(A)g'(x'_0).
$$

The number $g(A)$ fulfils both the conditions (9.2) and (9.3), and therefore $g'(A^T) = g(A)$.   □

9.3   Bounds for eigenvalues

We now use the theory of norms to obtain localization theorems for eigenvalues. First, however, we present several bounds for general matrices.

Theorem 9.3.1 (Gershgorin [29]) *Let $A = [a_{ik}]$ be a square matrix of order n (complex or real). Then all the eigenvalues of A are contained in the union $\bigcup_{i=1}^{n} K_i$ of the disks K_i in the complex plane with centers a_{ii} and radii $\sum_{j \neq i} |a_{ij}|$,*

$$
K_i = \left\{ z; |a_{ii} - z| \leq \sum_{j \neq i} |a_{ij}| \right\}.
$$

Moreover, the number of the eigenvalues of A in each component[2] of this union is equal to the number of disks that constitute this component. In particular, an isolated disk contains one and only one eigenvalue. The disk K_s is isolated if and only if

$$|a_{ss} - a_{tt}| > \sum_{i \neq s} |a_{si}| + \sum_{j \neq t} |a_{tj}|$$

holds for all indices $t \neq s$.

Example 9.3 For the complex matrix

$$A = \begin{bmatrix} 2 & -\frac{1}{2} & \frac{1}{4} & -\frac{1}{4} \\ \frac{1}{4} & 1+2i & 0 & \frac{1}{4} \\ -\frac{1}{2} & \frac{1}{4} & -1 & \frac{1}{2} \\ \frac{1}{4} & -\frac{1}{2} & \frac{1}{2} & -2-2i \end{bmatrix}$$

of order 4 we have (see Fig. 15)

$$K_1 = \{z; |z-2| \leq 1\}, \qquad K_3 = \{z; |z+1| \leq \tfrac{5}{4}\},$$
$$K_2 = \{z; |z-1-2i| \leq \tfrac{1}{2}\}, \quad K_4 = \{z; |z+2+2i| \leq \tfrac{5}{4}\}.$$

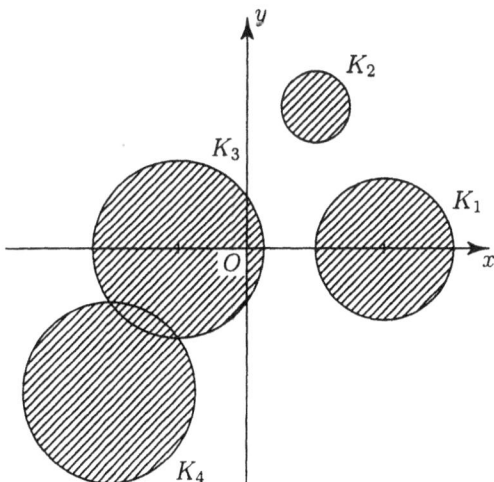

Fig. 15

---

[2] The component of a point set M in the plane is a maximal connected subset of M.

The disks K_1 and K_2 are isolated, the third component of the union $K_1 \cup K_2 \cup K_3 \cup K_4$ is $K_3 \cup K_4$.

Proof. To return to the proof of Theorem 9.3.1, let z be a complex number that does not belong to $\bigcup_{i=1}^n K_i$, i.e.,

$$|a_{ii} - z| > \sum_{j \neq i} |a_{ij}|, \quad i = 1, \dots, n.$$

Then $A - zI$ is uniformly diagonally dominant. It is thus a non-singular matrix by Theorem 5.3.4 and therefore z is not an eigenvalue.

In order to prove the second statement, we introduce a parameter $t \in [0, 1]$ and the matrix

$$A(t) = \begin{bmatrix} a_{11} & ta_{12} & \dots & ta_{1n} \\ ta_{21} & a_{22} & \dots & ta_{2n} \\ \vdots & \vdots & \ddots & \vdots \\ ta_{n1} & ta_{n2} & \dots & a_{nn} \end{bmatrix}.$$

$A(0)$ is the diagonal matrix $\mathrm{diag}(a_{11}, \dots, a_{nn})$ and $A(1) = A$. Put

$$K_i(t) = \left\{ z; |a_{ii} - z| \leq t \sum_{j \neq i} |a_{ij}| \right\}$$

for $i = 1, \dots, n$. Clearly $K_i(t) \subset K_i$ for any $t \in [0, 1]$. By the first statement of Gershgorin's theorem, all the eigenvalues of all the matrices $A(t)$ are contained in $\bigcup_{i=1}^n K_i$. If $C_1, \dots, C_r$ are all the components of $\bigcup_{i=1}^n K_i$ we employ the continuous dependence of the set of eigenvalues of the matrices $A(t)$ on the parameter t to find out that $A(0)$ has the same number of eigenvalues as $A(1)$ (that is, as A) in the component C_i. But the number of eigenvalues of $A(0)$ equals the number of the centers a_{kk} in the component C_i, i.e., the number of disks that constitute C_i. This proves the second statement.

The third statement follows readily from the well-known geometrical fact that two circles have no point in common if and only if the distance of their centers (in our case, the points a_{ss} and a_{tt}) is greater than the sum of their radii (in our case, $\sum_{i \neq s} |a_{si}| + \sum_{j \neq t} |a_{tj}|$). $\qquad \Box$

Using, in addition, Theorems 5.3.1 and 5.2.5, one can prove the next theorem in an analogous way.

Theorem 9.3.2 *Let $A = [a_{ik}]$ be a square matrix of order n and let the numbers $c_1, \ldots, c_n$ be such that the matrix*

$$\begin{bmatrix} c_1 & -|a_{12}| & \cdots & -|a_{1n}| \\ -|a_{21}| & c_2 & \cdots & -|a_{2n}| \\ \vdots & \vdots & \ddots & \vdots \\ -|a_{n1}| & -|a_{n2}| & \cdots & c_n \end{bmatrix}$$

is in the class K_0 (of Chapter 5). Then all the eigenvalues of A are contained in the union $\bigcup_{i=1}^{n} K_i$ of the disks $K_i = \{z; |a_{ii} - z| \leq c_i\}$. Moreover, the number of the eigenvalues contained in each component of the union is equal to the number of disks that constitute this component.

Remark. This theorem is a generalization of Gershgorin's theorem since for $c_i = \sum_{k \neq i} |a_{ik}|$ the corresponding matrix is in K_0. (See Exercise 9.10.)

Let us present, without proof, a not widely known bound for eigenvalues of a general matrix that follows from a theorem on roots of a polynomial equation [51].

Theorem 9.3.3 *Let A be a square (in general complex) matrix of order n, let α be a (complex) number for which $\operatorname{tr}(\alpha I - A)^{-1} \neq 0$. Then at least one eigenvalue of A lies in every closed disk that contains both the numbers α and $\tilde{\alpha}$, where*

$$\tilde{\alpha} = \alpha - \frac{n}{\operatorname{tr}(\alpha I - A)^{-1}}.$$

Further, if we put

$$r = \frac{n}{2|\operatorname{tr}(\alpha I - A)^{-1}|},$$

then at least one eigenvalue of A lies in the disk with center $\frac{1}{2}(\alpha + \tilde{\alpha})$ and radius r.

The following theorem, presented also without proof, can be used for the localization of a simple eigenvalue of a general matrix [13].

Theorem 9.3.4 *Let*

$$A = \begin{bmatrix} a_{11} & a_1^T \\ a_2 & A_{22} \end{bmatrix}$$

be a square matrix of order n partitioned in such a way that the block A_{22} is of order $n-1$. Let g be a norm in the space of $(n-1)$-component column vectors, let g' be the dual norm corresponding to the bilinear form $x'x$ in the space of $(n-1)$-component row vectors. Let there exist numbers β and γ such that

$$g(a_2)g'(a_1^T) \leq \beta, \qquad \hat{g}(A_{22} - a_{11}I_2) \geq \gamma > 2\sqrt{\beta}. \tag{9.20}$$

Then there is one and only one eigenvalue of A in the disk $|a_{11} - z| \leq \frac{1}{2}[\gamma - \sqrt{\gamma^2 - 4\beta}]$, and this eigenvalue is simple.

Remark. It is even possible to show that all the remaining eigenvalues λ of A satisfy the inequality

$$|a_{11} - \lambda| \geq \frac{1}{2}[\gamma + \sqrt{\gamma^2 - 4\beta}]$$

and that the eigenvalue $\hat{z}$ in the disk mentioned is the limit of the sequence $z_0 = a_{11}, z_{k+1} = a_{11} - a_1^T(A_{22} - z_k I)^{-1}a_2$. One may also obtain the bound $|z_k - z| \leq \frac{1}{2}\sigma^k[\gamma - \sqrt{\gamma^2 - 4\beta}]$, where $\sigma = [\gamma - \sqrt{\gamma^2 - 4\beta}]/[\gamma + \sqrt{\gamma^2 - 4\beta}]$.

Theorem 9.3.4 can be used if we know an approximation λ_1 to some eigenvalue. If the associated approximate eigenvector has a nonzero first component we can solve the system

$$\begin{bmatrix} a_{11} - \lambda_1 & a_1^T \\ a_2 & A_{22} - \lambda_1 I_2 \end{bmatrix} \begin{bmatrix} 1 \\ x^{(0)} \end{bmatrix} = \begin{bmatrix} \xi \\ 0 \end{bmatrix}$$

for an unknown vector $x^{(0)}$ and an unknown number ξ. If $\xi = 0$, then λ_1 is an eigenvalue. If $\xi \neq 0$ (and $x^{(0)}$ does exist), then the matrix A has the same eigenvalues as the matrix

$$\begin{bmatrix} \lambda_1 + \xi & a_1^T \\ -\xi x^{(0)} & A_{22} - x^{(0)}a_1^T \end{bmatrix} \tag{9.21}$$

since this matrix is of the form $T^{-1}AT$, where

$$T = \begin{bmatrix} 1 & 0 \\ x^{(0)} & I_2 \end{bmatrix}, \qquad T^{-1} = \begin{bmatrix} 1 & 0 \\ -x^{(0)} & I_2 \end{bmatrix}.$$

For a small ξ, the inequalities (9.20) are likely to be fulfilled and Theorem 9.3.4 can then be applied to the matrix (9.21).

We now turn to the localization of eigenvalues of symmetric (or Hermitian) matrices. The situation is much more favourable here, even in the case of multiple or close eigenvalues. First we consider the case when we know some approximation to an eigenvalue and an associated eigenvector.

Theorem 9.3.5 *Let A be a Hermitian or symmetric matrix. If ξ is a real number, c is a positive number, and $x \neq 0$ is a vector for which*

$$g_2(Ax - \xi x) \leq cg_2(x)$$

(where g_2 denotes the Euclidean norm), then there exists an eigenvalue λ of A such that

$$|\lambda - \xi| \leq c.$$

Proof. Put $B = A - \xi I$. Then $g_2(Bx) \leq cg_2(x)$, and thus

$$\frac{g_2(Bx)}{g_2(x)} \leq c.$$

By Theorem 9.2.3, the smallest in modulus eigenvalue β of the Hermitian (or symmetric) matrix B fulfils the relation

$$|\beta| = \hat{g}_2(B).$$

Therefore $|\beta| \leq g_2(Bx)/g_2(x)$ for any vector $x \neq 0$, and thus

$$|\beta| \leq c.$$

By Theorem 1.7.12, however, there exists an eigenvalue λ of A such that $\beta = \lambda - \xi$ and the theorem is proved. $\square$

If we put $x = [1, 0, \ldots, 0]^T$ and $\xi = a_{11}$ in Theorem 9.3.5, we readily obtain the following statement.

Corollary 7 *The Hermitian matrix*

$$\begin{bmatrix} a_{11} & a_1^* \\ a_1 & A_{22} \end{bmatrix}$$

has at least one eigenvalue in the interval $|a_{11} - x| \leq g_2(a_1)$.

We now prove a very useful theorem, which is due to Löwner [46] and is a special case of the Hoffman-Wielandt theorem [36].

Theorem 9.3.6 *Let $A = [a_{ik}]$ and $B = [b_{ik}]$ be Hermitian or symmetric matrices of order n. Let $a_1 \geq a_2 \geq \cdots \geq a_n$ be the eigenvalues of A and $b_1 \geq b_2 \geq \cdots \geq b_n$ the eigenvalues of B. Then*

$$\sum_{i=1}^{n}(a_i - b_i)^2 \leq \sum_{i,k=1}^{n} |a_{ik} - b_{ik}|^2. \tag{9.22}$$

Remark. The right-hand side of (9.22) can be rewritten as $N^2(A - B)$, where N is the Schur norm.

Proof. Put $A_0 = \text{diag}(a_1, \ldots, a_n)$ and $B_0 = \text{diag}(b_1, \ldots, b_n)$. Since $N(PU) = N(P) = N(UP)$ for any unitary matrix U and any square matrix P (see Exercise 9.1) and since $A = U_1 A_0 U_1^*$, $B = U_2 B_0 U_2^*$ with unitary matrices U_1 and U_2 we have

$$\begin{aligned} N^2(A - B) &= N^2(U_1 A_0 U_1^* - U_2 B_0 U_2^*) \\ &= N^2(A_0 - U_1^* U_2 B_0 U_2^* U_1) \\ &= N^2(A_0 - W B_0 W^*) \\ &= N^2(A_0 W - W B_0) \end{aligned}$$

for the unitary matrix $W = U_1^* U_2$.

Thus, it is sufficient to prove that

$$\min\{N^2(A_0 U - U B_0); U \text{ is unitary}\} = \sum_{i=1}^{n}(a_i - b_i)^2. \tag{9.23}$$

Denote the left-hand side of (9.23) by m. As

$$N^2(A_0 U - U B_0) = \sum_{i,k=1}^{n} (a_i - b_k)^2 |u_{ik}|^2 \quad \text{for } U = [u_{ik}],$$

we have

$$\begin{aligned} m &= \min\left\{ \sum_{i,k=1}^{n} (a_i - b_k)^2 |u_{ik}|^2; U \text{ is unitary} \right\} \\ &\geq \min \sum_{i,k=1}^{n} (a_i - b_k)^2 c_{ik}, \end{aligned} \tag{9.24}$$

where the second minimum is taken over all doubly stochastic matrices $C = [c_{ik}]$.

This minimum does exist since the set of doubly stochastic matrices is closed and bounded [10].

Let $D = [d_{ik}]$ be a doubly stochastic matrix for which this minimum is attained. If there are several such matrices, let D be the matrix possessing the least sum of off-diagonal entries. It is even possible that there were several such matrices. We shall, however, show that then $D = I$, i.e., that D is the identity matrix. Suppose $D \neq I$. Then there is a diagonal entry in D that is less than 1. Let p be the least index for which $d_{pp} < 1$. Then there exist both an index $q \neq p$ for which $d_{qp} > 0$ and an index $r \neq p$ for which $d_{pr} > 0$. Clearly $q > p$ (since $d_{qp} < 1$, too) and $r > p$ (since $d_{rr} < 1$). We put $t = \min(d_{qp}, d_{pr})$ and construct a new matrix $\tilde{D} = (\tilde{d}_{ik})$ from D in such a way that we subtract the number t from d_{qp} and from d_{pr} and add t to d_{pp} and d_{qr}. The matrix $\tilde{D}$ is again doubly stochastic. The new value of the sum $\sum_{i,k=1}^{n}(a_i - b_k)^2 \tilde{d}_{ik}$ is then

$$\sum_{i,k=1}^{n}(a_i - b_k)^2 d_{ik} +$$
$$[(a_p - b_p)^2 + (a_q - b_r)^2 - (a_q - b_p)^2 - (a_p - b_r)^2]t$$
$$= \sum_{i,k=1}^{n}(a_i - b_k)^2 d_{ik} - 2(a_p - a_q)(b_p - b_r)t.$$

Since $q > p$ and $r > p$ we have $a_p - a_q \geq 0$ and $b_p - b_r \geq 0$. The new sum is thus not greater than the previous one. Therefore, both the sums are identical but the sum of off-diagonal entries of the matrix D is by t less than that of $\tilde{D}$. This contradiction shows that $D = I$. By (9.24) then

$$m \geq \sum_{i=1}^{n}(a_i - b_i)^2;$$

however,

$$\sum_{i=1}^{n}(a_i - b_i)^2 = \sum_{i,k=1}^{n}(a_i - b_k)^2 |u_{ik}|^2$$

for $U = [u_{ik}] = I$, and thus also

$$m \le \sum_{i=1}^{n}(a_i - b_i)^2.$$

This proves (9.23).   □

We shall now be concerned with bounds for eigenvalues of Hermitian or symmetric matrices that are based on the use of diagonal entries. First, however, we prove an auxiliary statement [19], which is, after all, useful for the construction of special matrices with the given spectrum.

Theorem 9.3.7 *Let A be a Hermitian or symmetric matrix of order m with the eigenvalues $\alpha_1, \ldots, \alpha_m$ and B a Hermitian or symmetric matrix of order n with the eigenvalues $\beta_1, \ldots, \beta_n$. Let u be a normalized (i.e. of length 1) eigenvector of A associated with α_1 and v a normalized eigenvector of B associated with β_1.*

Then for any complex number η, the matrix

$$C = \begin{bmatrix} A & \eta u v^* \\ \bar{\eta} v u^* & B \end{bmatrix}$$

has the eigenvalues $\alpha_2, \ldots, \alpha_m, \beta_2, \ldots, \beta_n, \gamma_1, \gamma_2$, where the real numbers γ_1 and γ_2 are the eigenvalues of

$$\hat{C} = \begin{bmatrix} \alpha_1 & \eta \\ \bar{\eta} & \beta_1 \end{bmatrix}.$$

Proof. Let $u, u_2, \ldots, u_m$ be mutually orthogonal normalized eigenvectors of B associated with the eigenvalues $\beta_1, \beta_2, \ldots, \beta_n$, respectively. Form the $(m + n)$-component vectors

$$\tilde{u}_i = \begin{bmatrix} u_i \\ 0 \end{bmatrix}, \quad i = 2, \ldots, m,$$

$$\tilde{v}_j = \begin{bmatrix} 0 \\ v_j \end{bmatrix}, \quad j = 2, \ldots, n.$$

Multiplying the partitioned matrices, we obtain

$$C\tilde{u}_i = \begin{bmatrix} Au_i \\ \bar{\eta} v u^* u_i \end{bmatrix} = \alpha_i \tilde{u}_i, \quad i = 2, \ldots, m,$$

since $u^* u_i = 0$. Similarly,

$$C\tilde{v}_j = \begin{bmatrix} \eta u v^* v_j \\ B v_j \end{bmatrix} = \beta_j \tilde{v}_j, \qquad j = 2, \ldots, n.$$

Let $\hat{C}$ have orthogonal normalized eigenvectors $(r_1, s_1)^T$ and $(r_2, s_2)^T$ associated with the eigenvalues γ_1 and γ_2 from the statement of the theorem. The vectors

$$\tilde{w}_k = \begin{bmatrix} r_k u \\ s_k v \end{bmatrix}, \qquad k = 1, 2,$$

then fulfil the relations

$$\begin{aligned} C\tilde{w}_k &= \begin{bmatrix} r_k A u + \eta s_k u v^* v \\ \bar{\eta} r_k v u^* u + s_k B v \end{bmatrix} = \begin{bmatrix} (\alpha_1 r_k + \eta s_k) u \\ (\bar{\eta} r_k + \beta_1 s_k) v \end{bmatrix} = \begin{bmatrix} \gamma_k r_k u \\ \gamma_k s_k v \end{bmatrix} \\ &= \gamma_k \tilde{w}_k, \qquad k = 1, 2. \end{aligned}$$

We have thus found $m + n$ eigenvectors of C that are, as can be easily shown, mutually orthogonal and normalized and associated with the eigenvalues given in the theorem. □

Now we are ready to prove a theorem [19] that implies two important corollaries. The first of them presents inequalities for eigenvalues and eigenvectors of a Hermitian (or symmetric) matrix, the second characterizes pairs of n-component column vectors y and z to which there exists a doubly stochastic matrix D such that $y = Dz$.

Theorem 9.3.8 *Let* $y = [y_1, \ldots, y_n]^T$ *and* $z = [z_1, \ldots, z_n]^T$ *be real vectors,* $n \geq 1$. *Then the following properties are equivalent:*

1. *If* $y_{p_1} \geq y_{p_2} \geq \cdots \geq y_{p_n}$ *and* $z_{q_1} \geq z_{q_2} \geq \cdots \geq z_{q_n}$ *hold for the permutations* $(p_1, \ldots, p_n)$ *and* $(q_1, \ldots, q_n)$ *of the numbers* $1, \ldots, n$, *then*

$$\begin{aligned} y_{p_1} &\leq z_{q_1}, \\ y_{p_1} + y_{p_2} &\leq z_{q_1} + z_{q_2}, \\ &\vdots \quad \vdots \quad \vdots \\ y_{p_1} + y_{p_2} + \cdots + y_{p_{n-1}} &\leq z_{q_1} + z_{q_2} + \cdots + z_{q_{n-1}}, \\ y_1 + y_2 + \cdots + y_n &= z_1 + z_2 + \cdots + z_n. \end{aligned} \tag{9.25}$$

2. *There exists a real symmetric matrix of order n whose eigenvalues are $z_1, \ldots, z_n$ and whose diagonal entries are $y_1, \ldots,$
   y_n.*

3. *There exists a Hermitian matrix of order n whose eigenvalues
   are $z_1, \ldots, z_n$ and whose diagonal entries are $y_1, \ldots, y_n$.*

4. *There exists a real orthogonal matrix $C = [c_{ik}]$ of order n
   such that $y = Gz$, where $G = [c_{ik}^2]$.*

5. *There exists a unitary matrix $U = [u_{ik}]$ of order n such that
   $y = Hz$, where $H = [|u_{ik}|^2]$.*

6. *There exists a doubly stochastic matrix D of order n such
   that $y = Dz$.*

Proof. We shall prove the implications 1. $\to$ 2., 2. $\to$ 3., 3. $\to$ 5.,
5. $\to$ 6., 6. $\to$ 1., 2. $\to$ 4., and 4. $\to$ 5. Without loss of generality,
we may assume such a notation that

$$p_1 = q_1 = 1, \quad p_2 = q_2 = 2, \quad \ldots, \quad p_n = q_n = n,$$

that is $y_1 \geq \cdots \geq y_n$ and $z_1 \geq \cdots \geq z_n$.

1. $\to$ 2. This implication will be proved by induction on n. If
$n = 1$, then $y_1 = z_1$ and the statement is evident. If $n = 2$ we
have $y_1 \geq y_2$, $z_1 \geq z_2$, $y_1 \leq z_1$, $y_1 + y_2 = z_1 + z_2$. It can be easily
checked that for $w = \sqrt{(z_1 - y_1)(z_1 - y_2)}$ the real symmetric
matrix

$$\begin{bmatrix} y_1 & w \\ w & y_2 \end{bmatrix}$$

has the eigenvalues z_1 and z_2. Let now $n > 2$. We assume
that the implication 1. $\to$ 2. holds for any $2n'$ numbers whenever
$1 \leq n' < n$. Put

$$m = \min_{s=1,\ldots,n-1} \left(\sum_{i=1}^s z_i - \sum_{i=1}^s y_i \right);$$

let this minimum be attained for $s = t$, $1 \leq t \leq n - 1$.
Since $z_1 - y_1 \geq m$ and

$$\sum_{i=1}^{n-1} z_i - \sum_{i=1}^{n-1} y_i = -z_n + y_n \geq m,$$

we have $z_1 - z_n \geq 2m + y_1 - y_n$. This implies

$$z_1 - z_n \geq 2m. \qquad (9.26)$$

We introduce the numbers $z_1' = z_i - m$, $z_j' = z_j$, $j = 2, \ldots, n-1$, $z_n' = z_n + m$. Then

$$\sum_{i=1}^{s} y_i \leq \sum_{i=1}^{s} z_i', \quad s = 1, \ldots, t-1, \qquad (9.27)$$

$$\sum_{i=1}^{t} y_i = \sum_{i=1}^{t} z_i',$$

and further,

$$\sum_{i=t+1}^{s} y_i \leq \sum_{i=t+1}^{s} z_i', \quad s = t+1, \ldots, n-1, \qquad (9.28)$$

$$\sum_{i=t+1}^{n} y_i = \sum_{i=t+1}^{n} z_i'.$$

The relations (9.25) imply that the numbers $y_1, \ldots, y_t$ and $z_1', \ldots, z_t'$ fulfil the assumptions of 1. even when $z_1' \geq z_2' \geq \cdots \geq z_t'$ may not be true. By the induction hypothesis, there exists a real symmetric matrix A of order t whose eigenvalues are $z_1', \ldots, z_t'$ and whose diagonal entries are $y_1, \ldots, y_t$. In an analogous way, the existence of a real symmetric matrix A_2 of order $n - t$ whose eigenvalues are $z_{t+1}', \ldots, z_n'$ and whose diagonal entries are $y_{t+1}, \ldots, y_n$ follows from (9.28). Denote by u the real normalized eigenvector of A_1 associated with the eigenvalue z_1' and by v the real normalized eigenvector of A_2 associated with the eigenvalue z_n'. Since $z_1' \geq z_n'$ by (9.26) and $z_1' \leq z_1$, $z_1' + z_n' = z_1 + z_n$, by the statement proved for $n = 2$, there exists a real number w such that the matrix

$$\begin{bmatrix} z_1' & w \\ w & z_n' \end{bmatrix}$$

has the eigenvalues z_1 and z_n. By Theorem 9.3.7, the real symmetric matrix

$$\begin{bmatrix} A_1 & wuv^T \\ wvu^T & A_2 \end{bmatrix}$$

has then the eigenvalues $z_1, z_2, \ldots, z_n$ and the diagonal entries $y_1, y_2, \ldots, y_n$.

2. $\rightarrow$ 3. Obvious, since any real symmetric matrix is also Hermitian.

3. $\rightarrow$ 5. Let a Hermitian matrix $A = [a_{ik}]$ be such as required by 3. By Theorem 2.4.7, there exists a unitary matrix $U = [u_{ik}]$ such that $A = UZU^*$, where $Z = \mathrm{diag}(z_1, \ldots, z_n)$. Since $a_{ii} = y_1$, $i = 1, \ldots, n$, it follows that

$$y_i = a_{ii} = \sum_{k=1}^{n} |u_{ik}|^2 z_k$$

or

$$y = Hz, \quad H = [|u_{ik}|^2].$$

5. $\rightarrow$ 6. The implication follows from the fact that the matrix H of 5. is doubly stochastic (see Exercise 4.5).

6. $\rightarrow$ 1. Let $y = Dz$, where $D = [d_{ik}]$ is a doubly stochastic matrix. Then, first of all,

$$\sum_{i=1}^{n} y_i = \sum_{i=1}^{n} \sum_{k=1}^{n} d_{ik} z_k = \sum_{k=1}^{n} \left(\sum_{i=1}^{n} d_{ik} \right) z_k = \sum_{k=1}^{n} z_k.$$

Let now s be a fixed integer, $1 \leq s \leq n - 1$. We introduce numbers $v_1, \ldots, v_n$ as

$$v_k = \sum_{i=1}^{s} d_{ik}, \qquad k = 1, \ldots, n.$$

Thus $0 \leq v_k \leq 1$ and $\sum_{k=1}^{n} v_k = s$. Since $z_1 \geq \cdots \geq z_n$, we obtain

$$\begin{aligned}
\sum_{i=1}^{s} y_i &= \sum_{i=1}^{s} \sum_{k=1}^{n} d_{ik} z_k = \sum_{k=1}^{n} v_k z_k \\
&= \sum_{k=1}^{s-1} v_k z_k + \left(\sum_{k=s}^{n} v_k \right) z_s = \sum_{k=1}^{s-1} v_k z_k + \left(s - \sum_{k=1}^{s-1} v_k \right) z_s \\
&= \sum_{k=1}^{s-1} v_k (z_k - z_s) + s z_s \\
&\leq \sum_{k=1}^{s-1} (z_k - z_s) + s z_s = \sum_{i=1}^{s} z_i.
\end{aligned}$$

2. → 4. Let $A = [a_{ik}]$ be a real symmetric matrix satisfying the requirements of 2. By Theorem 2.4.6, there exists a real orthogonal matrix $C = [c_{ik}]$ such that $A = CZC^T$, where $Z = \text{diag}(z_1, \dots, z_n)$. Since $a_{ii} = y_i$, $i = 1, \dots, n$, we have

$$y_i = a_{ii} = \sum_{k=1}^{n} c_{ik}^2 z_k,$$

or

$$y = Gz, \quad G = [c_{ik}^2].$$

4. → 5. Obvious, since every real orthogonal matrix is also unitary. □

We shall now reformulate the equivalence of 1. and 2. (or 1. and 3.).

Corollary 8 *If the diagonal entries $a_{11}, \dots, a_{nn}$ of a Hermitian or symmetric matrix A of order n fulfil the inequalities $a_{11} \geq a_{22} \geq \cdots \geq a_{nn}$ and if $\lambda_1 \geq \lambda_2 \geq \cdots \geq \lambda_n$ are the eigenvalues of this matrix, then*

$$\begin{aligned}
\lambda_1 &\geq a_{11}, \\
\lambda_1 + \lambda_2 &\geq a_{11} + a_{22}, \\
&\vdots \quad \vdots \quad \vdots \\
\lambda_1 + \lambda_2 + \cdots + \lambda_{n-1} &\geq a_{11} + a_{22} + \cdots + a_{n-1,n-1}, \\
\lambda_1 + \lambda_2 + \cdots + \lambda_n &= a_{11} + a_{22} + \cdots + a_{nn}.
\end{aligned} \tag{9.29}$$

Conversely, if the inequalities (9.29) hold for the real numbers

$$a_{11} \geq a_{22} \geq \cdots \geq a_{nn} \text{ and } \lambda_1 \geq \lambda_2 \geq \cdots \geq \lambda_n,$$

then there exists a Hermitian or symmetric matrix A whose diagonal entries are $a_{11}, \dots, a_{nn}$ and whose eigenvalues are $\lambda_1, \dots, \lambda_n$.

Remark. The system (9.29) can clearly be written in the equivalent form

$$\lambda_n \le a_{nn},$$
$$\lambda_{n-1} + \lambda_n \le a_{n-1,n-1} + a_{nn},$$
$$\vdots \quad \vdots \quad \vdots$$
$$\lambda_2 + \cdots + \lambda_{n-1} + \lambda_n \le a_{22} + \cdots + a_{n-1,n-1} + a_{nn},$$
$$\lambda_1 + \lambda_2 + \cdots + \lambda_{n-1} + \lambda_n = a_{11} + a_{22} + \ldots$$
$$+ a_{n-1,n-1} + a_{nn}.$$

The equivalence of 1. and 6. shows that if y and z are vectors such that $y = Dz$ with a doubly stochastic matrix D, then the largest component of y is not greater than the largest component of z and the smallest component of y is not less than the smallest component of z. Analogous inequalities hold also for the sum of the two largest components and the sum of the two smallest components, etc.

We shall now present some bounds for eigenvalues of nonnegative matrices and matrices of the classes K and K_0 (M-matrices). These bounds will allow us to obtain bounds also for general matrices.

Theorem 9.3.9 *Let A be a nonnegative square matrix of order n and let $x = [x_1, \ldots, x_n]^T$ be a positive vector. Then*

$$\min_i \frac{(Ax)_i}{x_i} \le p(A) \le \max_i \frac{(Ax)_i}{x_i} \qquad (9.30)$$

are bounds for the Perron root $p(A)$.

Proof. Suppose

$$p(A) > \max_i \frac{(Ax)_i}{x_i};$$

then

$$p(A)x_i > (Ax)_i, \quad i = 1, \ldots, n,$$

or

$$Ax < p(A)x. \quad x > 0. \qquad (9.31)$$

By Theorem 4.3.2, there exists a nonnegative eigenvector y of A^T associated with $p(A)$,

$$A^T y = p(A)y, \quad y \geq 0, y \neq 0. \tag{9.32}$$

Left-multiplying (9.31) by y^T, we obtain $y^T A x < p(A)y^T x$; right-multiplying (9.32) by x, we have $y^T A x = p(A)y^T x$. This contradiction proves the right-hand inequality of (9.30). The proof of the left-hand inequality is similar. $\square$

Theorem 9.3.10 *Let $A \in K$, let $q(A)$ be the minimal eigenvalue of the matrix A (cf. Exercise 5.6), and let $y = [y_1, \ldots, y_n]^T$ be a vector such that $Ay > 0$ (and thus $y > 0$). Then*

$$\min_i \frac{(Ay)_i}{y_i} \leq q(A) \leq \max_i \frac{(Ay)_i}{y_i}. \tag{9.33}$$

Proof. Since $A^{-1} \geq 0$ by the property 11 of Theorem 5.1.1 and since, further,

$$q(A) = (p(A^{-1}))^{-1},$$

we apply Theorem 9.3.9 to the vector $x = Ay > 0$ and obtain

$$p(A^{-1}) \leq \max_i \frac{(A^{-1}x)_i}{x_i},$$

that is

$$q(A) = (p(A^{-1}))^{-1} \geq \frac{1}{\max_i(y_i/(Ay)_i)} = \min_i \frac{(Ay)_i}{y_i}.$$

The other inequality can be proved in an analogous way. $\square$

The formulae (9.33) and (5.17) can be also used to obtain a lower bound for the minimal (in modulus) eigenvalue of a diagonally dominant matrix.

Theorem 9.3.11 *Let $C = [c_{ik}]$ be a diagonally dominant square matrix of order n and let $\sigma = \min\{|\lambda|; \lambda \in S(C)\}$. Then*

$$\sigma \geq q(H(C)),$$

where $H(C) = [h_{ik}]$, $h_{ii} = |c_{ii}|$ and $h_{ik} = -|c_{ik}|$ for $i \neq k$, $i, k = 1, \ldots, n$.

Proof. The matrix C is nonsingular by Theorem 5.3.4. Clearly

$$\sigma = (\rho(C^{-1}))^{-1}.$$

By (5.17), however,

$$m(C^{-1}) \leq (H(C))^{-1}$$

and by Lemma 4.1.5 further

$$\rho(C^{-1}) \leq \rho(m(C^{-1})) \leq \rho((H(C))^{-1}).$$

We thus have

$$q(H(C)) = (\rho((H(C))^{-1}))^{-1} \leq (\rho(C^{-1}))^{-1} = \sigma.$$

$\square$

The presented bound together with the bound (9.33) allow us to obtain also a lower bound for the measure of nonsingularity $\hat{G}_2(A)$ of a symmetric diagonally dominant matrix with respect to the Euclidean norm g_2. Indeed, $\hat{g}_2(A) = \sigma$ for such a matrix by Theorem 9.2.3. For a non-symmetric matrix or for a norm different from g_2, the lower bound for the measure of nonsingularity cannot be obtained in this way. We can, however, directly estimate $\hat{g}_3(A)$ for uniformly diagonally dominant matrices (see Chapter 5) and $\hat{g}_1(A)$ for matrices whose transposes are uniformly diagonally dominant.

Theorem 9.3.12 *Let* $A = [a_{ik}]$ *be a uniformly diagonally dominant square matrix of order* n. *Then*

$$\hat{g}_3(A) \geq c = \min_{i=1,\ldots,n} \left(|a_{ii}| - \sum_{k \neq i} |a_{ik}| \right),$$

i.e., any eigenvalue λ *of* A *fulfils the inequality*

$$|\lambda| \geq c.$$

Proof. If c is the number introduced in the statement of the theorem, then

$$H(A)e \geq ce, \quad e = [1, 1, \ldots, 1]^T.$$

Since the matrix A is nonsingular by Theorem 5.3.3, we obtain by (5.17) that

$$m(A^{-1}) \leq (H(A))^{-1}.$$

Therefore

$$m(A^{-1})e \leq (H(A))^{-1}e \leq c^{-1}e.$$

But by Theorem 9.1.1 this means that $g_3(A^{-1}) \leq c^{-1}$. Hence, by Theorem 9.2.1,

$$\hat{g}_3(A) = (g_3(A^{-1}))^{-1} \geq c.$$

The last statement follows from Theorem 9.2.6. $\square$

In the conclusion we recall that the upper bounds for the spectral radius $\rho(A)$ of a matrix A which use norms (see Theorem 9.1.2) yield also bounds for the eigenvalues.

Exercises

9.1.* If N is the Schur norm and if U and V are unitary matrices of proper dimensions, prove that $N(UA) = N(A)$ and $N(AV) = N(A)$ for any matrix A.

9.2. Prove that the norm g_2 is dual to itself (with respect to the bilinear form $\langle x', x \rangle = x'^T x$).

9.3. Prove that the norm g_{YX} induced by the norms g_X and g_Y satisfies N1 to N4.

9.4. Prove that

$$N(A) \leq n^{1/2} g_i(A), \quad i = 1, 2, 3,$$

holds for a square matrix A of order n and each of the norms g_1, g_2, and g_3 of Example 9.1.

9.5. Assign to any square matrix A of order n the number

$$M(A) = n \max_{i,k} |a_{ik}|.$$

Prove that $M(A)$ fulfils N'1 to N'5.

9.6. Prove that $N(A) \leq M(A)$ holds for the norm $M(A)$ of Exercise 9.5 and the Schur norm $N(A)$.

9.7. Prove that

$$g_i(Ax) \leq M(A)g_i(x), \quad i = 1, 2, 3,$$

holds for each of the vector norms g_1, g_2 and g_3 of Example 9.1 and for the norm $M(A)$ of Exercise 9.5.

9.8. If T is a fixed nonsingular matrix of order n and g is a vector norm in the linear space of n-component column vectors, then the function $g_T(x) = g(Tx)$ is also a vector norm. Express the matrix norm induced by g_T in terms of the matrix norm induced by g.

9.9. Deduce from the result of the preceding exercise and from the result of Exercise 1.4 that to any matrix A there exists a vector norm g such that either $g(A) = \rho(A)$, or $g(A)$ is arbitrarily close to $\rho(A)$.

9.10.* Show that Theorem 9.3.2 generalizes Gershgorin's Theorem 9.3.1.

9.11.* If g is a norm and A a matrix for which $g(A) < 1$, prove that $I - A$ is nonsingular and $g((I - A)^{-1}) \leq (1 - g(A))^{-1}$.

9.12. Prove *Schur's inequality*: If $A = [a_{ik}]$ is a square matrix having the eigenvalues $\lambda_1, \ldots, \lambda_n$, then $\sum |\lambda_i|^2 \leq N^2(A)$. (Hint: Use Theorem 2.4.3.)

9.13. Prove that in Schur's inequality of the preceding exercise equality is valid if and only if the matrix A is normal.

9.14. If a diagonally dominant square matrix A has positive diagonal entries, prove that A is positive stable.

10

Direct Methods for Solving Linear Systems

In this chapter we study numerical methods for solving systems of n linear algebraic equations in n unknowns with a nonsingular matrix. The general case will be discussed in the conclusion of the chapter.

10.1 Nonsingular case

Solving a linear system with a nonsingular matrix, we have to take into account a number of factors, for example:

1. whether the matrix of the system in general or special, e.g. symmetric, positive definite, diagonally dominant, or banded;

2. whether the system is "well-" or "ill-conditioned" (see in what follows);

3. whether the system is large (consists of more than, say, 500 equations) and exceeds the capacity of the standard sub-routine;

4. whether we are going to solve only one system or more systems with the same matrix or with matrices of the same zero-nonzero structure (cf. Chapter 4);

5. whether the required accuracy of the result is likely to be reached in single precision or whether it is necessary to

carry out the computation in double precision or to refine
the result later.

A system of linear equations

$$Ax = b \tag{10.1}$$

with a nonsingular matrix A, and the matrix A itself as well, are
said to be *well-conditioned* if for a norm g (for example g_2 of
Example 9.1) the number

$$\kappa(A) = g(A)g(A^{-1}), \tag{10.2}$$

for which $\kappa(A) \geq 1$ always holds, is not substantially (say hun-
dred times or thousand times) greater than 1. Otherwise the
system as well as the matrix A are said to be *ill-conditioned*.

The number $\kappa(A)$ is called the *condition number* of A with
respect to the norm g. Its meaning can be expressed by the
following theorem which characterizes $\kappa(A)$ as the least upper
bound for the ratio of the relative error in the solution of (10.1)
and the relative error in an arbitrary right-hand side.

Theorem 10.1.1 *Let A be a nonsingular $n \times n$ matrix, let b_0 and
b_1 be n-component column vectors, $b_0 \neq 0$. If x_0 is the solution
of the system $Ax = b_0$ and x_1 the solution of the system $Ax = b_1$,
then*

$$\frac{g(x_1 - x_0)}{g(x_0)} \leq \kappa(A)\frac{g(b_1 - b_0)}{g(b_0)} \tag{10.3}$$

*holds for a vector norm g, where $\kappa(A)$ is the condition number of
A with respect to g.*

*Moreover, for the given matrix A there exist vectors $b_0 \neq 0$
and $b_1 \neq b_0$ such that equality takes place in (10.3).*

Proof. The properties of norms imply $0 < g(b_0) = g(A)g(x_0)$ and
$g(x_1 - x_0) = g(A^{-1}(b_1 - b_0)) \leq g(A^{-1})g(b_1 - b_0)$. Multiplying both
the inequalities, we easily obtain (10.3). By (9.3), there exists a
vector $x_0 \neq 0$ such that $g(Ax_0) = g(A)g(x_0)$ as well as a vector
$z_0 \neq 0$ such that $g(A^{-1}z_0) = g(A^{-1})g(z_0)$. Putting $b_0 = Ax_0$ and
$b_1 = b_0 + z_0$ we reach the equality in (10.3). $\qquad\square$

Large systems of linear equations whose matrix contains a relatively low number of nonzero entries (say 5%) are treated in Chapter 14.

Numerical methods for solving a "general" system (10.1) can be divided into two principal groups. The first group consists of methods that (at least theoretically) give an exact solution after a finite number of operations. They are called *direct methods* and we discuss them in this chapter. The other large group consists of methods that successively provide approximations to the solution. The computation is terminated if the approximation lies within an accuracy tolerance prescribed. These *iterative methods* will be treated in Chapter 11.

In Chapter 1, we claimed that the system (10.1) has a unique solution, which can be written as

$$x = A^{-1}b. \tag{10.4}$$

This formula had led to Cramer's rule, which can be also viewed as a direct method if we apply e.g. the expansion of the determinants by their columns (1.11). The value of Cramer's rule, however, is rather theoretical. It is suitable for numerical computation only if the order of the matrix does not exceed 3.

The fundamental idea of all other direct methods is to transform the system (10.1) into the equivalent form

$$Cx = d, \tag{10.5}$$

where C is a nonsingular upper (or lower) triangular matrix. Writing the system (10.5) as

$$
\begin{aligned}
c_{11}x_1 + c_{12}x_2 + \ldots \quad + c_{1,n-1}x_{n-1} \ + \ c_{1n}x_n &= d_1, \\
c_{22}x_2 + \ldots \quad + c_{2,n-1}x_{n-1} \ + \ c_{2n}x_n &= d_2, \\
\vdots \qquad \qquad \vdots \qquad \qquad \vdots \quad & \\
c_{n-1,n-1}x_{n-1} + c_{n-1,n}x_n &= d_{n-1}, \\
c_{nn}x_n &= d_n,
\end{aligned}
$$

we see that it can be easily solved by the so-called *back substitution*. From the last equation we calculate x_n (the matrix C is nonsingular and thus $c_{ii} \neq 0$ for $i = 1, \ldots, n$), substitute for x_n

into the last but one equation and calculate x_{n-1} from it, etc., until we calculate x_1 from the first equation.

The form (10.5) can be obtained e.g. by the well-known *method of Gaussian elimination* the description of which we present for the sake of completeness. Suppose that the coefficient a_{11} of the given system

$$
\begin{array}{ccccccccc}
a_{11}x_1 & + & a_{12}x_2 & + & \ldots & + & a_{1n}x_n & = & b_1, \\
a_{21}x_1 & + & a_{22}x_2 & + & \ldots & + & a_{2n}x_n & = & b_2, \\
\vdots & & \vdots & & \ddots & & \vdots & & \vdots \\
a_{n1}x_1 & + & a_{n2}x_2 & + & \ldots & + & a_{nn}x_n & = & b_n
\end{array}
\tag{10.6}
$$

is nonzero. If $x_1^{(0)}, \ldots, x_n^{(0)}$ is the solution of (10.6), then it is also the solution of the system

$$
\begin{array}{ccccccccc}
a_{11}x_1 & + & a_{12}x_2 & + & \ldots & + & a_{1n}x_n & = & b_1, \\
 & & a_{22}^{(1)}x_2 & + & \ldots & + & a_{2n}^{(1)}x_n & = & b_2^{(1)}, \\
 & & \vdots & & \ddots & & \vdots & & \vdots \\
 & & a_{n2}^{(1)}x_2 & + & \ldots & + & a_{nn}^{(1)}x_n & = & b_n^{(1)},
\end{array}
\tag{10.7}
$$

where the second equation is obtained by subtracting the first equation multiplied by a_{21}/a_{11} from the second equation of (10.6), the third equation of (10.7) is obtained by subtracting the first equation of (10.6) multiplied by a_{31}/a_{11} from the third equation of (10.6), and so forth. In general,

$$
a_{ik}^{(1)} = a_{ik} - a_{i1}a_{11}^{-1}a_{1k}, \qquad i,k = 2, \ldots, n,
\tag{10.8}
$$

and on the right-hand side we get

$$
b_i^{(1)} = b_i - a_{i1}a_{11}^{-1}b_1, \qquad i = 2, \ldots, n..
$$

If $a_{22}^{(1)} \neq 0$ in (10.7) we may similarly proceed further. The solution $x_1^{(0)}, x_2^{(0)}, \ldots, x_n^{(0)}$ satisfies also the system obtained from (10.7) by subtracting the second equation multiplied by $a_{32}^{(1)}/a_{22}^{(1)}$ from the third equation, subtracting the second equation multiplied by $a_{42}^{(1)}/a_{22}^{(1)}$ from the fourth equation, etc.

Here we have

$$
a_{ik}^{(2)} = a_{ik}^{(1)} - a_{i2}^{(1)}(a_{22}^{(1)})^{-1}a_{2k}^{(1)}, \qquad i,k = 3, \ldots, n,
$$

and on the right-hand side

$$b_i^{(2)} = b_i^{(1)} - a_{i2}^{(1)}(a_{22}^{(1)})^{-1}b_2^{(1)}, \qquad i = 3, \ldots, n.$$

Continuing the process in this way, we finally arrive at the system

$$
\begin{aligned}
a_{11}x_1 + a_{12}x_2 + \cdots + a_{1,n-1}x_{n-1} &&+ a_{1n}x_n &= b_1, \\
a_{22}^{(1)} + \cdots + a_{2,n-1}^{(1)}x_{n-1} &&+ a_{2n}^{(1)}x_n &= b_2^{(1)}, \\
+ \cdots + a_{3,n-1}^{(2)}x_{n-1} &&+ a_{3n}^{(2)}x_n &= b_3^{(2)}, \\
\ddots \qquad\quad\ \vdots\ && \vdots\qquad & \ \ \vdots \\
a_{n-1,n-1}^{(n-2)}a_{n-1} &+ a_{n-1,n}^{(n-2)}x_n &= b_{n-1}^{(n-2)}, \\
&& a_{nn}^{(n-1)}x_n &= b_n^{(n-1)}
\end{aligned}
$$

$$(10.9)$$

having an upper triangular matrix and possessing the same solution as the system (10.6). The whole process, however, is feasible only if

$$a_{11} \neq 0, \ a_{22}^{(1)} \neq 0, \ a_{33}^{(2)} \neq 0, \ \ldots, \ a_{n-1,n-1}^{(n-2)} \neq 0. \qquad (10.10)$$

Then we have (putting $a_{1i}^{(0)} = a_{1i}$)

$$a_{ij}^{(k)} = a_{ij}^{(k-1)} - a_{ik}^{(k-1)}(a_{kk}^{(k-1)})^{-1}a_{kj}^{(k-1)}, \ i,j = k+1, \ldots, n, \qquad (10.11)$$

$$b_i^{(k)} = b_i^{(k-1)} - a_{ik}^{(k-1)}(a_{kk}^{(k-1)})^{-1}b_k^{(k-1)}, \qquad (10.12)$$

$$i = k+1, \ldots, n, \ k = 1, \ldots, n-1.$$

We now introduce another notation. Put

$$\ell_{ik} = a_{ik}^{(k-1)}(a_{kk}^{(k-1)})^{-1}, \qquad i = k+1, \ldots, n, \qquad (10.13)$$

$$u_{kj} = a_{kj}^{(k-1)}, \quad j = k, k+1, \ldots, n, \ k = 1, \ldots, n. \qquad (10.14)$$

Then the system (10.9) is of the form (we have simplified the notation for the right-hand side)

$$
\begin{aligned}
u_{11}x_1 + u_{12}x_2 + \ldots + u_{1n}x_n &= \tilde{b}_1, \\
u_{22}x_2 + \ldots + u_{2n}x_n &= \tilde{b}_2, \\
\vdots \qquad\quad \ddots \qquad\ \vdots\ &\quad \vdots \\
u_{nn}x_n &= \tilde{b}_n.
\end{aligned}
$$

$$(10.15)$$

We shall first consider the conditions (10.10). Our only assumption was that the matrix A of the system (10.1) is nonsingular. The condition $a_{11} \neq 0$, however, need not be fulfilled. But
if $a_{11} = 0$, then some other coefficient a_{k1} is nonzero and changing the order of equations (which is an inessential operation) we
can obtain a new coefficient a_{11}, which is already nonzero. Our
considerations can similarly proceed further. The determinant
of A equals (by theorems on operations with determinants) the
determinant of the matrix of the system (10.7), which is equal to
the product of a_{11} and the determinant of the $(n-1) \times (n-1)$
matrix $[a_{ik}^{(1)}]$, $i, k = 2, \ldots, n$. This matrix is thus nonsingular as
well. (If some equations are interchanged, the determinant may
possibly change its sign.)

The assumptions (10.10) are therefore inessential if we take
possible changes in the order of equations into account. In the
s-th step, we choose the ith equation to "eliminate" the kth "unknown" if the corresponding coefficient $a_{ik}^{(s-1)}$ is nonzero. This
coefficient (or also its position in the matrix) is called the *pivot*
(or *pivotal element*) *of the sth step*. The problem how to choose
the pivot can be treated from various viewpoints. For example,
we may take such an entry of the matrix $a_{ik}^{(s-1)}$, $i, k = s, \ldots, n$,
which is the largest in modulus for pivot in the sth step (the
complete pivoting). Since we choose the pivot, in general, in a
column other than the sth we admit also an order of elimination
different from the elimination of the first unknown x_1 first, the
elimination of x_2 afterwards, etc., the last unknown eliminated
being x_{n-1}. The advantage of the complete pivoting is that the
accuracy is not lost in the division by a small number – at least in
the particular step. A disadvantage consists in the fact that the
search for the maximal entry through the whole matrix $a_{ik}^{(s-1)}$ is
rather time-consuming for a computer. Therefore, the so-called
partial pivoting is used more frequently. Here we choose for pivot
the largest in modulus entry of the first column of the matrix
$a_{ik}^{(s-1)}$, i.e., that entry which is equal to $\pm \max_{i=s,s+1,\ldots,n} |a_{is}^{(s-1)}|$.
Another possibility is to choose a positive number m, to fix some
order of the entries of the matrix (e.g. to consider the columns
successively and order their entries from above down), and to take
for pivot the first entry found whose modulus is greater than or

equal to m. (It may happen that no entry of the matrix satisfies this condition if we have chosen a too large "threshold" m. Then we can use e.g. $\frac{1}{2}m$ as a new threshold and repeat the whole procedure.) This method is particularly important for solving large linear systems with sparse matrices (Chapter 14), where, however, some further criteria (besides the threshold condition) are employed for the choice of the pivot.

We now turn back to the system (10.15) and conditions (10.13) and (10.14). We define a lower triangular matrix $L = [\ell_{ik}]$ of order n as follows: $\ell_{ii} = 1$, $i = 1, \ldots, n$; for $i > k$, the ℓ_{ik} are given in (10.2), and, of course, $\ell_{ik} = 0$ for $i < k$. The upper triangular matrix of the system (10.15) will be denoted by U. Therefore, $U = [u_{ik}]$, where $u_{ik} = 0$ for $i > k$. We shall show that

$$LU = A. \tag{10.16}$$

Let thus i and j be integers from $N = \{1, 2, \ldots, n\}$. First let $i \leq j$. We shall prove that then

$$a_{ij} = \sum_{k=1}^{i} \ell_{ik} u_{kj}. \tag{10.17}$$

This is true for $i = 1$ by (10.14). If $i > 1$ we sum up the relations (10.11) for $k = 1, \ldots, i - 1$. By (10.13) and (10.14) we then have

$$a_{ij}^{(i-1)} = a_{ij} - \sum_{k=1}^{i-1} \ell_{ik} u_{kj}.$$

Since the left-hand side is $\ell_{ii} u_{ij}$, (10.17) follows.

Let now $i > j$. We shall prove that

$$a_{ij} = \sum_{k=1}^{j} \ell_{ik} u_{kj}. \tag{10.18}$$

If $j = 1$ then (10.18) holds by (10.13) and (10.14). If $j > 1$ we sum up the relations (10.11) for $k = 1, \ldots, j - 1$ to obtain

$$a_{ij}^{(j-1)} = a_{ij} - \sum_{k=1}^{j-1} \ell_{ik} u_{kj}.$$

The left-hand side, however, equals $\ell_{ij}u_{jj}$ by (10.13) and (10.14), and thus (10.18) holds again. But (10.17) and (10.18) give (10.16) which was to be proved.

If we, in addition, write the matrix U as DC, where

$$D = \text{diag}(a_{11}, a_{22}^{(1)}, a_{33}^{(2)}, \ldots, a_{nn}^{(n-1)})$$

and C is an upper triangular matrix, then the result of Exercise 1.15 implies that the matrix A is strongly nonsingular and

$$a_{kk}^{(k-1)} = \frac{\det A(N_k)}{\det A(N_{k-1})}, \qquad k = 2, \ldots, n, \tag{10.19}$$

where $N_k = \{1, 2, \ldots, k\}$. We have thus proved the following theorem.

Theorem 10.1.2 *The system* (10.1) *can be solved by the method of Gaussian elimination with the pivots* $(1,1), (2,2), \ldots, (n-1, n-1)$ *if and only if the matrix A is strongly nonsingular. The resulting matrix U of the system* (10.15) *is identical with the matrix DC in the unique factorization $A = BDC$ of Theorem 1.4.3.*

Further we wish to remark that if we perform the partial pivoting columnwise, then it is important for the individual equations to be "numerically balanced". We should avoid the situation when the coefficients of one equation are of order 10^5 and all the coefficients of another equation are of order 10^{-3}. In such a case we would have chosen pivots from incomparable quantities. It is therefore convenient to scale the equations, e.g. to multiply them by a suitable power of 2 or 10 (according to the arithmetic of the computer used) so that all the equations have roughly the same largest (in modulus) entry on the left-hand side.

The method of Gaussian elimination can be performed also as a block method. In fact, if a_{ik} are matrices (a_{ii} are square matrices), x_i are column vectors and the number of the components of x_i equals the order of a_{ii}, then the whole procedure is feasible as far as a_{11} is nonsingular, $a_{22}^{(1)} = a_{22} - a_{21}a_{11}^{-1}a_{12}$ is nonsingular, $a_{33}^{(2)} = a_{33} - a_{32}(a_{22} - a_{21}a_{11}^{-1}a_{12})^{-1}a_{23}$ is nonsingular, etc. This means that the matrices

$$a_{11}, \begin{bmatrix} a_{11} & a_{12} \\ a_{21} & a_{22} \end{bmatrix}, \begin{bmatrix} a_{11} & a_{12} & a_{13} \\ a_{21} & a_{22} & a_{23} \\ a_{31} & a_{32} & a_{33} \end{bmatrix}, \ldots, A$$

are nonsingular. The resulting matrix U of the system (10.9) or (10.15) is then upper block triangular.

We now turn back to the point version of the method. Suppose that A is a Hermitian or real symmetric matrix. First of all, we then have the next theorem.

Theorem 10.1.3 *If $A = [a_{ik}]$ is a Hermitian (real symmetric) strongly nonsingular matrix, then for $s = 1, 2, \ldots, n - 1$ each matrix $A^{(s)} = [a_{ik}]^{(s)}$, $i, k = s + 1, \ldots, n$ is Hermitian (real symmetric) and strongly nonsingular.*

Proof. We use induction on s. Putting $A^{(0)} = A$, we find that the theorem holds for $s = 0$. Let $s \geq 1$ and let the theorem be true for the matrix $A^{(s-1)}$. Then

$$
\begin{aligned}
a_{ik}^{(s)} &= a_{ik}^{(s-1)} - a_{is}^{(s-1)}(a_{ss}^{(s-1)})^{-1}a_{sk}^{(s-1)} \\
&= \bar{a}_{ki}^{(s-1)} - \bar{a}_{ks}^{(s-1)}(\bar{a}_{ss}^{(s-1)})^{-1}\bar{a}_{si}^{(s-1)} \\
&= \bar{a}_{ki}^{(s)}
\end{aligned}
$$

and thus $A^{(s)}$ is Hermitian (real symmetric) as well. The strong nonsingularity of $A^{(s)}$ follows from Theorems 1.4.4 and 1.4.2 as for $N_s = \{1, 2, \ldots, s\}$ and $M_t = \{s + 1, \ldots, s + t\}$, $t \geq 1$, we have

$$
\det A^{(s)}(M_t) = \frac{\det A(N_s \cup M_t)}{\det A(N_s)},
$$

since $A^{(s)}(M_t) = [A(N_s \cup M_t)/A(N_s)]$ (see Chapter 1). □

The theorem implies that in the computation with a Hermitian (real symmetric) matrix we need only the entries $a_{ik}^{(s)}$ above and on the main diagonal. Even more suitable is then the *Choleski method* described in the following theorem.

Theorem 10.1.4 *Let $A = [a_{ik}]$ be a Hermitian (real symmetric) strongly nonsingular matrix of order n. Then there exists a unique upper triangular matrix $Z = [z_{ik}]$ having positive diagonal entries and a diagonal matrix $E = \mathrm{diag}(\varepsilon_1, \ldots, \varepsilon_n)$ with $\varepsilon_i = \pm 1$ such that*

$$
A = Z^* E Z. \tag{10.20}
$$

The entries z_{ik}, $i \geq k$, and the numbers ε_i can be recursively calculated from the formulae

$$z_{kk} = \sqrt{\left| a_{kk} - \sum_{j=1}^{k-1} \varepsilon_j |z_{jk}|^2 \right|}, \qquad (10.21)$$

$$\varepsilon_k = \text{sgn}\left[a_{kk} - \sum_{j=1}^{k-1} \varepsilon_j |z_{jk}|^2 \right], \qquad (10.22)$$

$$z_{ik} = \frac{1}{\varepsilon_i z_{ii}} \left[a_{ik} - \sum_{j=1}^{i-1} \varepsilon_j \bar{z}_{ji} z_{jk} \right], \qquad (10.23)$$

ordering the computation as follows:

$$z_{11}, \varepsilon_1, z_{12}, z_{22}, \varepsilon_2, z_{13}, z_{23}, z_{33}, \varepsilon_3, \ldots, z_{1n}, z_{2n}, \ldots, z_{nn}, \varepsilon_n.$$

Proof. Since A is strongly nonsingular Theorem 1.4.3 implies the existence of a unique factorization $A = BDC$, where B is a lower and C an upper triangular matrix, both having 1's on their diagonals, and D is a diagonal matrix, $D = \text{diag}(d_1, \ldots, d_n)$, $d_1 = \det A(N_1)$, $d_k = \dfrac{\det A(N_k)}{\det A(N_{k-1})}$ for $k = 2, \ldots, n$, $N_k = \{1, 2, \ldots, k\}$. Since $A^* = A$, the factorization $A = C^* D^* B^*$ has the same properties and $C = B^*$, $D = D^*$. Thus $A = BDB^*$ and D is real. Take $W = \text{diag}(\sqrt{|d_{ii}|})$, $E = \text{diag}(\text{sgn } d_i)$, and $Z = WB^*$. Then we have

$$A = BDB^* = BWEWB^* = Z^* EZ$$

and, moreover, Z is an upper triangular matrix with positive diagonal entries.

But if we suppose (10.20), $z_{ii} > 0$, and $\varepsilon_i = \pm 1, i = 1, \ldots, n$, then

$$a_{kk} = \varepsilon_k z_{kk}^2 + \sum_{j=1}^{k-1} \varepsilon_j |z_{jk}|^2, \qquad k = 1, \ldots, n,$$

$$a_{ik} = \varepsilon_i z_{ii} z_{ik} + \sum_{j=1}^{i-1} \varepsilon_j \bar{z}_{ji} z_{jk}, \qquad i > k, \ i, k = 1, \ldots, n.$$

Therefore (10.21), (10.22), and (10.23) necessarily hold and these relations determine Z and E uniquely since, as the theorem claims.

we can successively calculate the numbers z_{ik} and ε_k from (10.21) to (10.23) in the order shown above. □

Remark. For a Hermitian (real symmetric) strongly nonsingular matrix A, Theorem 10.1.4 and the algorithm presented can be used as an efficient numerical criterion for the positive definiteness of A (the case of $\varepsilon_1 = \cdots = \varepsilon_n = 1$) and as a method for the calculation of the inertia of A (since $\mathrm{In}\, A = \mathrm{In}\, E$).

10.2 General case

Another way how to solve the system (10.1) numerically, even in the case when A is not a square but an $m \times n$ matrix with $m \geq n$, is based on a theorem close to the Gram-Schmidt orthonormalization process 2.4.1. We present it without proof [11].

Theorem 10.2.1 *Let A be an $m \times n$ matrix, $m \geq n$, and let its columns be linearly independent. Then there exists a unique pair Q and R of matrices, Q being an $m \times n$ and R an $n \times n$ matrix such that*

$$Q^*Q = D = \mathrm{diag}(d_1, \ldots, d_n) \qquad (10.24)$$

is a diagonal matrix having positive diagonal entries and R is an upper triangular matrix having 1's on the diagonal, and, moreover,

$$A = QR. \qquad (10.25)$$

The matrix $Q = [Q_1, \ldots, Q_n]$ (and $R = [r_{ik}]$ as well) can be obtained as the matrix $A^{(n)}$ in the sequence $A^{(0)} = A$, $A^{(1)}, \ldots, A^{(n)}$, where

$$A^{(k)} = [Q_1, Q_2, \ldots, Q_k \cdot A^{(k)}_{k+1}, \ldots, A^{(k)}_n], \quad k = 1, \ldots, n,$$

is calculated from $A^{(k-1)} = [Q_1, Q_2, \ldots, Q_{k-1}, A^{(k-1)}_k, \ldots, A^{(k-1)}_n]$ according to the formulae

$$Q_k = A^{(k-1)}_k, \quad d_k = \langle Q_k, Q_k \rangle, \quad r_{kk} = 1, \qquad (10.26)$$

$$r_{ki} = \frac{\langle Q_k, A^{(k-1)}_i \rangle}{d_k}, \quad A^{(k)}_i = A^{(k-1)}_i - r_{ki} Q_k,$$

$$i = k+1, \ldots, n.$$

The factorization (10.25) can be used to solve both the system (10.1) with a nonsingular matrix A and the so-called *overdetermined system*

$$Ax = b$$

with an $m \times n$ matrix A, $m > n$, as soon as the columns of A are linearly independent.

Theorem 10.2.2 *Let the columns of the $m \times n$ matrix A of the system*

$$Ax = b \qquad (10.27)$$

be linearly independent. If $A = QR$ is the factorization of A of Theorem 10.2.1, then the solution x_0 of the system

$$Rx = D^{-1}Q^*b \qquad (10.28)$$

fulfils the equation

$$A^*(b - A\hat{x}) = 0; \qquad (10.29)$$

this relation characterizes vectors $\hat{x}$ such that among all vectors x they minimize the Euclidean norm of the residual $b - Ax$, i.e.,

$$g_2(b - A\hat{x}) \le g_2(b - Ax) \qquad (10.30)$$

for any vector x.

Proof. First we shall show that (10.29) implies (10.30). Indeed, for an arbitrary vector x and a vector $\hat{v}$ satisfying (10.29) we have

$$
\begin{aligned}
g_2^2(b - Ax) &= \langle b - Ax, b - Ax \rangle \\
&= \langle b - A\hat{x} + A(\hat{x} - x), b - A\hat{x} + A(\hat{x} - x) \rangle \\
&= \langle b - A\hat{x}, b - A\hat{x} \rangle + \langle b - A\hat{x}, A(\hat{x} - x) \rangle + \\
&\quad + \langle A(\hat{x} - x), b - A\hat{x} \rangle + \langle A(\hat{x} - x), A(\hat{x} - x) \rangle \\
&= g_2^2(b - A\hat{x}) + g_2^2(A(\hat{x} - x)) \\
&\ge g_2^2(b - A\hat{x})
\end{aligned}
$$

since $\langle A(\hat{x} - x), b - A\hat{x} \rangle = \langle \hat{x} - x, A^*(b - A\hat{x}) \rangle = 0$, and thus also

$$\langle b - A\hat{x}, A(\hat{x} - x) \rangle = 0.$$

Let now $Rx_0 = D^{-1}Q^*b$. Then

$$A^*(b - Ax_0) = R^*Q^*(b - QRx_0)$$
$$= R^*Q^*b - R^*Q^*QD^{-1}Q^*b = 0$$

as $Q^*QD^{-1} = I$. □

Remark. The equations (10.29) for $\hat{x}$ (having a nonsingular matrix!) are called *normal equations* corresponding to the (overdetermined or nonsingular) system (10.27). We have thus found a solution of the normal equations by the method described. This solution, however, is unique since the matrix A^*A is nonsingular. Indeed, if A^*A were singular, then by Theorem 1.5.3 there would exist a vector $y \neq 0$ such that $A^*Ay = 0$. But then $0 = y^*A^*Ay = \langle Ay, Ay \rangle$, i.e. $Ay = 0$. which contradicts the fact that the columns of A are linearly independent. Hence, x_0 of (10.28) is the unique solution of the system (10.29).

We saw in Theorem 2.7.1 that any $m \times n$ matrix A can be written as

$$A = U \begin{bmatrix} D & 0 \\ 0 & 0 \end{bmatrix} V^*, \tag{10.31}$$

where U and V are unitary matrices and D is a diagonal matrix (whose order equals the rank r of A).

Put

$$A^+ = V \begin{bmatrix} D^{-1} & 0 \\ 0 & 0 \end{bmatrix} U^*, \tag{10.32}$$

where A^+ is an $n \times m$ matrix.

Theorem 10.2.3 *If A is an $m \times n$ matrix written in the form (10.31), then the matrix A^+ of (10.32) fulfils the following relations:*

1. $AA^+A = A$.
2. $A^+AA^+ = A^+$.
3. $(AA^+)^* = AA^+$.
4. $(A^+A)^* = A^+A$.

Proof. By (10.31) and (10.32),

$$AA^+ = U \begin{bmatrix} I & 0 \\ 0 & 0 \end{bmatrix} U^*, \qquad A^+A = V \begin{bmatrix} I & 0 \\ 0 & 0 \end{bmatrix} V^*,$$

where AA^+ is an $m \times m$ matrix, A^+A an $n \times n$ matrix, and I the identity matrix of order r. Hence 3. and 4. hold. Right-multiplying the first equality by the matrix A from (10.31) we come to 1. Left-multiplying this equality by the matrix A^+ from (10.32), we obtain 2.   □

Remark. The matrix A^+ fulfilling 1. to 4. of Theorem 10.2.3 is called the *pseudoinverse* or the *Moore-Penrose inverse* of A. The pseudoinverse A^+ is uniquely determined by the conditions 1. to 4. and the matrix A.[1] Hence, even though the factorization (10.31) need not be unique, the resulting A^+ defined by (10.32) is always the same.

Theorem 10.2.4 *If the columns of a matrix A are linearly independent, then A^*A is nonsingular and $A^+ = (A^*A)^{-1}A^*$.*

Proof. The first statement was proved in the remark following Theorem 10.2.2. It thus suffices to verify that the matrix $(A^*A)^{-1}A^*$ satisfies 1. to 4. of Theorem 10.2.3.   □

Theorem 10.2.5 *If the columns of a matrix A are linearly independent, then the solution x_0 of the normal equation*

$$A^*(b - Ax) = 0$$

corresponding to the system

$$Ax = b$$

can be expressed as

$$x_0 = A^+b.$$

Proof. We have shown that under the hypothesis of the theorem the solution of the normal equations is unique. It is therefore sufficient to verify that $x_0 = A^+b$ fulfils the relation

$$A^*(b - Ax_0) = 0.$$

By Theorem 10.2.4, however, $A^+ = (A^*A)^{-1}A^*$ so that

$$A^*(b - AA^+b) = A^*b - A^*A(A^*A)^{-1}A^*b = 0.$$

□

---

[1]See [50].

Remark. The solution of the system $Ax = b$ can be expressed in the form $x = A^+b$ also in the case of an $m \times n$ matrix A with $m < n$, i.e. in the case of an *underdetermined* linear system. If the system in question has a solution, then A^+b is the (unique) solution of the system $Ax = b$ whose Euclidean norm $g_2(x)$ is minimal[2]. Here, A^+ is again the (unique) matrix fulfilling 1. to 4. of Theorem 10.2.3.

The QR factorization of Theorem 10.2.1 is suitable also for the computation of the pseudoinverse of a matrix having linearly independent columns.

Theorem 10.2.6 *Let the columns of A be linearly independent and let $A = QR$ be the factorization of Theorem 10.2.1. Then the pseudoinverse A^+ of A satisfies the relation*

$$A^+ = R^{-1}D^{-1}Q^*. \tag{10.33}$$

Proof. Let $A = QR$. Since $A^+ = (A^*A)^{-1}A^*$ by Theorem 10.2.4, we have

$$A^+ = (R^*Q^*QR)^{-1}R^*Q^*$$
$$R^{-1}D^{-1}(R^*)^{-1}R^*Q^* = R^{-1}D^{-1}Q^*.$$

$\square$

In the conclusion we mention several approaches to solving systems of linear algebraic equations having special matrices. The case of a band matrix was treated in Chapter 8. In Chapter 14 we shall consider this problem for a large sparse matrix, i.e. a matrix having relatively few nonzero entries. In addition, we now present the so-called *conjugate gradient method* for a positive definite matrix A without proof [53].

Theorem 10.2.7 *Let*

$$Ax = b$$

be a linear system with a positive definite matrix A of order n. Choosing a vector x_0 and putting $p_0 = r_0 = b - Ax_0$, we recur-

---

[2]See [44].

sively calculate

$$a_i = \frac{\langle r_i, r_i \rangle}{\langle Ap_i, p_i \rangle},$$
$$x_{i+1} = x_i + a_i p_i,$$
$$r_{i+1} = r_i - a_i A p_i,$$
$$b_i = \frac{\langle r_{i+1}, r_{i+1} \rangle}{\langle r_i, r_i \rangle},$$
$$p_{i+1} = r_{i+1} + b_i p_i$$

for $i = 0, 1, \ldots, n - 1$.

If $r_i \neq 0$ for all $i < n$, then x_n is the solution. If $r_i = 0$ occurs for some $i < n$, then x_i is the solution.

Exercises

10.1. Calculate the condition number $\kappa(A)$ (accurate to 3 decimal places) of the matrix

$$\begin{bmatrix} 0.001 & 1 \\ 1 & 999 \end{bmatrix}$$

a) with respect to the maximum norm,

b) with respect to the Euclidean norm.

10.2. Solve the linear system

$$x_1 + 1000x_2 = 1000,$$
$$x_1 + 0.001x_2 = 1$$

using two-digit floating-point arithmetic and apply

a) the partial pivoting,

b) the balancing and partial pivoting.

10.3. Construct and solve the normal equations corresponding to the system

$$\tfrac{1}{2}x_1 + \tfrac{1}{2}x_2 = 1,$$
$$x_1 \qquad = \tfrac{1}{2},$$
$$x_2 = \tfrac{1}{2}.$$

10.4. Calculate the pseudoinverse of the matrix

$$\begin{bmatrix} \frac{1}{2} & 1 & 0 \\ \frac{1}{2} & 0 & 1 \end{bmatrix}.$$

10.5. Find the pseudoinverse of the matrix

$$\begin{bmatrix} 1 & 2 & -3 \\ -2 & -4 & 6 \end{bmatrix}.$$

10.6. If U and V are unitary matrices prove that the relations $(UA)^* = A^+U^*$ and $(AV)^+ = V^*A^+$ hold for a matrix A such that the products exist.

10.7. If A^+ is the pseudoinverse of A prove that A is the pseudoinverse of A^+.

11

Iterative Methods for Solving Linear Systems

If we wish to solve an equation (or a system of equations) of the form $x = f(x)$, where x is an unknown quantity or vector and f a function or a vector of functions, we may use an *iterative method* and proceed as follows: We choose an initial approximation x_0 and calculate $x_1, x_2, \ldots$ successively from the formula

$$x_{k+1} = f(x_k), \quad k = 0, 1, \ldots, \tag{11.1}$$

If the sequence $\{x_k\}$ of numbers or vectors converges, then its limit is the solution of the problem given (if the function f is continuous in a neighborhood of the limit).

The iterative methods are advantageous from the viewpoint of computer programming and calculation. However, it is often a problem to find a transformation of the equation (or the system of equations) $g(x) = 0$ into the form $x = f(x)$ such that the presented iterative method (11.1) gives a converging sequence if the initial approximation x_0 is chosen within a rather wide range and not only in the close neighborhood of the unknown solution and, moreover, such that the method converges sufficiently rapidly.

11.1 General case

We shall be concerned with the case where x is an n-component column vector and the function $f(x)$ is of the form $Ax + b$ with a square matrix A of order n and a fixed vector b. Then we have the following theorem.

Theorem 11.1.1 *The iterative method*

$$x_{k+1} = Ax_k + b, \quad k = 0, 1, \ldots, \tag{11.2}$$

for solving the linear system

$$(I - A)x = b \tag{11.3}$$

converges for any initial vector x_0 and any vector b if and only if

$$\rho(A) < 1 \tag{11.4}$$

holds for the spectral radius $\rho(A)$. The limit $\hat{x}$ of the sequence $\{x_k\}$ is then the solution of (11.3).

A condition sufficient for (11.4) *is: For some norm $g(A)$ of the matrix A induced by a vector norm g, $g(A) < 1$.*

In this case, we have the estimates

$$g(\hat{x} - x_k) \leq [g(A)]^k g(x_0) + [g(A)]^k [1 - g(A)]^{-1} g(b), \tag{11.5}$$

$$g(\hat{x} - x_k) \leq g(A)[1 - g(A)]^{-1} g(x_k - x_{k-1}). \tag{11.6}$$

Proof. Let (11.4) be satisfied. Then the matrix $I - A$ is nonsingular since 1 is not an eigenvalue of A. By Theorem 1.7.13,

$$\lim_{k \to \infty} A^k = 0. \tag{11.7}$$

We now show by induction that

$$x_k = A^k x_0 + (I - A^k)(I - A)^{-1} b. \tag{11.8}$$

This is true for $k = 1$. Let thus $k > 1$ and suppose that (11.8) holds for $k - 1$. Then

$$\begin{aligned}
x_k &= Ax_{k-1} + b \\
&= A(A^{k-1} x_0 + (I - A^{k-1})(I - A)^{-1} b) + b \\
&= A^k x_0 + (A - A^k)(I - A^k)(I - A)^{-1} b + (I - A)(I - A)^{-1} b \\
&= A^k x_0 + (I - A^k)(I - A)^{-1} b,
\end{aligned}$$

which is (11.8). The relations (11.7) and (11.8) imply immediately that

$$\lim_{k \to \infty} x_k = \hat{x} = (I - A)^{-1} b, \tag{11.9}$$

which is, of course, a solution of the system (11.3) and, moreover, this solution is unique.

To prove (11.5), we employ the properties of norms from Chapter 9. By (11.8), (11.9), and the result of Exercise 9.11, we have

$$
\begin{aligned}
g(x_k - \hat{x}) &= g(A^k x_0 - A^k(I - A)^{-1}b) \\
&\leq g(A^k)g(x_0) + g(A^k)g((I - A)^{-1})g(b) \\
&\leq (g(A))^k g(x_0) + (g(A))^k(1 - g(A))^{-1})g(b).
\end{aligned}
$$

Further,

$$
\begin{aligned}
x_k - \hat{x} &= A(x_{k-1} - \hat{x}) \\
&= A(x_{k-1} - x_k + x_k - \hat{x}) \\
&= A(x_{k-1} - x_k) + A(x_k - \hat{x}).
\end{aligned}
$$

This implies

$$
g(x_k - \hat{x}) \leq g(A)g(x_{k-1} - x_k) + g(A)g(x_k - \hat{x})
$$

and, therefore,

$$
g(x_k - \hat{x}) \leq g(A)[1 - g(A)]^{-1}g(x_{k-1} - x_k),
$$

which proves (11.6).

Let now (11.2) converge for any x_0 and any b. Let α be an eigenvalue of A and u an associated eigenvector,

$$
Au = \alpha u.
$$

For $b = u$ and $x_0 = 0$ we then easily obtain by induction that

$$
x_k = (1 + \alpha + \cdots + \alpha^{k-1})u, \quad k = 1, 2, \ldots,
$$

and the condition for convergence gives

$$
|\alpha| < 1.
$$

This means that really $\rho(A) < 1$. □

11.2   The Jacobi method

We present one of the possible transformations of a general system of the form

$$Cx = y, \tag{11.10}$$

where $C = [c_{ik}]$, $i, k = 1, \ldots, n$, into the form (11.3). We write

$$C = D - \tilde{C}, \tag{11.11}$$

where $D = \operatorname{diag}(c_{11}, c_{22}, \ldots, c_{nn})$. If $c_{ii} \neq 0$ for all i, we put

$$A = D^{-1}\tilde{C}, \quad z = D^{-1}y.$$

The iterative method (11.2) is then called the *Jacobi method*. It can also be described as follows: To solve the system (11.10), we choose an initial vector $x^{(0)}$ with components $x_1^{(0)}, x_2^{(0)}, \ldots, x_n^{(0)}$. We substitute the numbers $x_2^{(0)}, \ldots, x_n^{(0)}$ for $x_2, \ldots, x_n$ into the first equation of the system and, calculating x_1, we obtain a new $x_1^{(1)}$. We substitute the numbers $x_1^{(0)}, x_3^{(0)}, \ldots, x_n^{(0)}$ for $x_1, x_3, \ldots, x_n$ into the second equation and calculate $x_1^{(1)}$. We proceed in this way until we calculate $x_n^{(1)}$ from the last equation. The whole procedure is then repeated, that is, we substitute $x_k^{(1)}$ and calculate $x_k^{(2)}$ and so forth.

This process can be expressed by the formula

$$Dx^{(k+1)} - \tilde{C}x^{(k)} = y, \tag{11.12}$$

or

$$x^{(k+1)} = D^{-1}\tilde{C}x^{(k)} + D^{-1}y.$$

Hence, by the general Theorem 11.1.1 the condition

$$\rho(D^{-1}\tilde{C}) < 1 \tag{11.13}$$

guarantees the convergence of the Jacobi method (11.12) for an arbitrary choice of the initial vector $x^{(0)}$ (and for any right-hand side y). The next theorem is concerned with a simpler condition for convergence.

Theorem 11.2.1 *If C is a diagonally dominant matrix (in the sense of Chapter 5), then the Jacobi method always converges.*

Proof. Let $C = [c_{ik}]$ be a diagonally dominant matrix of order n, i.e., let there exist positive numbers $h_1, \ldots, h_n$ such that

$$|c_{ii}|h_i > \sum_{k \neq i} |c_{ik}|h_k, \quad i = 1, \ldots, n. \tag{11.14}$$

As in (11.11), we put

$$C = D - \tilde{C},$$

where $D = \text{diag}(c_{11}, c_{22}, \ldots, c_{nn})$ is the "diagonal part" and $-\tilde{C}$ the "off-diagonal part" of C. We further put $H = \text{diag}(h_1, \ldots, h_n)$. The norm $g_3(Z) = \max_i \sum_k |z_{ik}|$ of the matrix $Z = H^{-1}D^{-1}\tilde{C}H$ is then less than 1. Indeed, by (11.14)

$$\sum_{k \neq i} |z_{ik}| = \frac{1}{|c_{ii}|h_i} \sum_{k \neq i} |c_{ik}|h_k < 1.$$

Hence, $\rho(Z) < 1$ by Theorem 9.1.2. But the matrices Z and $D^{-1}\tilde{C}$ are similar and, therefore, they have the same eigenvalues in virtue of Theorem 1.7.2. Thus

$$\rho(D^{-1}\tilde{C}) < 1$$

and the Jacobi method always converges. □

Theorem 11.2.2 *If C is a Hermitian (real symmetric) matrix and D a positive definite Hermitian (real symmetric) matrix, then the method*

$$Dx^{(k+1)} = (D - C)x^{(k)} + y \tag{11.15}$$

for the solution of the system (11.10) converges for any initial vector $x^{(0)}$ and any vector y if and only if both C and $2D - C$ are positive definite matrices.

Proof. Since D is a nonsingular matrix the iteration (11.15) is equivalent to the iteration

$$x^{(k+1)} = (I - D^{-1}C)x^{(k)} + D^{-1}y.$$

By Theorem 11.1.1, this iteration converges if and only if

$$\rho(I - D^{-1}C) < 1 \tag{11.16}$$

holds. By the result of Exercise 2.11, there exists a positive
definite matrix W such that $W^2 = D$. Now

$$W(I - D^{-1}C)W^{-1} = I - W^{-1}CW^{-1}$$

so that the matrix $I - D^{-1}C$ has the same eigenvalues as the Hermitian (symmetric) matrix $I - W^{-1}CW^{-1}$ whose all eigenvalues
are real by Theorem 2.4.5. The inequality (11.16) is thus satisfied
if and only if these eigenvalues belong to the open interval $(-1, 1)$,
i.e., if and only if the eigenvalues of the matrix $W^{-1}CW^{-1}$ are
contained in the interval $(0, 2)$. But this holds if and only if both
the matrices $W^{-1}CW^{-1}$ and $2I - W^{-1}CW^{-1}$ are positive definite. that is, in virtue of Theorem 2.5.9, if and only if both the
matrices C and $W(2I - W^{-1}CW^{-1})W = 2D - C$ are positive
definite.   □

Remark. Hence, if all the diagonal entries of a Hermitian (or
real symmetric) matrix C are positive and if D is the diagonal
part of C, then the Jacobi method converges for any initial vector
and any right-hand side if and only if C and $2D - C$ are positive
definite matrices.

11.3   The Gauss-Seidel method

Solving the system (11.10), we may proceed also in another way.
We calculate a new $x_1^{(1)}$ from the first equation as in the Jacobi
method. We then substitute this new $x_1^{(1)}$ and not the "old" $x_1^{(0)}$
into the second equation and calculate $x_2^{(1)}$. Into the third equation we already substitute new $x_1^{(1)}, x_2^{(1)}$, and old $x_4^{(0)}, \ldots, x_n^{(0)}$
and calculate $x_3^{(1)}$, etc. To employ matrix notation in the description of this process, put

$$C = D - L - U, \tag{11.17}$$

where D is the diagonal part of C as before, L is the part of C
below the diagonal, and U is the part above the diagonal, both
taken with the negative sign,

$$L = [\ell_{ik}], \quad \ell_{ik} = -c_{ik} \quad \text{for } i > k, \quad \ell_{ik} = 0 \quad \text{for } i \leq k,$$
$$U = [u_{ik}], \quad u_{ik} = -c_{ik} \quad \text{for } i < k, \quad u_{ik} = 0 \quad \text{for } i \geq k.$$

For the vector $x^{(k+1)}$ we then clearly have

$$(D - L)x^{(k+1)} - Ux^{(k)} = y, \tag{11.18}$$

or

$$x^{(k+1)} = (D - L)^{-1}Ux^{(k)} + (D - L)^{-1}y. \tag{11.19}$$

This is the *Gauss-Seidel method*. Its convergence is established in the following theorem. Especially the second statement of the theorem is important for practical computation.

Theorem 11.3.1 *The Gauss-Seidel method converges for any choice of the initial vector $x^{(0)}$ and any right-hand side y if*

$$\rho((D - L)^{-1}U) < 1. \tag{11.20}$$

This condition is satisfied if C is diagonally dominant.

Proof. The first statement follows from (11.19) by Theorem 11.1.1. Let now C be diagonally dominant, i.e., let (11.14) hold. Put $H = \mathrm{diag}(h_1, \dots, h_n)$. Express the matrix C as in (11.17) and write

$$\tilde{L} = H^{-1}LH, \qquad \tilde{U} = H^{-1}UH.$$

To establish the convergence of the Gauss-Seidel method, it is sufficient to prove (11.20), which is equivalent to

$$\rho((D - \tilde{L})^{-1}\tilde{U}) < 1.$$

By Theorem 9.1.2, this condition is satisfied if

$$g_3((D - \tilde{L})^{-1}\tilde{U}) < 1$$

holds for the norm g_3. To prove this last inequality, we use the fact that, by (9.3), there exists a vector $u \neq 0$ such that

$$g_3((D - \tilde{L})^{-1}\tilde{U}) = \frac{g_3((D - \tilde{L})^{-1}\tilde{U}u)}{g_3(u)}.$$

Denote $(D\tilde{L})^{-1}\tilde{U}u$ as $v = [v_1, \dots, v_n]^T$. Let $g_3(v) = \max_k |v_k| = |v_s|$. The equality

$$(D - \tilde{L})v = \tilde{U}u$$

then implies that

$$c_{ss}v_s = \sum_{i=1}^{s-1} c_{si}h_i h_s^{-1} v_i = - \sum_{i=s+1}^{n} c_{si}h_i h_s^{-1} u_i$$

for $u = [u_1, \ldots, u_n]^T$, from which it follows that

$$g_3(v) = |v_s| \leq \sum_{i=1}^{s-1} \frac{|c_{si}|h_i}{|c_{ss}|h_s}|v_i| + \sum_{i=s+1}^{n} \frac{|c_{si}|h_i}{|c_{ss}|h_s}|u_i|$$

$$\leq a_s g_3(v) + b_s g_3(u),$$

where

$$a_s = \frac{1}{|c_{ss}|h_s} \sum_{i=1}^{s-1} |c_{si}|h_i, \quad b_s = \frac{1}{|c_{ss}|h_s} \sum_{i=s+1}^{n} |c_{si}|h_i.$$

Since $a_s + b_s < 1$, the above inequality implies

$$g_3((D - \tilde{L})^{-1}\tilde{U}) = \frac{g_3(v)}{g_3(u)} < 1.$$

$\square$

The next important theorem is also concerned with the convergence of the Gauss-Seidel method.

Theorem 11.3.2 *The Gauss-Seidel method converges if the matrix C is positive definite.*

We do not prove this theorem as it is a particular case of a more general theorem.

Theorem 11.3.3 *Let C be a positive definite matrix. Let*

$$C = D - L - L^*, \tag{11.21}$$

where the matrix D is positive definite as well. Then the matrix $D - L$ is nonsingular and the iteration

$$(D - L)x^{(k+1)} = L^*x^{(k)} + y, \tag{11.22}$$

called the generalized Gauss-Seidel method, converges for any initial vector $x^{(0)}$ and any right-hand side y.

Proof. Suppose first that the matrix $D - L$ is singular. By Theorem 1.5.3, there exists a vector $z \neq 0$ such that

$$(D - L)z = 0. \tag{11.23}$$

But since

$$z^*(C + D)z = z^*Cz + z^*Dz > 0$$

and, on the other hand,

$$\begin{aligned}
z^*(C + D)z &= z^*(D - L - L^* + D)z \\
&= z^*(D - L^*)z + z^*(D - L)z \\
&= 0
\end{aligned}$$

by (11.23), we obtained a contradiction. The matrix $D - L$ is thus nonsingular and the iteration (11.22) is well defined. It can be rewritten as

$$x^{(k+1)} = (D - L)^{-1}L^*x^{(k)} + (D - L)^{-1}y.$$

By Theorem 11.1.1, it is sufficient to prove that $\rho((D - L)^{-1}L^*) < 1$. First we prove a lemma.

Lemma 11.3.4 *Let c be a real positive number and let s and ω be complex numbers such that*

$$s + \bar{s} + c > 0, \qquad \bar{s} = \omega(s + c).$$

Then $|\omega| < 1$.

Proof. If $s = 0$, then $\omega = 0$ and the statement holds. Let thus $s \neq 0$. Then $|\omega| \neq 0$ and

$$\begin{aligned}
|s|^2 &= s\bar{s} \\
&= \omega\bar{\omega}(s + c)(\bar{s} + c) \\
&= |\omega|^2[s\bar{s} + (s + \bar{s})c + c^2] \\
&= |\omega|^2[|s|^2 + c(s + \bar{s} + c)] \\
&> |\omega|^2|s|^2.
\end{aligned}$$

Therefore $|\omega| < 1$. □

We now employ the lemma. Let λ be the eigenvalue of the matrix $(D - L)^{-1}L^*$ such that $|\lambda| = \rho((D - L)^{-1}L^*)$. If u is an associated eigenvector then

$$(D - L)^{-1}L^*u = \lambda u,$$

or

$$L^*u = \lambda(D - L)u.$$

Left-multiplying this equality by the vector u^*, we obtain $u^*L^*u = \lambda(u^*Du - u^*Lu)$. Put $u^*Lu = -s$, $u^*Du = c$, and $\lambda = -\omega$. Then $c > 0$, $s + \bar{s} + c = u^*(D - L - L^*)u = u^*Cu > 0$, and also $\bar{s} = \omega(s + c)$. By the lemma, $|\lambda| = |\omega| < 1$ as was to be proved.   □

11.4   The SOR method

To solve the system (11.11) we may use – instead of the Gauss-Seidel iteration (11.18) – a more general iterative method depending on one real parameter ω,

$$(D - \omega L)x^{(k+1)} = ((1 - \omega)D + \omega U)x^{(k)} + \omega y, \qquad (11.24)$$

where D is again a diagonal matrix, L is a lower and U is an upper triangular matrix, and

$$C = D - L - U.$$

This procedure is called the *successive overrelaxation* (SOR) *method*. For $\omega = 1$ it is identical with the Gauss-Seidel method. (Strictly speaking, the term "overrelaxation" should be used only with $\omega > 1$.) Clearly, (11.24) is equivalent to

$$x^{(k+1)} = (D - \omega L)^{-1}((1 - \omega)D + \omega U)x^{(k)} + (D - \omega L)^{-1}\omega y \quad (11.25)$$

with the matrix $(D - \omega L)^{-1}((1 - \omega)D + \omega U)$ whose spectral radius determines whether the method converges.

Theorem 11.4.1 *A necessary condition for the method* (11.24) *to converge for any initial vector* $x^{(0)}$ *and any right-hand side* y *is that* $0 < \omega < 2$.

Proof. Let $\lambda_1, \lambda_2, \ldots, \lambda_n$ be the eigenvalues of the matrix

$$(D - \omega L)^{-1}((1 - \omega)D + \omega U)$$

(some of them, however, may coincide). The spectral radius ρ of this matrix is $\max_i |\lambda_i|$ and its determinant is

$$\det(D - \omega L)^{-1} \det((1 - \omega)D + \omega U) =$$
$$(\det D)^{-1}(1 - \omega)^n \det D = (1 - \omega)^n$$

since both $D - \omega L$ and $(1 - \omega)D + \omega U$ are triangular matrices. By (1.39) then

$$\lambda_1 \lambda_2 \ldots \lambda_n = (1 - \omega)^n$$

and, moreover,

$$\rho^n \geq |\lambda_1||\lambda_2| \ldots |\lambda_n| = |1 - \omega|^n,$$

which finally implies

$$\rho \geq |1 - \omega|.$$

Hence, in order to obtain $\rho < 1$ we must necessarily have $|1 - \omega| < 1$, that is $0 < \omega < 2$. $\square$

The converse of the theorem holds in some cases as we shall see in the next theorem where the matrix D from (11.24) may be more general than only diagonal.

Theorem 11.4.2 *Let C be a positive definite matrix, $C = D - L - U$, $U = L^*$, where D is positive definite as well. Then for any ω, $0 < \omega < 2$, the method (11.24) converges (for any $x^{(0)}$ and y).*

Proof. Let $0 < \omega < 2$. Put

$$L_1 = (1 - \omega)D + \omega L,$$
$$D_1 = (2 - \omega)D.$$

Therefore, D_1 is positive definite and

$$C_1 = D_1 - L_1 - L_1^* = \omega(D - L - L^*) = \omega C$$

is positive definite as well. By Theorems 11.3.3 and 11.3.1 then $\rho((D_1 - L_1)^{-1}L_1^*) < 1$. But

$$(D_1 - L_1)L_1^* = (D - \omega L)^{-1}((1 - \omega)D + \omega L^*)$$

and thus (11.24) really always converges.   □

Given a system $Cx = y$, the Gauss-Seidel method as well as, more generally, the successive overrelaxation method depend on the order in which the new components are substituted for the old in the equations whereas the Jacobi method is independent of this order. We suppose here that the kth component is calculated from the kth equation of the system.

The choice of this order is important since it allows us to influence the rate of convergence of the method. The rate of convergence is usually measured by the spectral radius of the corresponding matrix, i.e., by the number

$$\rho((D - L)^{-1}U)$$

in the Gauss-Seidel method and by the number

$$\rho((D - \omega L)^{-1}((1 - \omega)D + \omega U))$$

in the successive overrelaxation method with parameter ω. Clearly, the less the spectral radius of the method, the greater the rate of convergence becomes. It is interesting that in some cases, which rather frequently occur in practical computations, we are able to find even the optimal parameter of the successive overrelaxation method, i.e. the parameter ω_0 for which

$$\rho((D - \omega_0 L)^{-1}((1 - \omega_0)D + \omega_0 U)) \leq$$
$$\rho((D - \omega L)^{-1}((1 - \omega D + \omega U))$$

holds for all ω such that $0 < \omega < 2$, provided that we know the spectral radius $\rho(D^{-1}(L + U))$ of the Jacobi method. We shall see in Theorems 11.4.6 and 11.4.8 that this is true e.g. in the case of a symmetric matrix C from the class K that has the so-called *property* A,[1] which means that the digraph $\vec{G}(C)$ of C is symmetric and bipartite (its every polygon has an even number of edges).

By (3.11), the following theorem holds.

---

[1]See Young [62].

Theorem 11.4.3 *A symmetric or Hermitian matrix C has property A if and only if it may be transformed by a simultaneous permutation of rows and columns into the form*

$$\begin{bmatrix} D_1 & -L^* \\ -L & D_2 \end{bmatrix},$$

where D_1 and D_2 are diagonal matrices of order at least 1.

For the sake of brevity we introduce the following notion. The labelling $O(i)$ of the set of points V of the graph $G = (V, H)$ by numbers $1, 2, \ldots, |V|$ is called an *ordering* of G. The ordering $O(i)$ is said to be *consistent* if there exists another labelling $W(i)$ of V by positive integers (not necessarily distinct) such that

$$O(i) < O(k) \leftrightarrow W(i) = W(k) - 1 \qquad (11.26)$$

holds for each edge $(i, k) \in H$.

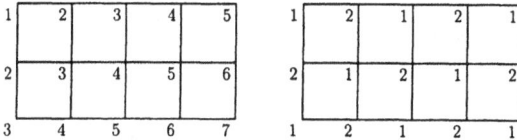
Fig. 16

For example, the ordering of the rectangular lattice $M_{3,5}$ "following the rows" and that "following the columns" are consistent, the corresponding labelling $W(k)$ for both cases being given in Fig. 16a. The checker-board ordering of the same rectangular lattice (we label all "white" squares (points) first and all "black" squares afterwards) is also consistent and the corresponding labelling $W(k)$ is shown in Fig. 16b (where the top left-hand corner point is "white").

Theorem 11.4.4 *The points of an undirected graph G may be consistently ordered if and only if G is a bipartite graph. In this case, we obtain one such ordering if we first label all points of one subset (of the two subsets which constitute the partition of the set of the points of G) and then all points of the other subset.*

Proof. Let the graph $G = (V, H)$ be consistently ordered. Let $W(i)$ be the corresponding labelling of V. Suppose that G contains a polygon $(u_1, u_2, \ldots, u_{2k+1}, u_1)$ of odd length $2k+1$. Summing up the equalities

$$
\begin{aligned}
W(u_1) \;-\; W(u_2) &= \; \varepsilon_1 \;\; (= \pm 1), \\
W(u_2) \;-\; W(u_3) &= \; \varepsilon_2 \;\; (= \pm 1), \\
\vdots \qquad\qquad \vdots \qquad & \;\;\; \vdots \qquad \vdots \\
W(u_{2k+1}) \;-\; W(u_1) &= \; \varepsilon_{2k+1} \;\; (= \pm 1),
\end{aligned}
$$

we obtain 0 on the left-hand side whereas the right-hand side equals an odd number, which is a contradiction. Therefore G is a bipartite graph.

If G is a bipartite graph, then it suffices to label the points in the way suggested in the theorem. The labelling W for which $W(k) = 1$ if k is in the first subset and $W(k) = 2$ if k is in the second subset clearly meets the requirement of (11.26). □

The matrix formulation of this fact can be stated as follows.

Theorem 11.4.5 *Let a Hermitian (real symmetric) matrix C have property A, let the graph $G(C)$ be consistently ordered, and let the corresponding labelling $W(i)$ assume r distinct values. Then there exists a simultaneous permutation of rows and columns of C such that the resulting matrix is block tridiagonal with r diagonal blocks which themselves are diagonal matrices.*

Conversely, if C is a block tridiagonal Hermitian (symmetric) matrix whose diagonal blocks are diagonal matrices, then C has property A and the ordering of $G(C)$ corresponding to the natural ordering of rows is consistent.

Proof. It is sufficient to start with those rows and columns that correspond to the smallest value of $W(i)$, then continue with the rows and columns that correspond to the nearest larger value of $W(i)$, etc. The resulting matrix is block tridiagonal since $c_{ik} \neq 0$ only if either $i = k$ or $|W(i) - W(k)| = 1$.

The converse follows from the fact that it is sufficient to assign to the kth point the number of the block row in which the kth row is contained. □

Now we can prove the following important theorem.

Theorem 11.4.6 *Let the matrix C of the system $Cx = y$ be positive definite and have property A. Then every consistent ordering of the graph $G(C)$ corresponds to an ordering of the equations of the system for which the spectral radius of the Gauss-Seidel method is ρ^2 where ρ is the spectral radius of the corresponding Jacobi method; moreover, $\rho < 1$.*

More generally, with the same ordering of equations the spectral radius of the successive overrelaxation method with parameter ω is

$$\rho_\omega = \frac{1}{4}\left\{\omega\rho + \sqrt{\omega^2\rho^2 + 4(1-\omega)}\right\}^2 \qquad (11.27)$$

$$\textit{for } \omega \le \omega_0,$$

$$\rho_\omega = \omega - 1 \qquad \textit{for } \omega > \omega_0, \qquad (11.28)$$

where

$$\omega_0 = \frac{2}{1 + \sqrt{1 - \rho^2}}. \qquad (11.29)$$

The radius is smallest for $\omega = \omega_0$. Then

$$\rho_{\omega_0} = \omega_0 - 1 = \frac{1 - \sqrt{1 - \rho^2}}{1 + \sqrt{1 - \rho^2}}. \qquad (11.30)$$

Remark. By the theorem, the spectral radius of neither the Gauss-Seidel method nor the successive overrelaxation method depends on the choice of the consistent ordering. Note that under the conditions of the theorem there may be several such orderings (cf. Figures 16a, b).

Proof. Let a consistent ordering of $G(C)$ be given. Without loss of generality we may assume that the order of the equations in the system $Cx = y$ already corresponds to this ordering so that

$$i < k, \ \ell_{ki} \ne 0 \longrightarrow W(i) = W(k) - 1 \qquad (11.31)$$

holds for the corresponding labelling $W(i)$ from (11.26). Let D be the diagonal part of C, L the part of C below the diagonal, and L^* the part above the diagonal, both taken with the negative sign, so that

$$C = D - L - L^*.$$

The spectral radius ρ_1 of the corresponding Gauss-Seidel method equals $\max |\lambda|$ where λ runs through all the roots of the equation

$$\det(xD - xL - L^*) = 0.$$

More generally, the spectral radius ρ_ω of the corresponding successive overrelaxation method with parameter ω, $0 < \omega < 2$, equals $\max |\lambda|$ where λ runs through all the roots of the equation

$$\det((x + \omega - 1)D - \omega xL - \omega L^*) = 0. \tag{11.32}$$

Since the Gauss-Seidel method is a particular case of the successive overrelaxation method for $\omega = 1$, we shall prove directly (11.28), from which the first statement clearly follows.

Let d_i, $i = 1, \ldots, n$, be the diagonal entries of D (they are all positive by 3. of Theorem 2.5.1) and ℓ_{ik}, $i > k$, the entries of L. Suppose first that $\rho_\omega = 0$. Then all the roots of the equation (11.32) are zero and therefore $1 - \omega = 0$. Hence (11.32) is of the form

$$\det(xD - xL - L^*) = 0. \tag{11.33}$$

By (1.5), the left-hand side of (11.33) can be rewritten as

$$d_1 d_2 \ldots d_n \left[x^n - \left(\sum_{i<k} \frac{|\ell_{ik}|^2}{d_i d_k} \right) x^{n-1} + \ldots \right].$$

Since all the roots of (11.33) are zero we have

$$\sum_{i<k} \frac{|\ell_{ik}|^2}{d_i d_k} = 0,$$

which implies $L = 0$. But then also all the roots of the equation

$$\det(xD - L - L^*) = 0 \tag{11.34}$$

are zero. The maximum of the moduli of these roots is ρ. Hence also $\rho = 0$ and (11.28) holds.

We shall thus assume $\rho_\omega \neq 0$ in what follows. For an arbitrary nonzero number y, let $Z(y)$ be the nonsingular diagonal matrix

$$Z(y) = \text{diag}(y^{W(1)}, y^{W(2)}, \ldots, y^{W(n)}). \tag{11.35}$$

Let now λ be a nonzero root of the equation (11.32) and let u be an (in general complex) root of the equation

$$x^2 - \lambda = 0.$$

Introducing the matrix

$$(Z(u))^{-1}((\lambda + \omega - 1)D - \omega\lambda L - \omega L^*)Z(u), \qquad (11.36)$$

we shall show that it is identical with the matrix

$$(\lambda + \omega - 1)D - \omega u L - \omega u L^*. \qquad (11.37)$$

Since $Z(u)$ is a diagonal matrix the diagonal entries of both the matrices (11.36) amd (11.37) coincide and both these matrices also have the same zero entries. If now the (i, k)-entry, $i > k$, is nonzero, that is $\ell_{ik} \neq 0$, then by (11.31) the entry of the matrix (11.36) in this position equals

$$-\omega\lambda u^{W(i)+W(k)}\ell_{ik} \; (= -\omega\lambda u^{-1}\ell_{ik}) = -\omega u \ell_{ik}.$$

If $i < k$ and $\ell_{ki} \neq 0$, then the (i, k)-entry of the matrix (11.36) equals

$$-\omega u^{-W(i)+W(k)}\bar{\ell}_{ki} = -\omega u \bar{\ell}_{ki}.$$

Hence, the matrices (11.36) and (11.37) are identical. Moreover, the matrix (11.36) is singular by the hypothesis on λ. Therefore, the matrix (11.37) is singular as well, i.e. $\det(\mu D - L - L^*) = 0$ where $\mu = (\lambda + \omega - 1)/(\omega u)$. Substituting, in addition, u^2 for λ, we obtain

$$u^2 - \mu\omega u + \omega - 1 = 0. \qquad (11.38)$$

But the converse statement is also true: If μ is a root of the equation (11.34) and u a nonzero root of the equation (11.38), then u^2 is a root of the equation (11.32).

Using (11.31), we may – as above – verify that

$$(Z(-1))^{-1}(xD - L - L^*)Z(-1) = -(-xD - L - L^*)$$

holds for the matrix $Z(y)$ defined by (11.35) and any x. Hence, if x is a root of (11.34), then $-x$ is also a root of this equation.

Moreover, every such root is real. Indeed, if we introduce a diagonal matrix W such that $W^2 = D$, then W is real nonsingular and

$$xD - L - L^* = -W(W^{-1}(L + L^*)W^{-1} - xI)W$$

holds for any x. Each root of the equation (11.34) is thus an eigenvalue of the Hermitian (real symmetric) matrix $W^{-1}(L + L^*)W^{-1}$ and hence, by Theorem 2.4.5, this root is real. This implies that ρ itself is a root of (11.34).

If, in particular, $\omega = 1$, then every nonzero root μ of the equation (11.34) is also a root of (11.38), i.e., μ^2 is a root of the equation (11.32) with $\omega = 1$, and conversely. The root ρ (i.e., the spectral radius of the Jacobi method) is thus associated with the spectral radius $\rho_1 = \rho^2$ of the Gauss-Seidel method. Since $\rho_1 < 1$ by Theorem 11.3.2, we have also $\rho < 1$. The first statement of the theorem has been proved.

We now turn back to the equation (11.38). Its discriminant is $\omega^2\mu^2 + 4(1 - \omega)$. Hence, if

$$\mu^2 \geq \frac{4(\omega - 1)}{\omega^2} \tag{11.39}$$

holds for $\mu \geq 0$, then there are real numbers u_1 and u_2 satisfying (11.38) and corresponding to the root μ of (11.34) (and also to the root $-\mu$). Therefore, $\lambda_1 = u_1^2$ and $\lambda_2 = u_2^2$ are real roots of the equation (11.32). Further, $u_1 = \frac{1}{2}\omega\mu + \frac{1}{2}\sqrt{\omega^2\mu^2 + 4(1 - \omega)}$, which is the largest of the two roots u_1 and u_2, is an increasing function of μ (for $\mu \geq 0$). This implies that if $\rho^2 \geq 4(\omega - 1)/\omega^2$, which is equivalent to the inequality $\omega \leq \omega_0 = 2/[1 + \sqrt{1 - \rho^2}]$, then

$$\hat{u} = \frac{1}{2}\omega\rho + \frac{1}{2}\sqrt{\omega^2\rho^2 + 4(1 - \omega)} \tag{11.40}$$

is the largest of all the numbers obtained in this way.

If a nonnegative root μ of the equation (11.34) does not satisfy the inequality (11.39), i.e., if

$$\mu^2 < \frac{4(\omega - 1)}{\omega^2}, \tag{11.41}$$

then $\omega > 1$ and the roots u_1 and u_2 of (11.38) are complex conjugates with the common modulus $|u_1| = |u_2| = \sqrt{\omega - 1}$.

By contradiction, however, we can immediately prove that for $\omega \le \omega_0$ this number does not exceed the value of $\hat{u}$ from (11.40). Thus $\rho_\omega = \hat{u}^2$ in this case and (11.28) has been proved.

If now $\rho^2 < 4(\omega - 1)/\omega^2$, which is equivalent to $\omega > \omega_0$, then (11.41) holds for all the roots μ of (11.34). Hence, all the roots of all the equations (11.38) have the same modulus $\sqrt{\omega - 1}$, i.e., all the nonzero roots of (11.32) also have the same modulus $\omega - 1$ and, moreover, none of them is zero for $\omega \ne 1$. (This is a very interesting property by itself!) Therefore $\rho_\omega = \omega - 1$, which is (11.28). □

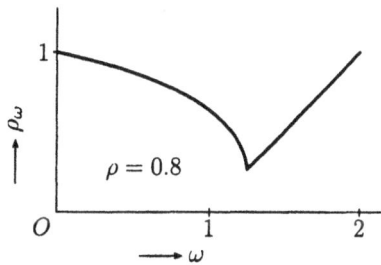

Fig. 17

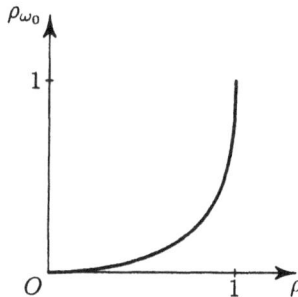

Fig. 18

Remark. The graph of the function ρ_ω from (11.28) and (11.28) is shown for $\rho = 0.8$ in Fig. 17. The value $\omega_0 = 1.25$ corresponds to the "tip". Fig. 18 displays the graph of the dependence of the optimal radius ρ_{ω_0} from (11.30) on ρ. It is apparent that for the values of ρ close to 1, which occur, in particular, in large systems arising from the solution of partial differential equations, the radius ρ_{ω_0} is substantially less than ρ. More accurate values are contained in the following table.

ρ	0.8	0.9	0.95	0.96	0.97	0.98	0.99
ρ_{ω_0}	0.25	0.393	0.524	0.562	0.609	0.668	0.753

ρ	0.995	0.997	0.998	0.999	0.9999	0.99999
ρ_{ω_0}	0.818	0.856	0.881	0.914	0.972	0.991

Asymptotically (as $n \to \infty$),

$$\rho_{\omega_0} \approx 1 - \frac{2\sqrt{2k}}{n} \quad \text{for} \quad \rho \approx 1 - \frac{k}{n^2},$$

where k is a suitable constant.

However, we must not omit the fact that in accordance with Fig. 17 the value of ρ_ω is very sensitive to the accuracy of ω_0 in the neighborhood of ω_0.

Further strong and useful theorems hold for the case of a monotone matrix C or a matrix C from the class K.

Theorem 11.4.7 *If the matrix C of the system $Cx = y$ is monotone and*

$$C = M - N,$$

where M is monotone and $N \geq 0$, then the iterative method

$$Mx^{(k+1)} = Nx^{(k)} + y \qquad (11.42)$$

converges for any initial vector and any right-hand side y.

Proof. Since (11.42) is equivalent to the iteration

$$x^{(k+1)} = M^{-1}Nx^{(k)} + M^{-1}y$$

the convergence is determined by the spectral radius $\rho(M^{-1}N) = \rho$. We have $M^{-1} \geq 0$ and thus $M^{-1}N \geq 0$. By Theorem 4.3.2, there exists a vector $z \neq 0$, $z \geq 0$, such that $M^{-1}Nz = \rho z$. Further

$$C^{-1}N = C^{-1}MM^{-1}N \; = C^{-1}(C+N)M^{-1}N$$
$$= (I + C^{-1}N)M^{-1}N.$$

from which

$$C^{-1}Nz = (I + C^{-1}N)M^{-1}Nz =$$
$$(I + C^{-1}N)\rho z = \rho z + C^{-1}N\rho z$$

follows. But the relations $C^{-1} \geq 0$ and $C^{-1}N \geq 0$ imply $\rho < 1$ since for $\rho \geq 1$ we should have $(\rho - 1)C^{-1}Nz + \rho z = 0$, which is impossible. □

In addition, we present a theorem without proof[2].

Theorem 11.4.8 *If the matrix C of the system $Cx = y$ belongs to the class K and has property A, then among all the Gauss-Seidel methods (which differ in the orderings of equations and unknowns) those corresponding to consistent ordering converge most rapidly.*

In the conclusion, note that we may also use the so-called *symmetric successive overrelaxation method* (SSOR): If $Cx = y$ is again a system with the matrix $C = D - L - U$, where D is the diagonal part of C, $-L$ the part of C below the diagonal, and $-U$ is the part above the diagonal, then this method can be viewed as a method whose each step consists of two "half-steps":

$$(D - \omega L)x^{(k+1/2)} = ((1 - \omega)D + \omega U)x^{(k)} + \omega y,$$
$$(D - \omega U)x^{(k+1)} = ((1 - \omega)D + \omega L)x^{(k+1/2)} + \omega y.$$

Therefore, the matrix realizing the calculation of $x^{(k+1)}$ from $x^{(k)}$ is

$$(D - \omega U)^{-1}((1 - \omega)D + \omega L)(D - \omega L)^{-1}((1 - \omega)D + \omega U)$$

and its spectral radius thus determines the convergence and the rate of convergence.

---

[2] See [59].

Exercises

11.1. Show that for the system of linear equations

$$3x_1 + 2x_2 + 2x_3 = 1,$$
$$2x_1 + 3x_2 + 2x_3 = 0,$$
$$2x_1 + 2x_2 + 3x_3 = -1$$

the Jacobi method does not converge whereas the Gauss-Seidel method does.

11.2. Show that for the system of linear equations

$$x_1 + x_2 \qquad\qquad = 1,$$
$$2(1 - \varepsilon)x_1 + x_2 + x_3 \qquad = 2,$$
$$x_3 + x_4 = -1,$$
$$-(1 - \varepsilon)^2 x_1 \qquad\qquad + x_4 = 5,$$

where ε is a small positive number (e.g. less than 0.1), the Jacobi method converges whereas the Gauss-Seidel method does not.

11.3. Find the spectral radius (accurate to 2 decimal places) of the successive overrelaxation method applied to the system of linear equations of Exercise 11.1 for the parameter a) $\omega = 1$; b) $\omega = 1.1$; c) $\omega = 0.9$; d) $\omega = 0.926$.

12

Matrix Inversion

In Chapter 1 we saw (cf. the proof of Theorem 1.3.4) that the problem of finding the inverse A^{-1} of a nonsingular $n \times n$ matrix A may be reduced to the solution of n linear systems with the same matrix A of coefficients:

If $Ax^{(i)} = e_i$, $e_i = [0, \ldots, 0, 1, 0, \ldots, 0]^T$ (where 1 is the i-th component), then A^{-1} can be constructed from the column vectors $x^{(i)}$ as the partitioned matrix $A^{-1} = [x^{(1)}, x^{(2)}, \ldots, x^{(n)}]$. To compute the inverse we can therefore use the methods of Chapter 10, especially the method of Gaussian elimination to find the factorization $A = BC$, where B and C are a lower and an upper triangular matrix, respectively (possible permutations are not taken into account). Then we have $A^{-1} = C^{-1}B^{-1}$ and the problem reduces to the inversion of the triangular matrices B and C. This problem, however, has been solved in (1.22) and (1.23).

12.1 Inversion of special matrices

We shall now show how the inverse of triangular matrices and some of its generalizations may be interpreted in graph-theoretical concepts. This is sometimes of importance for matrices containing only a small number of nonzero entries.

Theorem 12.1.1 *Let $A = [a_{ik}]$ be a nonsingular square matrix of order n which may be transformed into an upper triangular matrix by a simultaneous permutation of rows and columns. Then the digraph $\vec{G}(A)$ of the matrix A is acyclic (except for the loops) and the (i, k)-entry of $A^{-1} = [b_{ik}]$ is*

$$b_{ik} = \sum a_{ii}^{-1}(-a_{ii_1})a_{i_1 i_1}^{-1}(-a_{i_r k})a_{kk}^{-1}, \qquad (12.1)$$

where the summation is over all possible paths $(i, i_1, \ldots, i_r, k)$
from the vertex i *to the vertex* k *in* $\vec{G}(A)$ *(the subscript* r *is not,
of course, fixed here and may be also equal to zero). Moreover,
the same formula holds for the inverse of a partitioned matrix*
$A = [a_{ik}]$ *with square blocks* a_{ii} *which may be transformed into
an upper block triangular matrix by a simultaneous permutation
of block rows and block columns.*

Proof. Multiply A by the matrix $B = [b_{ik}]$, where the entries b_{ik}
are defined by (12.1). Putting $C = [c_{ik}] = AB$ we have

$$c_{ik} = \sum_j a_{ij} b_{jk}. \tag{12.2}$$

It is easy to see that $c_{ii} = \sum_j a_{ij} b_{ji} = a_{ii} b_{ii} = 1$ (or I_i in the
case of a partitioned matrix).

If $i \neq k$ we distinguish two cases: If there is no path from
the vertex i to the vertex k in $\vec{G}(A)$, then $a_{ik} = 0$. In addition,
$a_{ij} \neq 0$ for some $j \neq k$ implies $b_{jk} = 0$ since otherwise there
would be a path from i to k in $\vec{G}(A)$. Thus $c_{ik} = 0$ by (12.2). On
the other hand, if there is at least one path from i to k in $\vec{G}(A)$,
we have

$$c_{ik} = a_{ii} b_{ik} + \sum_{j \neq i} a_{ij} b_{jk} =$$
$$a_{ii} b_{ik} + \sum_{j \neq i} a_{ij} \sum a_{jj}^{-1}(-a_{jt_1}) a_{t_1 t_1}^{-1} \cdots (-a_{t_s k}) a_{kk}^{-1}$$

by (12.1) and (12.2). In the second summand, the index i does
not occur as t_p in any nonzero term as otherwise $\vec{G}(A)$ would
contain a cycle of length greater than 1. Therefore, the second
summand may be rewritten as

$$-a_{ii} \sum a_{ii}^{-1}(-a_{ij}) a_{jj}^{-1}(-a_{jt_1}) \cdots (-a_{t_s k}) a_{kk}^{-1},$$

where the summation is over all the paths from i to k in $\vec{G}(A)$.
Hence $c_{ik} = a_{ii} b_{ik} - a_{ii} b_{ik} = 0$, and $AB = I$. $\square$

We now consider the inverse of an almost acyclic matrix. A
matrix $A = [a_{ik}]$ is said to be *almost acyclic* if for all r-tuples of

distinct indices $i_1, i_2, \ldots, i_r$, $r \geq 3$, we have that at least one of the entries

$$a_{i_1 i_2}, a_{i_2 i_3}, a_{i_3 i_4}, \ldots, a_{i_r i_1}$$

is zero. (If we deal with a partitioned matrix $A = [a_{ik}]$, where the blocks a_{ii} are square, we speak of a *block almost acyclic matrix*.) In graph terminology, it means that the digraph $\vec{G}(A)$ does not contain a cycle of length greater than 2. The digraph of an irreducible almost acyclic matrix is symmetric and, moreover, the corresponding undirected graph is a tree. A symmetric matrix with this property was called acyclic in Chapter 8. An example of matrices of this type is an irreducible tridiagonal matrix. The following theorem is concerned with the inversion of almost acyclic matrices.

Theorem 12.1.2 *Let $A = [a_{ik}]$ be a nonsingular almost acyclic irreducible $n \times n$ matrix. Denote by $M_{i/j}$, $i \neq j$, the set consisting of the index i and all the indices k for which there exist indices $t_1, \ldots, t_s$, all distinct, such that*

$$a_{k t_1} \neq 0, a_{t_1 t_2} \neq 0, \ldots, a_{t_{s-1} t_s} \neq 0, a_{t_s i} \neq 0$$

and, in addition, $t_r \neq j$ for $r = 1, \ldots, s$. Further denote by H the set of all the pairs (i, k), $i \neq k$, $i, k - 1, \ldots, n$, such that $a_{ik} \neq 0$.

If $\det A(M_{i/j}) \neq 0$ whenever $a_{ij} \neq 0$, $i \neq j$, then the following holds:

1. *The system of equations*

$$x_{ii}^{-1} + \sum_{k,(i,k)\in H} [(x_{ii} - x_{ik} x_{kk}^{-1} x_{ki})^{-1} - x_{ii}^{-1}] = a_{ii}, \qquad (12.3)$$

$$i = 1, \ldots, n,$$

$$x_{ii}^{-1} x_{ik} (x_{kk} - x_{ki} x_{ii}^{-1} x_{ik})^{-1} = -a_{ik}, \quad (i, k) \in H, \quad (12.4)$$

in unknowns x_{ii}, $i = 1, \ldots, n$, and x_{ik}, $(i, k) \in H$, has a unique solution which gives the entries b_{ii}, $i = 1, \ldots, n$, and b_{ik}, $(i, k) \in H$, in the inverse $B = [b_{ik}]$ of the matrix A:

$$b_{ii} = x_{ii}, \qquad i = 1, \ldots, n, \qquad (12.5)$$

$$b_{ik} = x_{ik}, \qquad (i, k) \in H.$$

2. *The solution of* (12.3), (12.4) *can be found by the following
   algorithm: For each pair* $(i, j) \in H$ *we compute the number*
   $r_{i/j}$ *proceeding recursively according to the number of the
   elements of the set* $M_{i/j}$ *and using the relation*

$$r_{i/j} = a_{ii} - \sum_{k \neq j, (i,k) \in H} a_{ik} r_{k/i}^{-1} a_{ki}. \qquad (12.6)$$

Next we compute the numbers

$$r_i = a_{ii} - \sum_{k, (i,k) \in H} a_{ik} r_{k/i}^{-1} a_{ki} \qquad (12.7)$$

for $i = 1, \ldots, n$. *The solution of the system* (12.3), (12.4)
is then

$$x_{ii} = r_i^{-1}, \qquad (12.8)$$

$$x_{ik} = -r_i^{-1} a_{ik} r_{k/i}^{-1}. \qquad (12.9)$$

3. *The remaining entries* b_{pq} *in* B *(for* $p \neq q$, $(p, q) \notin H$) *are
   given by*

$$b_{pq} = b_{pt_1} b_{t_1 t_1}^{-1} b_{t_1 t_2} b_{t_2 t_2}^{-1} \ldots b_{t_{s-1} t_s} b_{t_s t_s}^{-1} b_{t_s q}, \qquad (12.10)$$

where $t_1, \ldots, t_s$ *are (uniquely determined) distinct indices
satisfying*

$$(p, t_1) \in H, \ (t_1, t_2) \in H, \ (t_2, t_3) \in H, \ \ldots, \ (t_s, q) \in H.$$

We omit the proof. In [15], the theorem is proved in a considerably more general form. In fact, the theorem also holds, for
example, for a partitioned matrix A with square diagonal blocks
a_{ii}. Then, of course, x_{ii} and x_{ik} are blocks, and $r_{i/j}$, r_i are blocks
as well (of the same dimensions as a_{ii}).

The following theorem is useful in matrix inversion.

Theorem 12.1.3 *If* A *and* B *are nonsingular* $n \times n$ *matrices and
the rank of their difference is* r, *then the rank of the difference of
their inverses is* r *as well.*

Proof. It follows from the expression $A^{-1} - B^{-1} = -B^{-1}(A - B) \times A^{-1}$ and from Theorem 1.5.4. $\qquad\square$

Theorem 12.1.3 may be used to compute B^{-1} if we know A^{-1} and if the difference can be expressed as a product of two rectangular matrices.

Theorem 12.1.4 *Let A be a nonsingular matrix. Let*

$$B = A - UQV, \qquad (12.11)$$

where U, Q, and V are matrices whose product is well defined. Then the following conditions are equivalent:
1. *B is nonsingular.*
2. $I - VA^{-1}UQ$ *is nonsingular.*
3. $I - QVA^{-1}U$ *is nonsingular.*

If any of these conditions is satisfied, then we have

$$B^{-1} = A^{-1} + A^{-1}UQ(I - VA^{-1}UQ)^{-1}VA^{-1}, \qquad (12.12)$$

$$B^{-1} = A^{-1} + A^{-1}U(I - QVA^{-1}U)^{-1}QVA^{-1}, \qquad (12.13)$$

Proof. Suppose that 1. holds and $I - VA^{-1}UQ$ is singular. Then there is a vector $y \neq 0$ such that

$$(I - VA^{-1}UQ)y = 0.$$

Put $A^{-1}UQy = x$ so that $Vx = y$. Clearly $x \neq 0$; furthermore, $Ax = UQy$, from which $Bx = Ax - UQVx = Ax - UQy = 0$. This contradicts the fact that B is nonsingular.

Conversely, suppose that 2. is true. Let $Bx = 0$, $x \neq 0$. Then $Ax = UQVx$, and for $y = Vx$ we get

$$A^{-1}UQy = x. \qquad (12.14)$$

Hence $(I - VA^{-1}UQ)y = y - Vx = 0$, from which $y = 0$, $Vx = 0$, and using (12.11) we have

$$Ax = Bx + UQVx = 0.$$

This contradicts the nonsingularity of A and the equivalence of 1. and 2. is proved. The proof of the equivalence of 1. and 3. is quite similar.

Suppose now that 1. through 3. hold. To prove the formula (12.12) we right-multiply the right-hand side of (12.12) by the matrix $A - UQV$. We obtain

$$I - A^{-1}UQV + A^{-1}UQ(I - VA^{-1}UQ)^{-1}V-$$
$$-A^{-1}UQ(I - VA^{-1}UQ)^{-1}VA^{-1}UQV$$
$$= I - A^{-1}UQV + A^{-1}UQ(I - VA^{-1}UQ)^{-1}(I - VA^{-1}UQ)V$$
$$= I.$$

The proof of (12.13) is similar; it is sufficient to left-multiply the right-hand side of (12.13) by $A - UQV$.   □

Remark. The formula (12.12) (and similarly (12.13)) can be obtained as the result of the following heuristic consideration. Let A be a nonsingular matrix and Y a square matrix of the same order as A. If the spectral radius $\rho(A^{-1}Y) < 1$ we may write

$$(A - Y)^{-1} = A^{-1} + A^{-1}YA^{-1} + A^{-1}YA^{-1}YA^{-1} + \dots \quad (12.15)$$

since $A - Y = A(I - A^{-1}Y)$ and by Theorem 1.7.14

$$(I - A^{-1}Y)^{-1} = I + A^{-1}Y + (A^{-1}Y)^2 + (A^{-1}Y)^3 + \dots.$$

Here we shall use (12.15), however, only formally. For $B = A - UQV$ we thus obtain (after formal manipulations again)

$$B^{-1} = A^{-1} + A^{-1}UQVA^{-1} + A^{-1}UQVA^{-1}UQVA^{-1} +$$
$$+A^{-1}UQVA^{-1}UQVA^{-1}UQVA^{-1} + \dots$$
$$= A^{-1} + A^{-1}UQ(I + VA^{-1}UQ + (VA^{-1}UQ)^2 + \dots)VA^{-1}$$
$$= A^{-1} + A^{-1}UQ(I - VA^{-1}UQ)^{-1}VA^{-1}.$$

Then, of course, it is necessary to perform the proof (if we do not suppose the series to be convergent, etc.).

In some cases, the formulae (12.12) and (12.13) may be used e.g. to shift ill condition from a large matrix to a small one. We present two particular cases [16].

Theorem 12.1.5 *Let* $A = [a_{ik}]$ *be a nonsingular square matrix of order* n. *Let* Z *be an* $n \times n$ *matrix of the form*

$$Z = \begin{bmatrix} c_1 & c_1 & \cdots & c_1 \\ c_2 & c_2 & \cdots & c_2 \\ \vdots & \vdots & \ddots & \vdots \\ c_n & c_n & \cdots & c_n \end{bmatrix}.$$

If $\gamma = 1 - \sum_{i,k=1}^{n} a_{ik}c_k \neq 0$, *then*

$$(A - Z)^{-1} = A^{-1} + \gamma^{-1}A^{-1}ZA^{-1}.$$

Proof. Apply Theorem 12.1.4 with $U = [c_1, c_2, \ldots, c_n]^T$, $Q = [1]$, and $V = [1, 1, \ldots, 1]$. □

The next theorem employs a method which resembles the aggregation in economic systems. Here, we suppose that the index set $N = \{1, 2, \ldots, n\}$ is decomposed into r disjoint subsets of consecutive integers, $N = N_1 \cup N_2 \cup \cdots \cup N_r$, so that this decomposition corresponds to partitioning $n \times n$ matrices into blocks.

With each $n \times n$ matrix $A = [a_{ik}]$ we then associate two matrices. First, the $r \times r$ matrix $c(A)$, called the *aggregation* of A, whose entry c_{pq} ($p \leq r$, $q \leq r$) equals

$$\frac{1}{n_p n_q} \sum_{i \in N_p, k \in N_q} a_{ik},$$

where $n_p = |N_p|$, $n_q = |N_q|$. In other words, if we partition A into the form $A = [A_{pq}]$ in accordance with the decomposition $N = N_1 \cup \cdots \cup N_r$, then c_{pq} is the arithmetic mean of the entries of A that belong to the block A_{pq}. The second matrix associated with A will be denoted by $s(A)$; its entry s_{pq} ($p \leq r$, $q \leq r$) equals the sum of all the entries in A_{pq}. Finally, we denote – for a moment – by $\mathscr{S}$ the set of all the square matrices $A = [A_{pq}]$ of order n that are "constant in each block A_{pq}", i.e., which possess the property that all entries in the block A_{pq} are equal, and this is true for all $p, q = 1, \ldots, r$. It is clear that for every square matrix P of order r there exists one and only one matrix $B \in \mathscr{S}$ such that $c(B) = P$. Now we can state the following.

Theorem 12.1.6 *Let A be a nonsingular $n \times n$ matrix and let $S \in \mathscr{S}$. Then the matrix $A - S$ is nonsingular if and only if the $r \times r$ matrix $I - s(A^{-1})c(S)$ is nonsingular. In this case we have*

$$(A - S)^{-1} = A^{-1} + A^{-1}MA^{-1}, \qquad (12.16)$$

where M is the matrix from $\mathscr{S}$ satisfying

$$c(M) = c(S)(I - s(A^{-1})c(S))^{-1}.$$

Proof. Denote by $J = [j_{pk}]$ the $r \times n$ matrix with entries

$$j_{pk} = \begin{cases} 1 & \text{for} \quad k \in N_p, \\ 0 & \text{for} \quad k \notin N_p, \end{cases}$$

$p = 1, \ldots, r$, $k = 1, \ldots, n$. Let D be the diagonal matrix of order r whose diagonal entries are

$$\frac{1}{n_1}, \frac{1}{n_2}, \ldots, \frac{1}{n_r},$$

where $n_k = |N_k|$, $k = 1, \ldots, r$.

Since $S \in \mathscr{S}$ we have

$$S = J^T D J S J^T D J = J^T c(S) J.$$

Using Theorem 12.4 with $U = J^T$, $Q = c(S)$, and $V = J$, we conclude from 2. that the matrix $A - S$ is nonsingular if and only if $I - J A^{-1} J^T c(S) = I - s(A)^{-1})c(S)$ is a nonsingular matrix. Formula (12.16) is an immediate consequence of (12.12):

$$\begin{aligned} (A - S)^{-1} &= A^{-1} + A^{-1} J^T c(S)(I - s(A^{-1})c(S))^{-1} J A^{-1} \\ &= A^{-1} + A^{-1} M A^{-1}. \end{aligned}$$

$\square$

12.2   The pseudoinverse

For completeness, we mention an algorithm[1] for computing the pseudoinverse defined in Chapter 10. If the given matrix is square and nonsingular we obtain an algorithm for computing the inverse.

---

[1]See [32].

Theorem 12.2.1 (Greville [32]) *Let* $A = [a_1, a_2, \ldots, a_n]$ *be an* $m \times n$ *matrix with columns* $a_1, \ldots, a_n$. *Denote by* A_k *the matrix* $A_k = [a_1, a_2, \ldots, a_k]$, $k = 1, \ldots, n$. *Then*

1.
$$A_1^+ = \begin{cases} (a_1^* a_1)^{-1} a_1^* & \text{for } a_1 \neq 0, \\ 0 & \text{for } a_1 = 0; \end{cases}$$

2. *for* $k \geq 2$ *we have*

$$A_k^+ = \begin{bmatrix} A_{k-1}^+ - d_k b_k \\ b_k \end{bmatrix},$$

where

$$d_k = A_{k-1}^+ a_k,$$
$$c_k = a_k - A_{k-1} d_k,$$
$$b_k = \begin{cases} (c_k^* c_k)^{-1} c_k^* & \text{for } c_k \neq 0, \\ (1 + d_k^* d_k)^{-1} d_k^* A_{k-1}^* & \text{for } c_k = 0; \end{cases}$$

3. $A^+ = A_n^+$.

These relations can thus be used to compute A^+ *recursively.*

Remark. If A is a nonsingular square matrix, then the first possibility takes place in both 1. and 2. ($a_1 \neq 0, c_k \neq 0$).

We conclude the chapter with a formula for improving the accuracy of the computed inverse. If B_0 is an approximate inverse of A, then – provided that the deviation of B_0 from A^{-1} is small – a more accurate approximation to A^{-1} is

$$B_1 = B_0 + B_0(I - AB_0). \tag{12.17}$$

Exercises

12.1. Given a nonsingular matrix partitioned in the form

$$\begin{bmatrix} A & B \\ C & D \end{bmatrix},$$

where A is nonsingular, show that its inverse can be expressed as

$$\begin{bmatrix} P & Q \\ R & S \end{bmatrix},$$

where

$$S = (D - CA^{-1}B)^{-1}, \quad Q = -A^{-1}BS, \quad R = -SCA^{-1},$$
$$P = A^{-1} - A^{-1}BR.$$

12.2. Let

$$\begin{bmatrix} A_{11} & A_{12} \\ A_{21} & A_{22} \end{bmatrix}$$

be a partitioned matrix with nonsingular blocks A_{11}, A_{22}. Show that each of the matrices

$$\begin{bmatrix} A_{11}^{-1} & -A_{11}^{-1}A_{12} \\ A_{21}A_{11}^{-1} & A_{22} - A_{21}A_{11}^{-1}A_{12} \end{bmatrix},$$

$$\begin{bmatrix} A_{11} - A_{12}A_{22}^{-1}A_{21} & A_{12}A_{22}^{-1} \\ -A_{22}^{-1}A_{21} & A_{22}^{-1} \end{bmatrix}$$

is the inverse of the other and prove that their determinants equal $\det A_{22}/\det A_{11}$, $\det A_{11}/\det A_{22}$.

12.3. Let A be a square matrix of order n. Denote by P the block tridiagonal matrix

$$P = \begin{bmatrix} A & -I & & & \\ -I & A & -I & & \\ \vdots & \vdots & \vdots & \ddots & \\ & & & A & -I \\ & & & -I & A \end{bmatrix}$$

having m block rows. If $S_k(x)$, $k = 0, 1, \ldots$, are the modified Chebyshev polynomials of second kind introduced in Exercise 7.3 prove that the following holds:

1. The matrix P is nonsingular if and only if $\det S_m(A) \neq 0$.

2. If $\det S_m(A) \neq 0$, then $P^{-1} = [Q_{ij}]$, $i, j = 1, \ldots, m$, where

$$Q_{ij} = \begin{cases} S_m^{-1}(A)S_{i-1}(A)S_{m-j}(A) & \text{for } j \geq i, \\ S_m^{-1}(A)S_{j-1}(A)S_{m-i}(A) & \text{for } j < i. \end{cases}$$

12.4. Let $B = [B_{ik}]$, where the blocks B_{ii} are nonsingular, be the inverse of a nonsingular block tridiagonal matrix. The formulae (12.10) then imply the relations

$$B_{ik} = B_{i,i+1}B_{i+1,i+1}^{-1}B_{i+1,i+2}B_{i+2,i+2}^{-1}\cdots B_{k-1,k-1}^{-1}B_{k-1,k}$$
$$\text{for } i < k,$$
$$B_{ik} = B_{i,i-1}B_{i-1,i-1}^{-1}B_{i-1,i-2}\cdots B_{k+1,k+1}^{-1}B_{k+1,k}$$
$$\text{for } i > k.$$

Verify the relations in the case of the matrix P^{-1} from Exercise 12.3.

13

Numerical Methods for Computing Eigenvalues of Matrices

The problem of the numerical calculation of eigenvalues can be considered from several points of view. First, it is possible that we need to know all the eigenvalues or, on the other hand, it is sufficient to compute only one or a few selected eigenvalues. Further, there is a great difference between the calculation of the eigenvalues of a symmetric (or Hermitian) matrix and of an unsymmetric matrix. Recently, it is often required to compute the eigenvalues of large (and usually sparse, cf. Chapter 14) matrices. The numerical methods chosen in case the matrix is sparse differ, as a rule, from those for dense matrices. Finally, special methods may be used in order to find one or all eigenvalues of a special matrix.

All these aspects cannot be considered in a single chapter. We thus confine ourselves to presenting a brief survey of the main methods.

13.1 Computation of selected eigenvalues

For determining the eigenvalue of largest modulus (this modulus is thus equal to the spectral radius) – provided this eigenvalue is unique and simple – the classical *power method* is applicable. The method is based on the following theorem.

Theorem 13.1.1 *If λ is the unique eigenvalue of A satisfying $|\lambda| = \rho(A)$ and if λ is simple, then the sequence of the matrices*

$(\lambda^{-1}A)^k$ *converges as* $k \to \infty$ *to the matrix* uv^T, *where* u *and* v *are the eigenvectors of* A *and* A^T, *respectively, associated with* λ *(Au* $= \lambda u, u \neq 0, A^T v = \lambda v, v \neq 0)$ *and normalized so that* $v^T u = 1$.

Proof. By Theorem 1.7.6, the matrix A is similar to a matrix in Jordan normal form whose block associated with λ is (since λ is assumed to be simple) of order 1; moreover, the moduli of the other eigenvalues are less than 1 under the hypothesis of the theorem. This means that there is a nonsingular matrix P such that

$$PAP^{-1} = A_J = \begin{bmatrix} \lambda & 0 \\ 0 & Z \end{bmatrix}, \quad \rho(Z) < |\lambda|.$$

Therefore,

$$P(\lambda^{-1}A)^k P^{-1} = (\lambda^{-1}A_J)^k = \begin{bmatrix} 1 & 0 \\ 0 & (\lambda^{-1}Z)^k \end{bmatrix},$$

and thus

$$\lim_{k\to\infty} (\lambda^{-1}A_J)^k = \begin{bmatrix} 1 & 0 \\ 0 & 0 \end{bmatrix} = e_1 e_1^T,$$

where $e_1 = [1, 0, \ldots, 0]^T$. Altogether we have

$$\lim_{k\to\infty} (\lambda^{-1}A)^k = (P^{-1}e_1)(e_1^T P). \tag{13.1}$$

But $Au = \lambda u$, $u \neq 0$, and, moreover, u is determined uniquely except for a nonzero factor. Further, $A_J e_1 = \lambda e_1$, from which $A(P^{-1}e_1) = P^{-1}(PAP^{-1})e_1 = P^{-1}A_J e_1 = P^{-1}(\lambda e_1) = \lambda(P^{-1}e_1)$. Hence,

$$u = \alpha P^{-1}e_1, \quad \alpha \neq 0. \tag{13.2}$$

Similarly we have $A^T v = \lambda v, v \neq 0$, and v is determined uniquely except for a nonzero factor. Now, $A_J^T e_1 = \lambda e_1$ and consequently $A^T(P^T e_1) = P^T((P^T)^{-1}A^T P^T)e_1 = P^T A_J^T e_1 = P^T(\lambda e_1) = \lambda(P^T e_1)$. Hence,

$$v = \beta P^T e_1, \quad \beta \neq 0. \tag{13.3}$$

By the assumption, $v^T u = 1$. Substituting from (13.2) and (13.3) we get

$$\beta(e_1^T P)\alpha(P^{-1}e_1) = 1,$$

and thus
$$\alpha\beta = 1$$
since $e_1^T e_1 = 1$.

Using (13.2) and (13.3), we find from (13.1) that
$$\lim_{k\to\infty} (\lambda^{-1}A)^k = (\alpha^{-1})u(\beta^{-1})v^T = uv^T,$$
as was to be proved. □

Now we are in a position to prove the next theorem.

Theorem 13.1.2 *Let A be a square matrix having a unique eigenvalue λ satisfying $|\lambda| = \rho(A)$; suppose that λ is simple.*

Let v be an eigenvector of A^T associated with λ. If z is an arbitrary vector such that $v^T z \neq 0$, then
$$\lim_{k\to\infty} (\lambda^{-1}A)^k z = y \neq 0$$
and y is an eigenvector of A associated with the eigenvalue λ.

Proof. By Theorem 13.1.1
$$\lim_{k\to\infty} (\lambda^{-1}A)^k z = (uv^T)z = u(v^T z),$$
where u is the eigenvector of A associated with the eigenvalue λ and satisfying $v^T u = 1$. Since $v^T z \neq 0$ the vector $y = (v^T z)u$ is an eigenvector of A associated with λ as well. □

Remark 1. The result of Theorem 13.1.2 implies that for a sufficiently large k the vectors $A^{k+1}z$ and $A^k z$ are approximately proportional:
$$A^{k+1}z \approx \mu A^k z;$$
the factor of proportionality μ is approximately equal to λ, and the vector $A^{k+1}z$ is approximately proportional to the eigenvector of A associated with λ. This can be used to determine λ and the eigenvector approximately.

Remark 2. The accuracy of the approximation to λ or to the eigenvector u obtained after a finite number of steps depends on the ratio $\xi = |\lambda_2|/|\lambda|$, where $|\lambda_2|$ is the maximal modulus of the remaining eigenvalues. If ξ is close to 1 the convergence is slow, if ξ is small the convergence is fast. The number k of necessary steps is, roughly speaking, such that ξ^k be of the same order of magnitude as the required relative accuracy of the result.

Applying the power method to the matrix A^{-1} we obtain a procedure for finding the eigenvalue of smallest modulus. Here, instead of computing A^{-1} we may solve the equation $Ax = (A^{-k})z$ in each step so that $x = A^{-(k+1)}z$.

A modification of the power method is particularly well-suited for the refinement of an approximately calculated eigenvalue. The procedure is based on the next theorem.

Theorem 13.1.3 *Let α be a simple eigenvalue of A and let α_0 be its approximation satisfying*

$$|\alpha_0 - \alpha| < |\alpha_0 - \lambda|$$

for all the other eigenvalues λ of the matrix A. Let v be an eigenvector of A^T associated with the eigenvalue α. If x_0 is an arbitrary vector such that $v^T x_0 \neq 0$, then the iteration

$$(A - \alpha_0 I)x_{k+1} = x_k$$

has the property that for large k the vector x_{k+1} is approximately proportional to x_k:

$$x_{k+1} \approx \sigma x_k.$$

Furthermore, $\sigma \approx 1/(\alpha - \alpha_0)$ and, on the other hand, x_{k+1} is approximately equal to an eigenvector of A associated with α.

Proof. It is sufficient to apply Theorem 13.1.2 to the matrix $(A - \alpha_0 I)^{-1}$.   □

Remark. The method described in Theorem 13.1.3 is called the *inverse iteration*. The number k of necessary iterations is again such that ξ^k, where

$$\xi = \max_{\lambda \neq \alpha} \left(\frac{|\alpha - \alpha_0|}{|\lambda - \alpha_0|} \right)$$

and the maximum is taken over all the eigenvalues of A different from α, is of the same order of magnitude as the relative accuracy required.

If A is symmetric (or Hermitian) and we employ the power method from Theorem 13.1.2 to compute λ, then a considerably more accurate approximation to λ is

$$\lambda \approx \frac{\langle A^{k+1}\tilde{z}, A^{k+1}\tilde{z}\rangle}{\langle A^{k+1}\tilde{z}, A^{k}\tilde{z}\rangle},$$

where $\tilde{z}$ is the initial vector or its nonzero multiple.

A modification of the power method can be used for computing two eigenvalues of largest (or smallest) modulus [61] provided that the moduli of all the other eigenvalues are smaller (or larger, respectively).

13.2 Computation of all the eigenvalues

One possible method for computing all the eigenvalues of a general $n \times n$ matrix with n less than 10 or 12 is to determine the characteristic polynomial $\varphi(y)$ of A and then solve $\varphi(z) = 0$. An appropriate method for finding the coefficients of the characteristic polynomial is the so-called *Krylov method* based on the Cayley-Hamilton Theorem 7.1.1.

Theorem 13.2.1 *Let A be a square matrix of order n and let u be an arbitrary vector, $u \neq 0$. If the system of n linear equations*

$$x_0 u + x_1 A u + x_2 A^2 u + \cdots + x_{n-1} A^{n-1} u = -A^n u \qquad (13.4)$$

in unknowns $x_0, x_1, \ldots, x_{n-1}$ has a unique solution $a_0, a_1, \ldots, a_{n-1}$, then

$$(-1)^n (z^n + a_{n-1} z^{n-1} + \cdots + a_1 z + a_0)$$

is the characteristic polynomial of A.

Proof. Suppose that $(-1)^n (z^n + a_{n-1} z^{n-1} + \cdots + a_1 z + a_0)$ is the characteristic polynomial of the matrix A. By the Cayley-Hamilton theorem we then have

$$A^n + a_{n-1} A^{n-1} + \cdots + a_1 A + a_0 I = 0$$

and thus also

$$A^n u + a_{n-1} A^{n-1} u + \cdots + a_1 A u + a_0 u = 0.$$

Therefore, if the system (13.4) has a unique solution, then $x_i = a_i$, $i = 0, 1, \ldots, n-1$. $\square$

Remark. This method is usually not recommendable for larger matrices since, in general, the zeros of the characteristic polynomial are much more sensitive to rounding errors in the coefficients a_i than to small changes in the entries of the original matrix A (see Exercise 13.1). In addition, for larger n the matrix of the system (13.4) is generally ill conditioned so that the computed values of a_i, $i = 0, 1, \ldots, n-1$, can be highly inaccurate. For these reasons the mostly used methods for determining the eigenvalues proceed in another way. They employ similarity transformations to transform the matrix A into a matrix B of special type such that the calculation of the eigenvalues of B is as simple as possible and the eigenvalue problem for B is not worse conditioned than that for A. The latter requirement and the requirement of numerical stability lead to the use of orthogonal transformations if A is real or unitary transformations if A is complex.

Theorem 13.2.2 (Householder [37]) *Let A be an arbitrary real (or complex) square matrix A of order n. Then there exists a sequence of $n-1$ orthogonal (or unitary) reflexion matrices (cf. Chapter 2) $W(w_k)$, $k = 1 \ldots, n-1$, such that*

$$B = W(w_{n-1}) \ldots W(w_1) A W(w_1) \ldots W(w_{n-1})$$

is a Hessenberg matrix (cf. Chapter 8), i.e. $B = [b_{ik}]$, where $b_{ik} = 0$ for $i - k > 1$. Moreover, the algorithm for determining the reflexion matrices $W(w_k)$ is carried out easily.

Proof. First we prove an auxiliary statement describing the construction of the matrix $W(w_1)$:

Put $u = [a_{31}, a_{41}, \ldots, a_{n1}]^T$, $|u| = \sqrt{(u^T u)}$. If $u \neq 0$ we introduce the vector w_1 by

$$w_1 = \begin{bmatrix} 0 \\ p \\ \xi u \end{bmatrix},$$

where p and ξ are defined by

$$\xi = \frac{1}{\sqrt{|u|^2 + [|a_{21}| + \sqrt{|a_{21}|^2 + |u|^2}]^2}}, \qquad (13.5)$$

$$p = \frac{a_{21}\xi[|a_{21}| + \sqrt{|a_{21}|^2 + |u|^2}]}{|a_{21}|} \quad \text{if} \quad a_{21} \neq 0,$$
$$p = \frac{\sqrt{2}}{2} \qquad\qquad\qquad\qquad \text{if} \quad a_{21} = 0. \qquad (13.6)$$

Then, $|w_1| = 1$ and the matrix

$$A_1 = W(w_1)AW(w_1),$$

where

$$W(w_1) = I - 2w_1 w_1^*,$$

is orthogonally (or unitarily) similar to A. In addition, the entries of $A_1 = [a_{ik}^{(1)}]$ are

$$a_{11}^{(1)} = a_{11},$$
$$a_{21}^{(1)} = \tfrac{1}{2}a_{21}\left[1 - \sqrt{1 + \frac{|u|^2}{|a_{21}|^2}}\right] \quad \text{if} \quad a_{21} \neq 0,$$
$$a_{21}^{(1)} = -|u|\frac{\sqrt{2}}{2} \qquad\qquad\qquad \text{if} \quad a_{21} = 0,$$
$$a_{31}^{(1)} = a_{41}^{(1)} = \ldots a_{n1}^{(1)} = 0.$$

In order to prove the statement we first note that the numbers ξ and p are chosen in such a way that we have (for both $a_{21} \neq 0$ and $a_{21} = 0$)

$$|p|^2 + \xi^2 |u|^2 = 1, \qquad (13.7)$$

$$2\xi(\bar{p}a_{21} + \bar{\xi}|u|^2) = 1. \qquad (13.8)$$

The relation (13.7) means that $|w_1| = 1$ so that the matrix $W(w_1)$ is orthogonal (or unitary) and also symmetric (or Hermitian). We now compute A_1 as follows:

$$A_1 = (I - 2w_1 w_1^*)A(I - 2w_1 w_1^*)$$
$$= A - 2w_1 w_1^* A + (-2Aw_1 + 4(w_1^* Aw_1)w_1)w_1^*.$$

The first column of the third term is zero since the first component of w_1^* is zero. Next, we find the first column of the sum of the first two terms. If $a_{21} \neq 0$, then we have

$$
\begin{bmatrix} a_{11} \\ a_{21} \\ u \end{bmatrix} - 2 \begin{bmatrix} 0 \\ p \\ \xi u \end{bmatrix} [0, \; \bar{p} \; \bar{\xi} u^*] \begin{bmatrix} a_{11} \\ a_{21} \\ u \end{bmatrix} =
$$

$$
= \begin{bmatrix} a_{11} \\ a_{21} - 2p(\bar{p}a_{21} + \bar{\xi}|u|^2) \\ u[1 - 2\xi(\bar{p}a_{21} + \bar{\xi}|u|^2)] \end{bmatrix}
$$

$$
= \begin{bmatrix} a_{11} \\ a_{21} - p/\xi \\ 0 \end{bmatrix}
$$

$$
= \begin{bmatrix} a_{11} \\ \frac{1}{2}a_{21}[1 - \sqrt{1 + |u|^2/|a_{21}|^2}] \\ 0 \end{bmatrix}
$$

by (13.8). If $a_{21} = 0$, then $p = (\sqrt{2})/2$ and $a_{21}^{(1)} = -p/\xi = -|u|$ as $\xi = \sqrt{2}/(2|u|)$.

The auxiliary statement implies that the first column of A_1 has the Hessenberg form. Put now

$$
w_2 = \begin{bmatrix} 0 \\ 0 \\ p_1 \\ \xi_1 u_1 \end{bmatrix},
$$

where p_1, ξ_1 are numbers and u_1 a vector constructed analogously as in the auxiliary statement but for the principal submatrix of A_1 obtained by deleting the first row and column. Then, the first two columns of the matrix $A_2 = W(w_2)A_1 W(w_2) = W(w_2)W(w_1)AW(w_1)W(w_2)$ will be in Hessenberg form (since the first column does not change). We proceed in this way until we obtain the desired Hessenberg matrix B. $\square$

Remark. If the matrix A from Theorem 13.2.2 is real symmetric or Hermitian, then B will be a real symmetric or Hermitian matrix as well. Hence, B will be tridiagonal. The computation of the eigenvalues of a tridiagonal matrix was considered in Chapter 8. For a Hessenberg matrix we may either use a method which recursively evaluates the top left principal minors of $B - \lambda I$ and eventually gives the value of the characteristic polynomial or employ the so-called QR *algorithm* described – for a general matrix – below.

In Chapter 10 (Theorem 10.2.1) we saw that each nonsingular square matrix A can be uniquely expressed in the form

$$A = QR, \tag{13.9}$$

where Q is a unitary (or orthogonal if A is real) matrix and R is an upper triangular matrix whose diagonal entries are real and positive.

The fundamental idea of the QR algorithm (Francis [25], Kublanovskaya [43]) and related algorithms is to put

$$A = Q_0 R_0$$

in accordance with (13.9), to define

$$A_1 = R_0 Q_0$$

and continue in forming the sequence of matrices

$$A_k = Q_k R_k, \qquad A_{k+1} = R_k Q_k, \qquad k = 1, 2, \ldots, \tag{13.10}$$

which is constructed by repeated application of (13.9). Since

$$A_{k+1} = R_k Q_k = Q_k^{-1}(Q_k R_k)Q_k = Q_k^{-1}A_k Q_k,$$

all the matrices A_k are similar to A and therefore have the same eigenvalues as follows from Theorem 1.7.2 (actually, we have not used the fact that Q_k are unitary and R_k upper triangular matrices).

Suppose now, for instance, that the sequence $\{Q_k\}$ converges to the identity matrix. Then the sequence $\{A_k\}$ converges to an upper triangular matrix and, moreover, the diagonal entries of this matrix are the eigenvalues of A. In fact, this situation occurs only in special cases (for example, the eigenvalues of A

must be real and positive). The convergence of the algorithm in the case of a general matrix A is studied e.g. in [61].

One can also apply the so-called *shift technique* and calculate not the factorization $A_k = Q_k R_k$, but

$$A_k - \sigma_k I = \tilde{Q}_k \tilde{R}_k. \tag{13.11}$$

This technique shifts the original spectrum of A by σ_k (if λ_j are the eigenvalues of A_k, then the eigenvalues of $A_k - \sigma_k I$ are $\lambda_j - \sigma_k$). Therefore, if the numbers σ_k are stored the spectrum of A may be readily found provided we know the spectrum of A_k. In addition, since (for the method without shifts)

$$A_k = Q_{k-1}^{-1} Q_{k-2}^{-1} \cdots Q_0^{-1} A Q_0 Q_1 \cdots Q_{k-1} \tag{13.12}$$

we have

$$A = Q_0 Q_1 \cdots Q_{k-1} A_k Q_{k-1}^* \cdots Q_0^*,$$

and the eigenvectors y of A are related to the eigenvectors z of A_k as follows:

$$y = Q_0 Q_1 \cdots Q_{k-1} z. \tag{13.13}$$

The same formula (with Q_j replaced by $\tilde{Q}_j$) holds for the method with shifts since the shift technique does not change the eigenvectors.

For a special shift strategy, the convergence is even cubic in certain important cases (it means, roughly speaking, that the number of correct decimal places increases approximately three times in each step).

The following theorem describes the application of the QR algorithm to band matrices.

Theorem 13.2.3 *If the matrix A is $(n-1, q)$-banded (cf. Chapter 8), then each of the matrices A_k given by (13.10) is $(n-1, q)$-banded as well. The same is true of the method with shifts using (13.11).*

Proof. Clearly, for $q = n - 1$ the theorem holds. Thus, assume $q < n - 1$. Let $A = [a_{ik}]$ be $(n-1, q)$-banded, i.e.,

$$a_{ik} = 0 \qquad \text{for} \quad i - k > q.$$

Let $A = QR$ where R is an upper triangular matrix with positive diagonal entries. Write

$$Q = Q_{-(n-1)} + Q_{-(n-2)} + \cdots + Q_{-1} + Q_0 + Q_1 + \cdots + Q_{n-1},$$
$$(13.14)$$

where Q_j contains those entries q_{uv} of $Q = [q_{ik}]$ for which $v - u = j$, its other entries being zero. Splitting the matrix R similarly, we get $R = R_0 + R_1 + \cdots + R_{n-1}$ where R_0 is a diagonal matrix with positive diagonal entries. Using now the assumption on the band form of A we have

$$Q_{-(n-1)}R_0 = 0,$$
$$Q_{-(n-1)}R_1 + Q_{-(n-2)}R_0 = 0,$$
$$\vdots$$
$$Q_{-(n-1)}R_{n-q-2} + Q_{-(n-2)}R_{n-q-3} + \cdots + Q_{-(q+1)}R_0 = 0,$$
$$(13.15)$$

since multiplying matrices U and V written in the form (13.14)

$$U = \sum_{i=-(n-1)}^{n-1} U_i, \qquad V = \sum_{i=-(n-1)}^{n-1} V_i,$$

we find that the product $W = UV$ may be expressed as

$$W = \sum_{k=-(n-1)}^{n-1} W_k,$$

where

$$W_k = \sum_{t=k-(n-1)}^{n-1} U_t V_{k-t} \quad \text{for} \quad k \geq 0,$$

$$W_k = \sum_{t=-(n-1)}^{k+(n-1)} U_t V_{k-t} \quad \text{for} \quad k < 0.$$

Since R_0 is nonsingular, (13.15) implies that

$$Q_{-(n-1)} = Q_{-(n-2)} = \cdots = Q_{-(q+1)} = 0.$$

Hence,

$$A_1 = RQ = (R_0 + \cdots + R_{n-1})(Q_{-q} + \cdots + Q_0 + \cdots + Q_{n-1})$$

and it follows that the entries $a_{ik}^{(1)}$ of A_1 satisfy $a_{ik}^{(1)} = 0$ for $i - k > q$. The same argument can be applied to all the matrices A_k and therefore the theorem holds. □

Theorem 13.2.4 *If A is a Hessenberg matrix, then each matrix A_k defined in (13.10) has Hessenberg form. If A is a real symmetric (or complex Hermitian) tridiagonal matrix, then each matrix A_k is real symmetric (or complex Hermitian), and tridiagonal as well. The same holds for the method with shifts.*

Proof. The first statement is an immediate consequence of Theorem 13.2.3 for $q = 1$. If, in addition, A is real and symmetric (or Hermitian), then each matrix A_k has the same property by (13.12). It follows that A_k is tridiagonal. Since the shift technique does not change the band structure of the matrix, the theorem holds also for the method with shifts. □

Remark. In the proof of Theorem 13.2.2, we reduced the matrix to Hessenberg form by the *Householder method* using the elementary reflexion matrices. Another possibility is to employ the *Givens method* [31] where the transformation is constructed by means of the plane rotation matrices (see Theorem 2.3.4 and (13.16) below). A comparison with Householder method shows that Givens method requires about twice as many operations.

Nevertheless, the plane rotations find their use in an iterative method reducing the given real symmetric (or complex Hermitian) matrix to a matrix which is arbitrarily close to a diagonal matrix (the *Jacobi method*). For simplicity we restrict ourselves to the most important case of a real symmetric matrix of order n. For each ordered pair of integers (i, k), $i \neq k$, $i, k = 1, \ldots, n$, and any real number α, we denote by $R_{ik}(\alpha)$ the square matrix which differs from the identity matrix of order n by entries $\sin \alpha$

in the position (i, k), $-\sin\alpha$ in the position (k, i), and $\cos\alpha$ in the positions (i, i) and (k, k). For $i < k$ we thus put

$$R_{ik}(\alpha) = \tag{13.16}$$

$$
\begin{bmatrix}
1 & & & & & & & & & \\
 & \ddots & & & & & & & & \\
 & & 1 & & & & & & & \\
 & & & \cos\alpha & \cdots & \cdots & \cdots & \sin\alpha & & \\
 & & & \vdots & 1 & & & \vdots & & \\
 & & & \vdots & & \ddots & & \vdots & & \\
 & & & \vdots & & & 1 & \vdots & & \\
 & & & -\sin\alpha & \cdots & \cdots & \cdots & \cos\alpha & & \\
 & & & & & & & & 1 & \\
 & & & & & & & & & \ddots & \\
 & & & & & & & & & & 1
\end{bmatrix}
\begin{matrix}
\\ \\ \\ \leftarrow i \\ \\ \\ \\ \leftarrow k \\ \\ \\
\end{matrix}
$$

It is easy to see that the matrix $R_{ik}(\alpha)$ is orthogonal and, moreover,

$$R_{ik}^T(\alpha) = R_{ik}(-\alpha) = R_{ki}(\alpha). \tag{13.17}$$

The principle of the Jacobi method [39] is described in the next theorem. Its verificaqtion is left to the reader (Exercise 13.7).

Theorem 13.2.5 *Let* $A = [a_{ik}]$ *be a real* $n \times n$ *symmetric matrix. If* $a_{pq} \neq 0$ *and if the angle* α_{pq} *satisfies*

$$\cot\alpha_{pq} = \frac{a_{pp} - a_{qq}}{2a_{pq}},$$

then the entry $b_{pq}(= b_{qp})$ *of the matrix*

$$B = R_{pq}(-\alpha_{pq})AR_{pq}(\alpha_{pq}) \tag{13.18}$$

is zero and, in addition,

$$\sum_{i,k,i\neq k} b_{ik}^2 = \sum_{i,k,i\neq k} a_{ik}^2 - 2a_{pq}^2. \tag{13.19}$$

Moreover, the angle α_{pq} *can be chosen such that*

$$|\alpha_{pq}| \leq \frac{\pi}{4}; \tag{13.20}$$

putting

$$t = \tan \alpha_{pq}, \quad c = \cos \alpha_{pq}, \quad s = \sin \alpha_{pq}, \qquad (13.21)$$
$$r = \frac{s}{1+c},$$

we then have

$$b_{pp} = a_{pp} + t a_{pq}, \qquad (13.22)$$
$$b_{qq} = a_{qq} - t a_{pq},$$
$$b_{ip} = b_{pi} = a_{ip} + s(a_{iq} + r a_{ip}),$$
$$b_{iq} = b_{qi} = a_{iq} - s(a_{ip} - r a_{iq}), \quad i = 1, \ldots, n, \quad q \neq i \neq p,$$
$$b_{ij} = a_{ij}, \qquad i \neq p, \; j \neq q.$$

The Jacobi method applies the theorem repeatingly to form a sequence of matrices similar to A until we arrive at a matrix close to diagonal. The particular versions of the Jacobi method differ in the way how the indices p and q are chosen in the individual iteration steps.

In the *classical Jacobi method*, we choose the pair (p, q) of indices such that

$$|a_{pq}| = \max\{|a_{ik}|; i, k = 1, \ldots, n, i \neq k\}.$$

For machine computation the search for the off-diagonal entry of greatest modulus is rather costly in time. To avoid this search the *threshold method* is often used. We first fix a threshold value $b > 0$ and then choose for p and q the indices of the first off-diagonal entry whose modulus is greater than b, the off-diagonal entries being scanned in a preassigned order. If there is no such entry we reduce the threshold value e.g. to $b/2$. On the other hand, it is also possible to use the *cyclic Jacobi method*, where p, q run rowwise through all the superdiagonal positions of the matrix, the entry $a_{n,n-1}$ being followed by a_{12} again.

Exercises

13.1.* Consider the polynomial

$$(x - 1)(x - 2) \ldots (x - 10) = x^{10} - 55x^9 + \cdots + 10!$$

and replace the coefficient -55 of x^9 by $-55(1+\varepsilon)$, where $\varepsilon = 10^{-4}$. Find approximately the relative changes caused in the roots 10 and 1 by this perturbation.

13.2.* Given the 10×10 matrix

$$
\begin{bmatrix}
0 & 1 & 0 & 0 & \cdots & 0 & 0 \\
0 & 0 & 1 & 0 & \cdots & 0 & 0 \\
\vdots & \vdots & \vdots & \vdots & \ddots & \vdots & \vdots \\
0 & 0 & 0 & 0 & \cdots & 0 & 1 \\
\varepsilon & 0 & 0 & 0 & \cdots & 0 & 0
\end{bmatrix},
$$

investigate the changes in the eigenvalues if ε changes from 0 to 0.001.

13.3. Define the distance $d(S(A), S(B))$ of the spectra $S(A) = \{a_1, \ldots, a_n\}$, $S(B) = \{b_1, \ldots, b_n\}$ of two $n \times n$ matrices A and B as

$$
d(S(A), S(B)) = \min_{P \in \Pi_n} \max_i |a_i - b_{P_i}|
$$

where Π_n means the set of all permutations. Using Exercise 13.2, show that one cannot expect an estimate of the form

$$
d(S(A), S(B)) \leq K_n M^{1-k} N^k (A - B) \tag{13.23}
$$

if $k > 1/n$, M being $\max(N(A), N(B))$ and $N(X)$ being the Schur norm from Chapter 9. (K_n depends only on n.) Compare this with Theorem 9.3.6 and with a result from [3] where it is shown that (13.23) is true for $k = 1/n$ and $K_n = 2(2n - 1)$.

13.4. Show how to apply Theorem 9.3.6 as a stopping criterion for the Jacobi method (in the real case) when the required accuracy ε of the computed eigenvalues is given.

13.5. Use the Krylov method to find the characteristic polynomial and the eigenvalues of the matrix

$$
\begin{bmatrix}
-5 & 8 & -4 & 8 & 4 \\
-8 & 12 & -5 & 10 & 5 \\
8 & -10 & 5 & -8 & -4 \\
16 & -20 & 8 & -12 & -5 \\
-16 & 20 & -8 & 10 & 3
\end{bmatrix}.
$$

Choose, for example, $u = [0, 0, 0, 0, 1]^T$.

13.6. Find the eigenvalues of the matrix P from Exercise 12.3 provided you know the eigenvalues of A.

13.7. Prove Theorem 13.2.5.

14

Sparse Matrices

In this chapter the solution of linear systems having a nonsingular, large, and sparse matrix is treated in detail. We have already mentioned in Chapter 10 that a "large" matrix means a matrix of order 500 or more and that a matrix is *sparse* if it has at most 5 % nonzero entries or if it has only several (say, less than 10) nonzero entries in each row and column. Recently, systems of this kind occur more and more frequently, for example, in the analysis of mutual economic relations (cf. Example 5.1) or in engineering, where they arise from structural analysis, from solving ordinary and partial differential equations, etc.

The numerical solution of systems with a large sparse matrix has its own specific problems (cf. [57]). In fact, the primary task here is to use the internal storage of the computer efficiently when the data are loaded into the machine as well as during the whole computation.

14.1 Storing. Elimination ordering

It is disadvantageous to store a large sparse matrix in the computer in the usual way, i.e. all entries row by row or column by column. It is much more efficient to leave out the zero entries and, in turn, to store the position of each nonzero entry. We shall present two such procedures for an $m \times n$ matrix $A = [a_{ik}]$. We store the sequence

$$m, n, i_1, k_1, a_{i_1 k_1}, i_2, k_2, a_{i_2 k_2}, \ldots, i_N, k_N, a_{i_N, k_N},$$

where N is the number of nonzero entries of A. Constructing this sequence, we search for the non-zeros through the rows or through the columns of A. The number of data stored is thus $3N+2$. The other way consists in storing m and n again, then the number r of nonzero entries in the first row, $k_1, a_{1k_1}, k_2, a_{1k_2}, \ldots, k_r, a_{1k_r}$, the number s of nonzero entries in the second row, $\ell_1, a_{2\ell_1}, \ell_2, a_{2\ell_2}, \ldots,$ $\ell_s, a_{2\ell_s}$, etc., and finally the number t of nonzero entries in the mth row, $p_1, a_{mp_1}, \ldots, p_t, a_{mp_t}$. Similarly we can store a matrix searching for the non-zeros through its columns. The number of data stored is $2 + m + 2N$ here. If we have for example $m = n = 200$, $N = 2000$ (i.e., 5% nonzero entries), then the number of locations required is 6002 in the first case and 4202 in the second case. Storing the matrix in the usual way, we would need about 40,000 locations. The presented methods for storing, however, need a special software for matrix operations.

It is impossible to answer unambiguously the question whether direct or iterative methods are more advantageous for solving systems with large sparse matrices. The advantage of iterative methods consists in the fact that they – at least in the simplest cases – use the original matrix in each step and the rounding errors thus do not accumulate as they do in the case when we "get farther and farther away" from the initial data. It is even possible that the further development of computers, in particular parallel computers, will influence the answer to the above question.

We shall be concerned mostly with the application of the method of Gaussian elimination in what follows. The principal problem here is the choice of pivots. As we saw in Chapter 10 there are various approaches to the choice of pivotal elements. A further fact that has to be taken into account when we study large sparse matrices is the number of nonzero entries which arise during the elimination and which have to be stored.

Example 14.1 Let the system $Ax = b$ have a matrix A of the form

$$
\begin{bmatrix}
\times & \times & \times & \ldots & \times \\
\times & \times & & & \\
\times & & \times & & \\
\vdots & & & \ddots & \\
\times & & & & \times
\end{bmatrix},
$$

where each cross denotes a nonzero and zero entries are not

marked. Choosing the entry in position $(1, 1)$ for the pivotal element of the first step, we come to a "filled" matrix $A^{(1)}$, i.e., all the entries of $A^{(1)}$, possibly except for some diagonal entries, are nonzero. On the other hand, the sequence of pivots $(n, n), (n - 1, n - 1), \ldots, (2, 2)$ corresponds to the natural order of pivots $(1, 1)(2, 2), \ldots, (n - 1, n - 1)$ for the permuted matrix

$$
\begin{bmatrix}
\times & & & & \times \\
& \times & & & \times \\
& & \ddots & & \vdots \\
& & & \times & \times \\
\times & \times & \ldots & \times & \times
\end{bmatrix}.
$$

Clearly, no new nonzero then arises in the course of elimination.

We shall first consider the case of a symmetric positive definite or diagonally dominant matrix A of the system $Ax = b$. Then (and, as a rule, also in the case of a symmetric but not positive definite matrix A) the diagonal entries are always taken for pivotal elements. The only question is the order of pivots. To study this subject in more detail we shall use graph theory, which in similar problems of combinatorial character usually is an appropriate means.

We first introduce the concept of the Schur complement in a digraph [21]. Let $\vec{G} = (V, H)$ be a digraph whose set of vertices is V and set of arcs is H. Let $W \subset V$. The *Schur complement* of the set of vertices W in $\vec{G}$ is the digraph $\vec{G}[W] = (V \backslash W, \hat{H})$ such that $(i, k) \in \hat{H}$ $(i, k \in V \backslash W)$ if and only if there is a directed path from i to k in $\vec{G}$ whose all interior vertices (i.e. vertices of the path except for i and k) belong to W.

Example 14.2 If $\vec{G}$ is the digraph of Fig. 19a and $W = \{1, 2, 3\}$, then $\vec{G}[W]$ is the digraph displayed in Fig. 19b.

The following theorem presents an important property of the Schur complement.

Theorem 14.1.1 *Let $\vec{G} = (V, H)$ be a digraph, let $W_1 \subset V$, $W_2 \subset V$, and $W_1 \cap W_2 = \emptyset$. Put $\vec{G}_1 = \vec{G}[W_1]$. Then*

$$
\vec{G}_1[W_2] = \vec{G}[W_1 \cup W_2].
$$

Proof is left to the reader (Exercise 14.1).

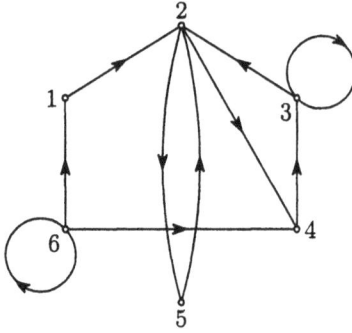
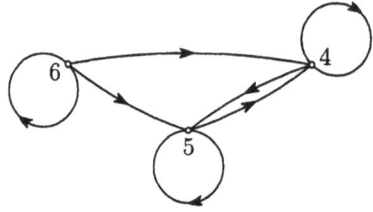

Fig. 19a   Fig. 19b

Next theorems show how the Schur complement in a digraph is connected with the Schur complement of a submatrix (cf. Chapter 1) and with the elimination. We first establish this relation for a matrix of class K (see Chapter 5).

Theorem 14.1.2 *Let a square matrix A of order n belong to the class K and let $M = \{1, 2, \ldots, m\} \subset N = \{1, 2, \ldots, n\}$. For the digraph $\vec{G}(B)$ of the Schur complement $B = [A/A(M)]$ of the principal submatrix $A(M)$ of A we then have*

$$\vec{G}(B) = \vec{G}_1[M],$$

where $\vec{G}_1 = \vec{G}(A)$ is the digraph of A. The digraph of the (matrix) Schur complement B is thus equal to the Schur complement of the corresponding set of vertices in the digraph of A.

Remark. The theorem claims that for a matrix of class K the combinatorial properties of the zero-nonzero structure of the (matrix) Schur complement depend only on the combinatorial properties of the zero-nonzero structure of the matrix given.

Proof. We prove the theorem by induction on the number $m = |M|$ of elements of the set M. If $m = |M| = 1$ and $M = \{1\}$, then $B = [b_{ik}]$, $i, k \in N\backslash\{1\}$, $b_{ik} = a_{ik} - a_{i1}a_{11}^{-1}a_{1k}$. Moreover, $B \in K$ by Theorem 5.2.11. For $i = k$ thus $b_{ii} > 0$. If $i \neq k$, then $a_{ik} \leq 0$, $a_{i1} \leq 0$, $a_{1k} \leq 0$, and $a_{11} > 0$ so that $b_{ik} \leq 0$ and, moreover, $b_{ik} = 0$ if and only if $a_{ik} = 0$ and $a_{i1}a_{1k} = 0$. Hence, (i, k) is an arc in the digraph $\vec{G}(B)$ if and only if at least one of the following two statements holds:

1. (i, k) is an arc in $\vec{G}(A)$.
2. Both $(i, 1)$ and $(1, k)$ are arcs in $\vec{G}(A)$.

In other words, (i, k) is an arc in $\vec{G}(B)$ if and only if there is a path from i to k in $\vec{G}(A)$ whose interior point (if any) belongs to $\{1\}$. The statement therefore holds.

Suppose now that $m = |M| > 1$ and that the statement holds for all matrices in K and all subsets M' for which $|M'| < |M|$. Put $M' = M\backslash\{m\}$, $C = [A/A(M')]$, and $\tilde{C} = [A(M)/A(M')]$. By the induction hypothesis we have $\vec{G}(C) = \vec{G}_1[M']$. Denote this digraph by $\vec{G}_2$ and put $M'' = \{m\}$. By Theorem 14.1.1, we obtain $\vec{G}_2[M''] = \vec{G}_1[M' \cup M''] = \vec{G}_1[M]$. On the other hand, $\vec{G}_2[M'']$ is the digraph of the Schur complement $[C/\tilde{C}]$, which follows from the statement we have already proved for $m = 1$. Using now Theorem 1.4.4, we come to the relation $[A/A(M)] = [C/\tilde{C}]$. Hence, $\vec{G}(B) = \vec{G}_2[M''] = \vec{G}_1[M]$ holds for the digraphs of both the sides, which was to be proved. $\qquad\square$

An analogous theorem holds also for the factorization of A into a product of a lower and an upper triangular matrices.

Theorem 14.1.3 *Let $A \in K$ be a matrix of order n. Then in every factorization $A = LU$ into a product of a lower triangular matrix L and an upper triangular matrix U (which always exists), the zero-nonzero structure of L and U is completely determined by the zero-nonzero structure of A. In more detail, if $\vec{G}(A)$ is the digraph of A, then $\vec{G}(L) = (N, H_1)$, where $N = \{1, 2, \dots, n\}$ and $(i, k) \in H_1$ if and only if $i \geq k$ and, moreover, in $\vec{G}(A)$ there exists a path from i to k whose all interior vertices are labelled by integers less than k. Similarly, $\vec{G}(U) = (N, H_2)$, where $(i, k) \in H_2$ if and only if $i \leq k$ and in $\vec{G}(A)$ there exists a path from i to k whose all interior vertices are labelled by integers less than i.*

Proof. The existence of the factorization $A = LU$ follows from 13. of Theorem 5.1.1. By Theorem 1.4.3, the difference between any two such factorizations consists in right-multiplying L by a diagonal nonsingular matrix and left-multiplying U by the inverse of this diagonal matrix. Of course, this does not change the zero-nonzero structure of L and U. We can thus suppose that $\ell_{ii} = 1$ for $i = 1, \dots, n$. By (10.9), the ith row of the part of

U above the diagonal (together with the diagonal part) concides with the first row of $A^{(i-1)} = [a_{k\ell}^{(i-1)}]$, $k, \ell = 1, \ldots, n$ (the notation here is consistent with the notation introduced for the method of Gaussian elimination in Chapter 10). Hence $u_{ik} \neq 0$ for $i \leq k$ if and only if $a_{ik}^{(i-1)} \neq 0$. By Theorems 14.1.2 and 10.1.3, this occurs if and only if there is a path in $\vec{G}(A)$ from i to k whose all interior points are labelled by integers from the set $\{1, 2, \ldots, i-1\}$, i.e., by integers less than i. An analogous statement holds for the entries of L as well, which follows e.g. from the application of the statement already proved to the matrix A^T: In fact, $A^T = U^T L^T$ is an analogous factorization and $\vec{G}(L^T)$ thus fulfils the condition proved (however, we then have to reverse the direction of each arc, see Exercise 14.2).   □

Remark. Labelling the vertices of an arbitrary digraph $\vec{G} = (V, H)$ by integers $1, 2, \ldots, |V|$, we introduce the *lower triangular part* of $\vec{G}$ as the digraph $\vec{G}_\ell = (V, H_1)$, where $(i, k) \in H_1$ if and only if $i \geq k$ and in $\vec{G}$ there is a path from i to k whose all interior vertices are labelled by integers less than k. The definition of the *upper triangular part* $\vec{G}_u$ of $\vec{G}$ is analogous. Theorem 14.1.3 can then be restated with the help of the lower and upper triangular parts of $\vec{G}(A)$ if we assume the natural ordering of vertices (corresponding to the order of the rows of A).

Now we are ready to prove the fundamental theorem on the zero-nonzero structure of the matrices $A^{(k)}$, L, and U for a general system $Ax = b$ with a strongly nonsingular matrix A.

Theorem 14.1.4 *Let the matrix A of the system $Ax = b$ be strongly nonsingular. Let $\vec{G}$ be the digraph of A whose vertices are labelled by the indices of rows (and columns). Then the digraph of the matrix $A^{(k)}$, $k = 1, \ldots, n - 1$, of the method of Gaussian elimination performed with the natural choice of pivotal elements is contained in the Schur complement $\vec{G}[N_k]$, where $N_k = \{1, 2, \ldots, k\}$, and the digraph of the lower triangular matrix L (or the upper triangular matrix U) in the factorization $A = LU$ is contained in the lower (or upper) triangular part of $\vec{G}$.*

In addition, if all the diagonal entries of A are nonzero, then there exists a (strongly nonsingular) matrix of the same zero-

nonzero structure as A for which the equality takes place in all the above inclusions.

Proof. We prove $\vec{G}(A^{(k)}) \subset \vec{G}[N_k]$ by induction on k. This is true for $k = 0$ if we put $A^{(0)} = A$ and $\vec{G}[N_0] = \vec{G}(A)$.

Thus let $k \geq 1$ and assume $\vec{G}(A^{(k-1)}) \subset \vec{G}[N_{k-1}]$. The matrix $A^{(k)}$ is obtained from $A^{(k-1)}$ according to the formula (10.11). But this means that if $a_{ij}^{(k)} \neq 0$, then at least one of the two following cases takes place: 1. $a_{ij}^{(k-1)} \neq 0$; 2. $a_{ik}^{(k-1)} \neq 0$ and $a_{kj}^{(k-1)} \neq 0$ hold simultaneously. Hence, if $(i,j) \in \vec{G}(A^{(k)})$, then in $\vec{G}[N_{k-1}]$ there is a path from i to j possibly containing a single interior vertex k. Thus $\vec{G}(A^{(k)})$ is contained in $\vec{G}_1[k]$ for $\vec{G}_1 = \vec{G}[N_{k-1}]$. By Theorem 14.1.1 applied to $W_1 = N_{k-1}$ and $W_2 = \{k\}$, however, we have $\vec{G}_1[k] = \vec{G}[N_k]$. Therefore, $\vec{G}(A^{(k)}) \subset \vec{G}[N_k]$ and the first part of the theorem has been proved. The statement on the digraphs of L and U follows from the statement just proved by the same argument as in the proof of Theorem 14.1.3. The last assertion then follows from the fact that for any matrix $A = [a_{ik}]$ having nonzero diagonal entries, there exists a matrix $A_0 \in K$ which has the same zero-nonzero structure as A. Indeed, it is sufficient to set $A_0 = I - \varepsilon m(A)$, where $m(A) = [|a_{ik}|]$ is the modulus of A and ε is a sufficiently small positive number (e.g., $\varepsilon < (g_1(A))^{-1}$, where g_1 is the norm of A from Theorem 9.2.5).   □

Remark. In applications we have, in the vast majority of cases, $\vec{G}(A^{(k)}) = \vec{G}[N_k]$, $\vec{G}(L) = \vec{G}_\ell$, and $\vec{G}(U) = \vec{G}_u$ as soon as the diagonal entries of A are nonzero since, as a rule, no cancellation takes place in the formula (10.11).

If we solve several systems with matrices of the same zero-nonzero structure it is advantageous to try to find an elimination order in which relatively few new nonzeros arise since – provided that all the diagonal entries of A are nonzero – the zero-nonzero structure of A essentially determines the zero-nonzero structures of L and U.

It is very important that all the presented facts on the zero-nonzero structure of the matrices $A^{(k)}$, L, and U are true also in the case of a partitioned matrix A whose diagonal blocks are

square. The assumption of strong nonsingularity then means
that all the partitioned submatrices

$$A_{11}, \quad \begin{bmatrix} A_{11} & A_{12} \\ A_{21} & A_{22} \end{bmatrix}, \quad \text{etc.}$$

are nonsingular. The digraphs, however, are digraphs of parti-
tioned matrices (e.g. (i,k) is an arc in $\vec{G}(A)$ if $A_{ik} \neq 0$). In this
sense, Theorem 14.1.5 on the structure of the matrices L and U
in the case of a point (block) almost acyclic matrix A (Chapter
12) finds much wider applications.

Theorem 14.1.5 *Let A be an irreducible almost acyclic $n \times n$
matrix which is either positive definite or diagonally dominant.
Then:*

1. *For a fixed $r \in \{1, 2, \ldots, n\}$, there exists a permutation matrix P such that the matrix $B = P^T A P$ has the rth row of
   A as its last (nth) row (with entries, of course, permuted)
   and, furthermore, the digraph $\vec{G}(B)$ of the matrix B possesses the following property: If the path from the vertex i
   to the vertex n in $\vec{G}(B)$ passes through the vertex j, then
   $i < j$.*

2. *Suppose the matrix $B = [b_{ik}]$ of 1. is factored into the product
   LU, where $L = [\ell_{ij}]$ is a lower triangular matrix with unit
   diagonal entries and $U = [u_{ij}]$ is upper triangular. Then
   $u_{ik} \neq 0$ if and only if $i \leq k$ and $b_{ik} \neq 0$, and $\ell_{ik} \neq 0$ if and
   only if $i \geq k$ and $b_{ik} \neq 0$. In more detail, we have*

$$u_{ik} = b_{ik} \quad \text{for } i < k, \; i, k = 1, \ldots, n, \quad (14.1)$$

$$\ell_{ik} = b_{ik} u_{kk}^{-1} \text{ for } i > k, \; i, k = 1, \ldots, n, \quad (14.2)$$

$$u_{ii} = b_{ii} - \sum_{k=1}^{i-1} b_{ik} u_{kk}^{-1} b_{ki}, \; i = 1, \ldots, n. \quad (14.3)$$

Proof. Since A is irreducible $\vec{G}(A)$ is strongly connected by Theorem 3.2.2. In addition, $\vec{G}(A)$ contains no cycle of length greater
than 2, which clearly implies that $\vec{G}(A)$ is a symmetric digraph
such that the corresponding undirected graph is a tree. We

now use the numbering φ given by Exercise 3.5 and satisfying $\varphi(r) = 1$. The reverse numbering $\hat{\varphi} = n + 1 - \varphi$ then defines the desired permutation.

To prove 2. we use Theorem 14.1.4. If $B = LU$, then the digraph $\vec{G}(L)$ is included in the lower triangular part of $\vec{G}(B)$. We shall show that the lower triangular part of $\vec{G}(B)$ contains the arc (i, k) if and only if $i \geq k$ and (i, k) is an arc of $\vec{G}(B)$. Indeed, assume that (i, k), $i > k$ is an arc in $\vec{G}(L)$ and that $D_1 = (i, j_1, \ldots, j_s, k)$ is a path satisfying $j_t < k$, $t = 1, \ldots, s$. We prove that $s \geq 1$ is impossible, Since there is a path from i to n whose vertices are labelled by numbers not less than k, we conclude by Theorem 3.1.1 that there exists a path D_2 from i to k whose vertices are labelled by numbers not less than k. The paths D_2 and D_1 have thus precisely the two vertices i and k in common and, therefore, D_2 along with the path reverse to D_1 form a cycle of length at least $s + 2$ contained in $\vec{G}(B)$. Hence $s = 0$ and the arc (i, k) itself belongs to $\vec{G}(B)$, i.e., $\ell_{ik} \neq 0$ only if $i \geq k$ and $b_{ik} \neq 0$. For the same reason we have $u_{ik} \neq 0$ only if $i \leq k$ and $b_{ik} \neq 0$.

From (10.17) and (10.18) we obtain

$$b_{ij} = \sum_{k=1}^{\min(i,j)} \ell_{ik} u_{kj}; \qquad (14.4)$$

in addition, we assume $\ell_{ii} = 1$, $i = 1, \ldots, n$.

If for some $b_{ij} \neq 0$, $i \neq j$, the right-hand side of (14.4) contained a nonzero term with $k \neq \min(i,j)$, say with $k = k_0, k_0 \neq i, k_0 \neq j$, then (i, k_0, j) would be a cycle in $\vec{G}(B)$ of length 3 since by the preceding part of the proof we would have $b_{ik_0} \neq 0$ as well as $b_{k_0 j} \neq 0$. Hence, when $b_{ij} \neq 0$ and $i < j$ then $b_{ij} = u_{ij}$, and when $b_{ij} \neq 0$ for $i > j$ then $b_{ij} = \ell_{ij} u_{jj}$. We have thus shown that (14.2) and (14.3) hold. Substituting from these formulae into

$$b_{ii} = \sum_{k=1}^{i} \ell_{ik} u_{ki},$$

we obtain

$$u_{ii} = b_{ii} - \sum_{k=1}^{i-1} b_{ik} u_{kk}^{-1} b_{ki}, \quad i = 1, \ldots, n,$$

which is (14.3).   □

A positive definite matrix A can be transformed to block almost acyclic form e.g. by the following procedure:

Suppose that A is irreducible (otherwise we apply the procedure to each maximal irreducible principal submatrix separately). The (undirected) graph G of A is thus connected. Choose now a subset U_1 of the set of vertices of G (the set U_1 should be chosen such that a relatively small number of edges go from the vertices in U_1 to vertices outside U_1). Denote by $G_{11}, G_{12}, \ldots, G_{1s}$ the components of the graph obtained by deleting from G all the vertices in U_1 and all edges going from U_1. Let $U_{11}, \ldots, U_{1s}$ be the sets of those vertices in $G_{11}, \ldots, G_{1s}$, respectively, that are connected by an edge in G with at least one vertex of U_1. We proceed similarly in G_{11} using the set U_{11} etc., until no vertex remains. The sets

$$U_1, U_{11}, \ldots, U_{1s}, U_{111}, \ldots, U_{11r_1}, \ldots, U_{1sr_s}, \ldots$$

form a partition of the set of vertices of G and the condensation of G with respect to the partition (cf.Chapter 3) is a tree. We now order the rows (and columns) of A in such a way that we start with the rows corresponding to U_1, then we take the rows corresponding to $U_{11}, \ldots, U_{1s}$, etc. Finally, reversing this ordering we obtain a block almost acyclic matrix whose blocks are numbered as it was required in Theorem 14.1.5.

Next, we present a useful technique for investigating nonzero entries in the process of Gaussian elimination.

14.2   Envelopes. Profile

Let $A = [a_{ik}]$ be a square matrix of order n. For $i, k = 1, \ldots, n$ we define $p_i = \min(\min\{j; a_{ij} \neq 0\}, i)$ and $q_k = \min(\min\{j; a_{jk} \neq 0\}, k)$. Obviously p_i is just the column index of the first nonzero in row i (and q_k the row index of the first nonzero in column k),

$i, k = 1, 2, \ldots, n$, provided, of course, that these indices do not exceed i (or k, respectively). Then we define the *lower envelope of A* to be the set of the positions (i, p_i), $i = 1, \ldots, n$, whereas the *upper envelope of A* is the set of the positions (q_k, k), $k = 1, \ldots, n$.

Example 14.3 In the matrix

$$\begin{bmatrix} \hat{2} & \underline{-1} & 0 & 0 \\ 0 & \hat{3} & \underline{1} & 0 \\ \hat{4} & 0 & 0 & 0 \\ 0 & \hat{2} & 0 & \underline{1} \end{bmatrix},$$

the entries in the upper envelope are underlined, those in the lower envelope endowed by a hat.

We now prove an important theorem, in which a *full lower triangular matrix* with a given lower envelope means a matrix whose all entries in each row lying between the envelope entry and the diagonal entry are nonzero (analogously, we define a *full upper triangular matrix* requiring the nonzeros to be in columns).

Theorem 14.2.1 (George and Liu [28]). *Let A be a strongly nonsingular matrix and let $A = LU$ be some of its (always existing) factorizations into a product of a lower triangular matrix L and an upper triangular matrix U. Then the lower envelope of L and the upper envelope of U are the same as those of A. Moreover, if $A \in K$ and A is such that $p_i < i$ for $i = 2, \ldots, n$ and $q_k < k$ for $k = 2, \ldots, n$, then L is a full lower triangular matrix whose lower envelope is identical with that of A, and U is a full upper triangular matrix whose upper envelope is identical with that of A. The same is true even if A does not belong to K but all its diagonal entries are nonzero and if there is no cancellation in (10.11) during the calculation.*

Proof. Let $A = LU$, where $L = [\ell_{ik}]$, $U = [u_{ik}]$, and $\ell_{ik} = 0$ if $i < k$, $u_{ik} = 0$ if $i > k$. Then instead of (10.17) and (10.18) we obtain

$$a_{ij} = \sum_{k=1}^{\min(i,j)} \ell_{ik} u_{kj}. \tag{14.5}$$

As we have seen, we may choose the diagonal entries in L to be arbitrary nonzero numbers without affecting the zero-nonzero structures of L and U. We choose $\ell_{ii} = 1$, $i = 1, \ldots, n$. Then we have the recurrence relations

$$u_{ij} = a_{ij} - \sum_{k=1}^{i-1} \ell_{ik} u_{kj} \qquad \text{for } i \leq j, \qquad (14.6)$$

$$\ell_{ij} = \frac{1}{u_{jj}} \left(a_{ij} - \sum_{k=1}^{j-1} \ell_{ik} u_{kj} \right) \qquad \text{for } i > j. \qquad (14.7)$$

We shall use (14.6) and (14.7) to prove, by induction on $i+j$, that for $j < p_i$ we have $\ell_{ij} = 0$, for $j = p_i$ we have $\ell_{ij} \neq 0$, for $i < q_j$ we have $u_{ij} = 0$ and for $i = q_j$ we have $u_{ij} \neq 0$.

If $i + j = 3, i < j$, i.e. $i = 1, j = 2$, then $u_{12} = a_{12}$ and the assertion holds. Similarly if $i > j$, then $i = 2, j = 1$, we have $\ell_{21} = a_{21}/u_{11}$, and the assertion again holds. Let now $i + j > 3$ and suppose that the assertion holds for all i', j' satisfying $i'+j' < i+j$. If $j < p_i$, i.e. $j < i$, then (14.7) implies $\ell_{ij} = 0$ since $\ell_{ik} = 0$ for $k < j$, i.e. $k < p_i$ by the induction hypothesis $(i+k < i+j)$ and $a_{ij} = 0$. The same relation and argument give $\ell_{ij} = a_{ij}/u_{jj} \neq 0$ for $j = p_i$. The proof of the assertion on u_{ij} for $i < q_j$ and $i = q_j$ is similar; it only uses (14.6) instead of (14.7).

Suppose now that $A \in K$ and $p_i < i, q_k < k$ for $i, k = 2, \ldots, n$. By property 14. of Theorem 5.1.1 and Theorem 1.4.3, we can see that if $\ell_{ii} > 0$, $i = 1, 2, \ldots, n$, then $L \in K$ and $U \in K$ as well. We now use induction on $i + j$ to prove that we have $\ell_{ij} < 0$ for $i > j > p_i$ and $u_{ij} < 0$ for $j > i > q_j$. If $i+j = 5$ (there is nothing to prove in case $i + j < 5$) and $i > j$ it remains to investigate the case $i = 3, j = 2$, and $p_3 = 1$. Then $\ell_{32} = a_{32} - \ell_{31} u_{12} < 0$ since we have $a_{32} \leq 0$, $\ell_{31} < 0$, and $u_{12} < 0$ using the assertion we have proved and the assumption $q_2 < 2$. Similarly, the assertion holds for $j > i$, i.e., $j = 3, i = 2$, and $q_3 = 1$. Let now $i + j > 5$ and suppose that the assertion holds for all i', j' satisfying $i' + j' < i + j$. If $i > j$, then formula (14.7) and the fact that we have $L \in K$ and $U \in K$ imply that the sum on the right-hand side contains only nonnegative terms and, moreover, $\ell_{i,j-1} u_{j-1,j} > 0$ since for $j = p_i + 1$ we have $\ell_{ip_i} < 0$ by the preceding part of the proof, and $u_{j-1,j} < 0$ by the induction hypothesis or by the preceding part of the proof if $q_j = j - 1$; if $j > p_i + 1$, then the same is true of $\ell_{i,j-1}$ by the induction hypothesis. The proof of

$u_{ij} < 0$ for $j > i > q_j$ is analogous. The conclusion then follows by an argument similar to that used in the proof of Theorem 14.1.5. □

Theorem 14.2.1 shows that if A is a square matrix of order n, then the number

$$\Pi(A) = n + \sum_{i=1}^{n}(i - p_i) + \sum_{j=1}^{n}(j - q_j), \qquad (14.8)$$

called the *profile* of A (with natural ordering), plays an important role in the considerations regarding the occurence of additional nonzeros which are created during factorization and must be stored in the computer. The first summand on the right-hand side of (14.8) is the number of diagonal entries, the second is the number of entries below the diagonal and on the right of the lower envelope, the third above the diagonal and below the upper envelope.

Simultaneous permutations of the rows and columns of A change the profile as a rule and when dealing with large sparse matrices a very important task is to find a permutation matrix P_0 representing the permutation which minimizes the profile:

$$\Pi(P_0 A P_0^T) = \min_{P} \Pi(P A P^T).$$

This task belongs – like that of minimizing the bandwidth treated in Chapter 8 – to algorithmically most complex (NP-complete) problems. Also here, the Reverse CM algorithm and the GPSD algorithm from Chapter 8 give good results.

We now turn to the case when the matrix is neither positive definite nor diagonally dominant. Then, of course, there is no reason to take only diagonal entries for pivots.

Although we have more possibilities in this case, the problem of minimizing the profile over all possible admissible permutation matrices P, Q,

$$\Pi(P_0 A Q_0) =$$
$$= \min\{\Pi(PAQ); P, Q \text{ permutation matrices,}$$
$$PAQ \text{ strongly nonsingular}\},$$

is again very complex algorithmically. Here, too, graphs may be used.

As we saw, however, in Chapter 3, the bigraph representation is appropriate for the study of the matrix properties which do not depend on independent permutations of rows and columns.

We first note the properties of the bigraph $\vec{B}(A)$ of a nonsingular matrix A [21]. A bigraph $\vec{B} = (V, W, H)$ is called *linear* if one and only one arc begins at each vertex of V and one and only one arc ends at each vertex of W. For linear bigraphs, we clearly have $|V| = |W|$ and the arcs define a one-to-one mapping of the set V onto W.

Theorem 14.2.2 *Let $A = [a_{ik}]$ be a nonsingular matrix. Then its bigraph $\vec{B}(A)$ contains (as a subgraph) a linear bigraph with the same sets of vertices. Conversely, if $\vec{B} = (V, W, H)$ is a bigraph that contains a linear bigraph with the same sets of vertices, then there is a nonsingular matrix A whose bigraph equals $\vec{B}$.*

Proof. If the matrix $A = [a_{ik}]$ is nonsingular, then $\det A \neq 0$ so that the expression (1.5) for the determinant contains at least one nonzero term. This term defines a linear bigraph which is a subgraph of $\vec{B}(A)$.

Conversely, if $\vec{B} = (V, W, H)$ is a bigraph which contains a linear bigraph $\vec{B}_1$ as a subgraph having the same sets of vertices, then it suffices to take for A the matrix having ones in the positions that correspond to the arcs of $\vec{B}_1$, ε in the positions corresponding to the other arcs of $\vec{B}$, and zeros elsewhere. For a sufficiently small $|\varepsilon|$ (for example, $|\varepsilon| < 1/n$ where $n = |V| = |W|$), the matrix A is nonsingular since the matrix $P^T A$, where P is the permutation matrix obtained by setting $\varepsilon = 0$ in A, is diagonally dominant.   □

Remark. A bigraph $\vec{B}$ is said to be *nonsingular* if it contains a subgraph which is a linear bigraph having the same sets of vertices as $\vec{B}$. A bigraph $\vec{B}_1$ is called the *inverse* of a nonsingular bigraph $\vec{B} = (V, W, H)$ if $\vec{B}_1 = (W, V, H_1)$ and $(i, k) \in H_1$ if and only if $i \in W, k \in V$ and, moreover, (k, i) belongs to a linear bigraph (with the same sets of vertices) of $(V, W, H \cup \{(k, i)\})$.

It may be shown[1] that whenever A is a nonsingular matrix such that $\vec{B}(A) = \vec{B}$, then $\vec{B}(A^{-1})$ is included in $\vec{B}_1$.

If we now perform the elimination with the pivots $(1,1)$, $(2,2), \ldots, (n,n)$ it holds that (provided no cancellation takes place during the elimination) the bigraph of $A^{(k)}$ can be obtained in the following way: Let $\vec{L}$ be a linear subbigraph of $(N_k, N_k, H \cap N_k \times N_k)$, $N_k = \{1, 2, \ldots, k\}$, where $\vec{B}(A) = (N_n, N_n, H)$; the bigraph $\vec{L}$ does exist as the matrix in question is nonsingular. Then $\vec{B}(A^{(k)}) = (M_k, M_k, H_k)$ where $M_k = \{k+1, \ldots, n\}$ and, furthermore, $(i,j) \in H_k$, $i, j \in M_k$, if and only if in the digraph $\vec{B}(A) \cup \vec{L}^T$, where $\vec{L}^T$ is obtained from the bigraph $\vec{L}$ by reversing the direction of the arcs, there exists a path from the vertex i (in the first set) to j (in the second set) whose all vertices (in the two sets) are labelled by numbers not exceeding k. Here, the symbol $\vec{B}(A) \cup \vec{L}^T$ means the digraph with $2n$ vertices whose set of arcs is the union of the arcs of $\vec{B}(A)$ and $\vec{L}^T$.

Exercises

14.1. Prove Theorem 14.1.1.

14.2.* Prove that the digraph $\vec{G}(A^T)$ of the matrix A^T can be obtained from $\vec{G}(A)$ by reversing the direction of each arc.

14.3. Find the digraphs $\vec{G}(A^{(k)})$, $k = 1, 2, \ldots, 5$, $\vec{G}(L)$, and $\vec{G}(U)$ provided A is the matrix of the system

$$
\begin{array}{rcrcrcrcrcrcl}
x_1 & - & x_2 & & & & & & & & & = & 2, \\
& & 2x_2 & & & - & x_4 & & & - & x_6 & = & -3, \\
& - & x_2 & + & 2x_3 & & & & & & & = & 1, \\
& & & & & & 3x_4 & - & x_5 & & & = & 0, \\
-x_1 & & & & & & & + & x_5 & & & = & -5, \\
& & & & & - & x_4 & & & + & 2x_6 & = & -1.
\end{array}
$$

Verify your results by using Gaussian elimination to solve the system. Show also that A belongs to the class K introduced in Chapter 5.

---

[1]See [21].

Bibliography

[1] Bellman, R.: Introduction to Matrix Analysis. New York, Mc Graw-Hill Book Co, Inc., 1960.

[2] Berman, A. – Plemmons, R. J.: Nonnegative Matrices in the Mathematical Sciences. New York, Academic Press 1979.

[3] Bhatia, R. – Friedland, S.: Variation of Grassman Powers and Spectra. *Linear Algebra Appl.*, **40**, 1981, 1–18.

[4] Birkhoff, G.: Tres observaciones sobre el algebra lineal. *Univ. Nac. Tucumán Rev. ser. A*, **5**, 1946, 147–150.

[5] Birkhoff, G. - Mac Lane, S.: A Survey of Modern Algebra. New York, The MacMillan Comp., 1965.

[6] Carlson, D. – Datta, B. N.: On the effective computation of the inertia of a non-hermitian matrix. *Num. Math.*, **33**, 1979, 315–332.

[7] Cottle, R. W. – Dantzig, G. B.: Complementary pivot theory of mathematical programming. *Linear Algebra Appl.*, **1**, 1968, 103–125.

[8] Crabtree, D. – Haynsworth, E. V.: An identity for the Schur complement of a matrix. *Proc. AMS*, **22**, 1969, 364–366.

[9] Cuthill, E. – McKee, J.: Reducing the bandwidth of sparse symmetric matrices. In: Proc. 24th ACM National Conference, New York, N. Y., 1969, 157–172.

[10] Dieudonné, J.: Foundations of Modern Analysis. New York, Academic, 1961.

[11] Dahlquist, G. – Björck, A.: Numerical Methods. Englewood Cliffs, N. J., Prentice-Hall, 1974.

[12] Donoghue, W. F.: Monotone Matrix Functions and Analytic Continuation. Berlin, Springer-Verlag, 1974.

[13] Fiedler, M.: Some estimates of spectra of matrices. In: Symp. PICC, Roma 1960, 33–36.

[14] Fiedler, M. – Pták, V.: On matrices with non-positive off-diagonal elements and positive principal minors. *Czech. Math. J.*, (87) **12**, 1962, 382–400.

[15] Fiedler, M.: On inverting partitioned matrices. *Czech. Math. J.*, (88) **13**, 1963, 574–586.

[16] Fiedler, M. – Pták, V.: On aggregation in matrix theory and its application to numerical inverting of large matrices. *Bull. Acad. Polon., ser. math. astr. et phys.*, **11**, 1963, 757–759.

[17] Fiedler, M. – Pták, V.: Some generalizations of positive definiteness and monotonicity. *Num. Math.*, **9**, 1966, 163–172.

[18] Fiedler, M.: Additive compound matrices and an inequality for eigenvalues of symmetric stochastic matrices. *Czech. Math. J.*, (99) **24**, 1974, 392–402.

[19] Fiedler, M.: On a theorem by A. Horn. In: Math. Structures – Comp. Math. – Math Modelling. Sofia 1975, 251–255.

[20] Fiedler, M.: Eigenvectors of acyclic matrices. *Czech. Math. J.*, (100) **25**, 1975, 607–618.

[21] Fiedler, M.: Inversion of bigraphs and connections with the Gauss elimination. In: Graphs, Hypergraphs and Block Systems. Zielona Góra, 1976, 57–68.

[22] Fiedler, M.: A deflation formula for tridiagonal matrices. *Aplikace mat.*, **25**, 1980, 348–357.

[23] Fiedler, M.: Quasidirect decompositions of Hankel and Toeplitz matrices. *Linear Algebra Appl.*, **61**, 1984, 155–174.

[24] Fiedler, M.: Hankel and Loewner matrices. *Linear Algebra Appl.*, **58**, 1984, 75–95.

[25] Francis, J. G. F.: The QR transformation, I., II. *Computer J.*, **4**, 1961, 265–271, 332–345.

[26] Frobenius, G.: Über Matrizen aus nicht negativen Elementen. Berlin, S.-B. Preuss. Acad. Wiss. 1912, 456–477.

[27] Gantmacher, F. R.: Matrix Theory. New York, Chelsea 1959.

[28] George, A. – Liu, J. W. H.: A note on fill for sparse matrices. *SIAM J. Numer. Anal.*, **12**, 1975, 452–455.

[29] Gerschgorin, S.: Über die Abgrenzung der Eigenwerte einer Matrix. *Izv. Akad. Nauk SSSR Ser. Mat.*, **7**, 1931, 749–754.

[30] Gibbs, N. E. – Poole, W. G. – Stockmeyer, P. K.: An algorithm for reducing the bandwidth and profile of a sparse matrix. *SIAM J. Numer. Anal.*, **13**, 1976, 236–250.

[31] Givens, W.: Numerical computation of the characteristic values of a real symmetric matrix. Oak Ridge Nat. Lab., ORNL – 1574, 1954.

[32] Greville, T. N. E.: Some applications of the pseudoinverse of a matrix. *SIAM Rev.*, **11**, 1960, 15–22.

[33] Harary, F.: Graph Theory. Reading, Addison Wesley, 1969.

[34] Haynsworth, E. V.: Determination of the inertia of a partitioned Hermitian matrix. *Linear Algebra Appl.*, **1**, 1968, 73–81.

[35] Heinig, G. – Rost, K.: Invertierung einiger Klassen von Matrizen und Operatoren. I. Endliche Toeplitzmatrizen und ihre Verallgemeinerungen. Wiss. Inform. 12, 1979, TH Karl-Marx-Stadt, Sekt. Math.

[36] Hoffman, A. J. – Wielandt, H. W.: The variation of the spectrum of a normal matrix. *Duke Math. J.*, **20**, 1953, 37–39.

[37] Householder, A. S.: The approximate solution of matrix problems. *J. Ass. Comp. Math.*, **5**, 1958, 205–243.

[38] Householder, A. S.: The Theory of Matrices in Numerical Analysis. New York-Toronto-London, Blaisdell Publ. Co., 1964.

[39] Jacobi, C. G. J.: Über ein leichtes Verfahren die in der Theorie der Säkulärstörungen vorkommenden Gleichungen numerisch aufzulösen. *Crelle's J.*, **30**, 1846, 51–94.

[40] König, D.: Theorie der endlichen und unendlichen Graphen. Leipzig, Akad. Verlag, 1936.

[41] Koteljanskij, D. M.: Some properties of matrices with positive elements (in Russian). *Mat. Sb.*, **31** (73), 1952, 497–506.

[42] Krein, M. G. – Naimark, M. A.: The method of symmetric and Hermitian forms in the theory of the separation of the roots of algebraic equations. Charkov 1936. Transl. *Lin. and Multilin. Alg.*, **10**, 1981, 265–308.

[43] Kublanovskaya, V. N.: On some algorithms for solving the complete eigenvalue problem (in Russian). *Ž. Vyčisl. Mat. i Mat. Fiz.*, **1**, 1961, 555–570.

[44] Lancaster, P.: Theory of Matrices. New York – London, Academic Press, 1969.

[45] Lander, F. J.: The Bézoutiant and inversion of Hankel and Toeplitz matrices (in Russian). *Mat. Issled. Kishinev IX*, **2** (32), 1974, 69–87.

[46] Löwner, K.: Über monotone Matrixfunktionen. *Math. Zeitschrift*, **38**, 1934, 177–216.

[47] Marcus, M. – Minc, H.: A Survey of Matrix Theory and Matrix Inequalities. Boston, Allyn & Bacon, 1964.

[48] Oldenburger, R.: Infinite powers of matrices and characteristic roots. *Duke Math. J.*, **6**, 1940, 357–361.

[49] Ostrowski, A. M.: Über die Determinanten mit überwiegender Hauptdiagonale. *Comment. Math. Helv.*, **10**, 1937, 69–96.

[50] Penrose, R.: A generalized inverse for matrices. *Proc. Cambridge Phil. Soc.*, **51**, 1955, 406–413.

[51] Pólya, G. – Szegö, G.: Aufgaben und Lehrsätze aus der Analysis. II. Berlin, Springer-Verlag, 1925.

[52] Perron, O.: Zur Theorie der Matrizen. *Math. Ann.*, **64**, 1907, 248–263.

[53] Ralston, A.: A First Course in Numerical Analysis. New York, Mc Graw-Hill Book Co., 1965.

[54] Romanovsky, V.: Recherches sur les chains de Markoff. *Acta Math.*, **66**, 1936, 147–251.

[55] Schur, I.: Über die charakteristischen Wurzeln einer linearen Substitution mit einer Anwendung auf die Theorie der Integralgleichungen. *Math. Ann.*, **66**, 1909, 488–510.

[56] Taussky – Todd, O.: A generalization of a theorem of Lyapunov. *SIAM J.*, **9**, 1961, 640–643.

[57] Tewarson, R. P.: Sparse Matrices. New York-London, Academic Press 1973.

[58] Varga, R. S.: Orderings of the successive overrelaxation scheme. *Pac. J. Math.*, **9**, 1959, 925–939.

[59] Varga, R. S.: Matrix Iterative Analysis. Englewood Cliffs, N. J., Prentice-Hall 1962.

[60] Waerden, B. L. v. d.: Algebra. Grundlehren Math. Wiss., Berlin-Göttingen-Heidelberg, Springer-Verlag, 1955.

[61] Wilkinson, J. H.: The Algebraic Eigenvalue Problem. Oxford, Clarendon Press, 1965.

[62] Young, D. M.: Iterative methods for solving partial difference equations of elliptic type. *Trans. AMS*, **76**, 1954, 92–111.

[63] Young, D. M.: Iterative Solution of Large Linear Systems. New York, Academic Press, 1971.

Index

A CATALOG OF SELECTED
DOVER BOOKS
IN SCIENCE AND MATHEMATICS

Astronomy

CHARIOTS FOR APOLLO: The NASA History of Manned Lunar Spacecraft to 1969, Courtney G. Brooks, James M. Grimwood, and Loyd S. Swenson, Jr. This illustrated history by a trio of experts is the definitive reference on the Apollo spacecraft and lunar modules. It traces the vehicles' design, development, and operation in space. More than 100 photographs and illustrations. 576pp. 6 3/4 x 9 1/4. 0-486-46756-2

EXPLORING THE MOON THROUGH BINOCULARS AND SMALL TELESCOPES, Ernest H. Cherrington, Jr. Informative, profusely illustrated guide to locating and identifying craters, rills, seas, mountains, other lunar features. Newly revised and updated with special section of new photos. Over 100 photos and diagrams. 240pp. 8 1/4 x 11. 0-486-24491-1

WHERE NO MAN HAS GONE BEFORE: A History of NASA's Apollo Lunar Expeditions, William David Compton. Introduction by Paul Dickson. This official NASA history traces behind-the-scenes conflicts and cooperation between scientists and engineers. The first half concerns preparations for the Moon landings, and the second half documents the flights that followed Apollo 11. 1989 edition. 432pp. 7 x 10.
0-486-47888-2

APOLLO EXPEDITIONS TO THE MOON: The NASA History, Edited by Edgar M. Cortright. Official NASA publication marks the 40th anniversary of the first lunar landing and features essays by project participants recalling engineering and administrative challenges. Accessible, jargon-free accounts, highlighted by numerous illustrations. 336pp. 8 3/8 x 10 7/8. 0-486-47175-6

ON MARS: Exploration of the Red Planet, 1958-1978--The NASA History, Edward Clinton Ezell and Linda Neuman Ezell. NASA's official history chronicles the start of our explorations of our planetary neighbor. It recounts cooperation among government, industry, and academia, and it features dozens of photos from Viking cameras. 560pp. 6 3/4 x 9 1/4. 0-486-46757-0

ARISTARCHUS OF SAMOS: The Ancient Copernicus, Sir Thomas Heath. Heath's history of astronomy ranges from Homer and Hesiod to Aristarchus and includes quotes from numerous thinkers, compilers, and scholasticists from Thales and Anaximander through Pythagoras, Plato, Aristotle, and Heraclides. 34 figures. 448pp. 5 3/8 x 8 1/2.
0-486-43886-4

AN INTRODUCTION TO CELESTIAL MECHANICS, Forest Ray Moulton. Classic text still unsurpassed in presentation of fundamental principles. Covers rectilinear motion, central forces, problems of two and three bodies, much more. Includes over 200 problems, some with answers. 437pp. 5 3/8 x 8 1/2. 0-486-64687-4

BEYOND THE ATMOSPHERE: Early Years of Space Science, Homer E. Newell. This exciting survey is the work of a top NASA administrator who chronicles technological advances, the relationship of space science to general science, and the space program's social, political, and economic contexts. 528pp. 6 3/4 x 9 1/4.
0-486-47464-X

STAR LORE: Myths, Legends, and Facts, William Tyler Olcott. Captivating retellings of the origins and histories of ancient star groups include Pegasus, Ursa Major, Pleiades, signs of the zodiac, and other constellations. "Classic." – *Sky & Telescope.* 58 illustrations. 544pp. 5 3/8 x 8 1/2. 0-486-43581-4

A COMPLETE MANUAL OF AMATEUR ASTRONOMY: Tools and Techniques for Astronomical Observations, P. Clay Sherrod with Thomas L. Koed. Concise, highly readable book discusses the selection, set-up, and maintenance of a telescope; amateur studies of the sun; lunar topography and occultations; and more. 124 figures. 26 halftones. 37 tables. 335pp. 6 1/2 x 9 1/4. 0-486-42820-6

Browse over 9,000 books at www.doverpublications.com

Chemistry

MOLECULAR COLLISION THEORY, M. S. Child. This high-level monograph offers an analytical treatment of classical scattering by a central force, quantum scattering by a central force, elastic scattering phase shifts, and semi-classical elastic scattering. 1974 edition. 310pp. 5 3/8 x 8 1/2.   0-486-69437-2

HANDBOOK OF COMPUTATIONAL QUANTUM CHEMISTRY, David B. Cook. This comprehensive text provides upper-level undergraduates and graduate students with an accessible introduction to the implementation of quantum ideas in molecular modeling, exploring practical applications alongside theoretical explanations. 1998 edition. 832pp. 5 3/8 x 8 1/2.   0-486-44307-8

RADIOACTIVE SUBSTANCES, Marie Curie. The celebrated scientist's thesis, which directly preceded her 1903 Nobel Prize, discusses establishing atomic character of radioactivity; extraction from pitchblende of polonium and radium; isolation of pure radium chloride; more. 96pp. 5 3/8 x 8 1/2.   0-486-42550-9

CHEMICAL MAGIC, Leonard A. Ford. Classic guide provides intriguing entertainment while elucidating sound scientific principles, with more than 100 unusual stunts: cold fire, dust explosions, a nylon rope trick, a disappearing beaker, much more. 128pp. 5 3/8 x 8 1/2.   0-486-67628-5

ALCHEMY, E. J. Holmyard. Classic study by noted authority covers 2,000 years of alchemical history: religious, mystical overtones; apparatus; signs, symbols, and secret terms; advent of scientific method, much more. Illustrated. 320pp. 5 3/8 x 8 1/2.
                    0-486-26298-7

CHEMICAL KINETICS AND REACTION DYNAMICS, Paul L. Houston. This text teaches the principles underlying modern chemical kinetics in a clear, direct fashion, using several examples to enhance basic understanding. Solutions to selected problems. 2001 edition. 352pp. 8 3/8 x 11.   0-486-45334-0

PROBLEMS AND SOLUTIONS IN QUANTUM CHEMISTRY AND PHYSICS, Charles S. Johnson and Lee G. Pedersen. Unusually varied problems, with detailed solutions, cover of quantum mechanics, wave mechanics, angular momentum, molecular spectroscopy, scattering theory, more. 280 problems, plus 139 supplementary exercises. 430pp. 6 1/2 x 9 1/4.   0-486-65236-X

ELEMENTS OF CHEMISTRY, Antoine Lavoisier. Monumental classic by the founder of modern chemistry features first explicit statement of law of conservation of matter in chemical change, and more. Facsimile reprint of original (1790) Kerr translation. 539pp. 5 3/8 x 8 1/2.   0-486-64624-6

MAGNETISM AND TRANSITION METAL COMPLEXES, F. E. Mabbs and D. J. Machin. A detailed view of the calculation methods involved in the magnetic properties of transition metal complexes, this volume offers sufficient background for original work in the field. 1973 edition. 240pp. 5 3/8 x 8 1/2.   0-486-46284-6

GENERAL CHEMISTRY, Linus Pauling. Revised third edition of classic first-year text by Nobel laureate. Atomic and molecular structure, quantum mechanics, statistical mechanics, thermodynamics correlated with descriptive chemistry. Problems. 992pp. 5 3/8 x 8 1/2.   0-486-65622-5

ELECTROLYTE SOLUTIONS: Second Revised Edition, R. A. Robinson and R. H. Stokes. Classic text deals primarily with measurement, interpretation of conductance, chemical potential, and diffusion in electrolyte solutions. Detailed theoretical interpretations, plus extensive tables of thermodynamic and transport properties. 1970 edition. 590pp. 5 3/8 x 8 1/2.   0-486-42225-9

Browse over 9,000 books at www.doverpublications.com

Engineering

FUNDAMENTALS OF ASTRODYNAMICS, Roger R. Bate, Donald D. Mueller, and Jerry E. White. Teaching text developed by U.S. Air Force Academy develops the basic two-body and n-body equations of motion; orbit determination; classical orbital elements, coordinate transformations; differential correction; more. 1971 edition. 455pp. 5 3/8 x 8 1/2.   0-486-60061-0

INTRODUCTION TO CONTINUUM MECHANICS FOR ENGINEERS: Revised Edition, Ray M. Bowen. This self-contained text introduces classical continuum models within a modern framework. Its numerous exercises illustrate the governing principles, linearizations, and other approximations that constitute classical continuum models. 2007 edition. 320pp. 6 1/8 x 9 1/4.   0-486-47460-7

ENGINEERING MECHANICS FOR STRUCTURES, Louis L. Bucciarelli. This text explores the mechanics of solids and statics as well as the strength of materials and elasticity theory. Its many design exercises encourage creative initiative and systems thinking. 2009 edition. 320pp. 6 1/8 x 9 1/4.   0-486-46855-0

FEEDBACK CONTROL THEORY, John C. Doyle, Bruce A. Francis and Allen R. Tannenbaum. This excellent introduction to feedback control system design offers a theoretical approach that captures the essential issues and can be applied to a wide range of practical problems. 1992 edition. 224pp. 6 1/2 x 9 1/4.   0-486-46933-6

THE FORCES OF MATTER, Michael Faraday. These lectures by a famous inventor offer an easy-to-understand introduction to the interactions of the universe's physical forces. Six essays explore gravitation, cohesion, chemical affinity, heat, magnetism, and electricity. 1993 edition. 96pp. 5 3/8 x 8 1/2.   0-486-47482-8

DYNAMICS, Lawrence E. Goodman and William H. Warner. Beginning engineering text introduces calculus of vectors, particle motion, dynamics of particle systems and plane rigid bodies, technical applications in plane motions, and more. Exercises and answers in every chapter. 619pp. 5 3/8 x 8 1/2.   0-486-42006-X

ADAPTIVE FILTERING PREDICTION AND CONTROL, Graham C. Goodwin and Kwai Sang Sin. This unified survey focuses on linear discrete-time systems and explores natural extensions to nonlinear systems. It emphasizes discrete-time systems, summarizing theoretical and practical aspects of a large class of adaptive algorithms. 1984 edition. 560pp. 6 1/2 x 9 1/4.   0-486-46932-8

INDUCTANCE CALCULATIONS, Frederick W. Grover. This authoritative reference enables the design of virtually every type of inductor. It features a single simple formula for each type of inductor, together with tables containing essential numerical factors. 1946 edition. 304pp. 5 3/8 x 8 1/2.   0-486-47440-2

THERMODYNAMICS: Foundations and Applications, Elias P. Gyftopoulos and Gian Paolo Beretta. Designed by two MIT professors, this authoritative text discusses basic concepts and applications in detail, emphasizing generality, definitions, and logical consistency. More than 300 solved problems cover realistic energy systems and processes. 800pp. 6 1/8 x 9 1/4.   0-486-43932-1

THE FINITE ELEMENT METHOD: Linear Static and Dynamic Finite Element Analysis, Thomas J. R. Hughes. Text for students without in-depth mathematical training, this text includes a comprehensive presentation and analysis of algorithms of time-dependent phenomena plus beam, plate, and shell theories. Solution guide available upon request. 672pp. 6 1/2 x 9 1/4.   0-486-41181-8

Browse over 9,000 books at www.doverpublications.com

HELICOPTER THEORY, Wayne Johnson. Monumental engineering text covers vertical flight, forward flight, performance, mathematics of rotating systems, rotary wing dynamics and aerodynamics, aeroelasticity, stability and control, stall, noise, and more. 189 illustrations. 1980 edition. 1089pp. 5 5/8 x 8 1/4.   0-486-68230-7

MATHEMATICAL HANDBOOK FOR SCIENTISTS AND ENGINEERS: Definitions, Theorems, and Formulas for Reference and Review, Granino A. Korn and Theresa M. Korn. Convenient access to information from every area of mathematics: Fourier transforms, Z transforms, linear and nonlinear programming, calculus of variations, random-process theory, special functions, combinatorial analysis, game theory, much more. 1152pp. 5 3/8 x 8 1/2.   0-486-41147-8

A HEAT TRANSFER TEXTBOOK: Fourth Edition, John H. Lienhard V and John H. Lienhard IV. This introduction to heat and mass transfer for engineering students features worked examples and end-of-chapter exercises. Worked examples and end-of-chapter exercises appear throughout the book, along with well-drawn, illuminating figures. 768pp. 7 x 9 1/4.   0-486-47931-5

BASIC ELECTRICITY, U.S. Bureau of Naval Personnel. Originally a training course; best nontechnical coverage. Topics include batteries, circuits, conductors, AC and DC, inductance and capacitance, generators, motors, transformers, amplifiers, etc. Many questions with answers. 349 illustrations. 1969 edition. 448pp. 6 1/2 x 9 1/4.

     0-486-20973-3

BASIC ELECTRONICS, U.S. Bureau of Naval Personnel. Clear, well-illustrated introduction to electronic equipment covers numerous essential topics: electron tubes, semiconductors, electronic power supplies, tuned circuits, amplifiers, receivers, ranging and navigation systems, computers, antennas, more. 560 illustrations. 567pp. 6 1/2 x 9 1/4.   0-486-21076-6

BASIC WING AND AIRFOIL THEORY, Alan Pope. This self-contained treatment by a pioneer in the study of wind effects covers flow functions, airfoil construction and pressure distribution, finite and monoplane wings, and many other subjects. 1951 edition. 320pp. 5 3/8 x 8 1/2.   0-486-47188-8

SYNTHETIC FUELS, Ronald F. Probstein and R. Edwin Hicks. This unified presentation examines the methods and processes for converting coal, oil, shale, tar sands, and various forms of biomass into liquid, gaseous, and clean solid fuels. 1982 edition. 512pp. 6 1/8 x 9 1/4.   0-486-44977-7

THEORY OF ELASTIC STABILITY, Stephen P. Timoshenko and James M. Gere. Written by world-renowned authorities on mechanics, this classic ranges from theoretical explanations of 2- and 3-D stress and strain to practical applications such as torsion, bending, and thermal stress. 1961 edition. 560pp. 5 3/8 x 8 1/2.   0-486-47207-8

PRINCIPLES OF DIGITAL COMMUNICATION AND CODING, Andrew J. Viterbi and Jim K. Omura. This classic by two digital communications experts is geared toward students of communications theory and to designers of channels, links, terminals, modems, or networks used to transmit and receive digital messages. 1979 edition. 576pp. 6 1/8 x 9 1/4.   0-486-46901-8

LINEAR SYSTEM THEORY: The State Space Approach, Lotfi A. Zadeh and Charles A. Desoer. Written by two pioneers in the field, this exploration of the state space approach focuses on problems of stability and control, plus connections between this approach and classical techniques. 1963 edition. 656pp. 6 1/8 x 9 1/4.

     0-486-46663-9

Browse over 9,000 books at www.doverpublications.com